Parts Selection and Management

Parts Selection and Management

Edited by

MICHAEL G. PECHT

A JOHN WILEY & SONS, INC., PUBLICATION

Published by John Wiley & Sons, Inc., Hoboken, New Jersey.
Published simultaneously in Canada.

For general information on our other products and services please contact our Customer Care Department within the U.S. at 877-762-2974, outside the U.S. at 317-572-3993 or fax 317-572-4002.

Wiley also publishes its books in a variety of electronic formats. Some content that appears in print, however, may not be available in electronic format.

Library of Congress Cataloging-in-Publication Data is available.

ISBN 0-471-47605-6

Printed in the United States of America.

10 9 8 7 6 5 4 3 2 1

Contents

Preface

Product differentiation often motivates a company to adopt new technologies and insert them into its products. Decisions regarding when, where, and how a technology will be used differentiate the market winners from the losers. Few companies have failed because the right technology was not available; far more have failed when a technology was not effectively selected and managed.

This book provides an "eyes-on, hands-off" approach to parts selection and management, which enables companies to:

- employ risk assessment and mitigation techniques to address technology insertion
- organize and conduct the fact-finding processes necessary to improve part quality, integrity, application-specific reliability, and cost effectiveness
- make an informed companywide decision about parts selection and management based upon company resources, philosophy, and goals and customer demands
- choose parts to fit the functionality of the product, to satisfy system assembly and design level constraints, and to match subsequent manufacturing and handling requirements
- evaluate the part's actual "micro-environment" within a system and then choose the correct technique to fit the part to its intended environmental requirements
- maximize system supportability by preparing for (in design) and meeting the challenge of part obsolescence during system life
- improve supply chain interactions and communications with customers and regulatory agencies in order to minimize time-to-profit

Who this guidebook is for

This book presents a process aimed at increasing company profitability and reducing the time-to-profit. It should be used as a guide in the development of a part selection and management team and in the execution of daily business operations related to parts selection and management. Members of product development teams, product designers, supply chain managers, marketing professionals, business development professionals, contract negotiators, and proposal writers will find the guidance provided in this book useful.

What this guidebook contains

This guidebook presents information and a process guide for parts selection and management. It addresses risk assessment, decision-making steps, and subsequent management activities. The goal is to provide solutions that enable flexibility, innovation, and creativity in product development while ensure that the risks associated with part use are and continue to be acceptable.

Motivation for a Parts Selection and Management Program: Chapter 1 provides the reasons that necessitate the presence of a part selection and management process in any industry using electronic parts. The chapter explains how the parts selection and management process should be used and maintained to keep pace with technology advances, electronics marketplace changes, and the dynamics of the electronics industry supply chain.

Methodology for Parts Selection and Management: Chapter 2 presents the composition and responsibilities of the parts selection and management team and the infrastructure needed to support the process. The chapter continues with summaries of each of the processes that comprise parts selection and management, which are explained in detail in following chapters.

Product Requirements, Constraints, and Specifications: Chapter 3 addresses the translation of customer and market requirements into product-level requirements and constraints that are necessary to arrive at a preliminary specification. The reader is taught how to synchronize the technology insertion initiatives of a company with the product-specific design process through requirement tracking and technology road mapping.

Using the Part Datasheet: Chapter 4 explains how to read electronics part documents and datasheets, including information about the status of a particular datasheet or part, the general format of part numbers, ratings, thermal and electrical characteristics, and reliability information. This information will aid the development of part acceptance and rejection criteria for a product team's decision-making process.

Candidate Parts Selection: Making the First Cut: Chapter 5 addresses the candidate part identification process, procurement specifications, product analysis, preliminary design, and availability issues.

Manufacturer Assessment Procedure and Criteria: Chapter 6 details the part manufacturer assessment procedure, including criteria that determine if the manufacturer's policies and procedures are consistent with producing quality and reliable parts. The criteria include process control, handling, storage and shipping controls, corrective and preventive action, part traceability, and change notification processes.

Part Assessment Guidelines and Criteria: Chapter 7 addresses part assessment. A definition is given for a part group or family, which is used when data are not explicitly available for the candidate part. Criteria are presented by which the quality and integrity of a part group may be assessed. The criteria include average outgoing quality, process capability index, integrity monitoring tests, and assembly guidelines.

Electronic Part Distribution and Distributor Assessment: Chapter 8 presents the role of distributors in the electronic supply chain and discusses their relevance to part manufacturers (suppliers) and customers. The criteria by which the distributor may be assessed are outlined, and a discussion is presented on the trends in the electronics industry that affect distributors.

Tracking Part Changes Through the Part Supply Chain: Chapter 9 discusses the relevant standards and industry practices for part change notification and control. Examples of commonly made changes and the underlying reasons for them are also provided.

Parts Selection and Management to Avoid Counterfeit Electronic Parts: Chapter 10 introduces the counterfeiting problem with respect to electronic parts through examples and financial implications. Initiatives at organizational, industry, and governmental levels against counterfeiting activity are provided, along with specific recommendations for part and equipment manufacturers.

Equipment Supplier Intervention Techniques: Chapter 11 provides guidance for equipment supplier intervention in regard to supply chain members. Case studies are provided with example cost information.

Determination of the Life Cycle Environment: Chapter 12 marks the beginning of the application-specific tasks of parts selection and management. The focus of this chapter is on the different parameters that characterize a life cycle environment. Methods and tools used to determine the life cycle environment of an electronic part are provided.

Performance: Chapter 13 discusses part performance assessment. Methods of characterizing the thermal profile and making parts compliant to environmental requirements are presented. Three methodologies by which a part may be used beyond the manufacturer's specified temperature limits are introduced. Economic aspects of ensuring part conformance to environmental requirements are also presented.

Reliability Assessment: Chapter 14 presents the methodology for assessing part reliability. The chapter outlines the information necessary to conduct a reliability assessment, addresses the situations in which part manufacturer testing may be used to assess reliability, and presents guidance for reliability assessment through virtual qualification and accelerated testing.

Assembly Requirements and Constraints Assessment: Chapter 15 discusses how to select parts based on assembly requirements and limitations. The focus is on determining the manufacturability associated with including a candidate part within an assembly. Various assembly, routing compatibility, and test and rework acceptability issues are presented, and their impact on the cost and yield of the product is discussed.

Obsolescence Prediction and Management: Chapter 16 focuses on minimizing the impact of part obsolescence by providing a methodology to evaluate part and product life cycles, assess their life cycle mismatch, and then select the part based on mismatch assessment. The chapter also identifies obsolescence mitigation strategies, evaluates the pros and cons of the various obsolescence mitigation options, and provides a decision support methodology for selecting the best alternative.

Part Acceptance and Risk Management: Chapter 17 guides the reader through the risk management functions associated with parts selection. Risk assessment, mitigation, and management are discussed. Unmanaged risks and their economic impact are presented.

Environmental and Legislative Issues: Chapter 18 explains the issues associated with new materials that are used as substitutes for lead and halogenated flame retardants in electronic parts. The legislative framework driving the changes, as well as new materials and their implications for assembly, reliability, and the risks of patent infringement, are discussed. Industry groups working toward "green" electronics and eco-labeling strategies are identified.

Legal Liabilities: Chapter 19 explores the legal issues associated with the parts selection and management process. Both international and U.S. laws are discussed. Prudent business and engineering practices that may reduce the implications of a potential legal action are included. Particular attention is devoted to the use of parts beyond the manufacturer's temperature specifications.

Acknowledgments

This book was written and reviewed by experts with firsthand knowledge of parts selection and management issues. Additional research for this book was performed at the CALCE Electronics Products and Systems Center of the University of Maryland. The Center provides a knowledge and resource base to support the development of competitive electronic components, products, and systems. The CALCE Center is supported by more than 100 electronic products and systems companies from all sectors including telecommunications, computer, avionics, automotive, and military manufacturers. The following individuals contributed to the development of this guidebook by reviewing material and offering suggestions:

Angelotti, Phil (Hamilton Hallmark)

Anissipour, Amir (Boeing)

Audette, Pierre (Nortel)

Banks, Linda (HCFA)

Bardhan, Samuel (JDS Uniphase)

Blake, William (Alliant TechSystems)

Boru, Murat (CALCE EPSC)

Boullié, Julien (Aerospatiale)

Brennom, Tom (Rockwell Collins)

Charpenel, Philippe (Aerospatiale)

Condra, Lloyd (Boeing)

Cooper, Mark (Litton)

Corboy, Justine (Sarnoff Laboratories)

Davis, Don (Hamilton Sundstrand)

Day, Virginia (Mitre Corporation)

Easter, Rick (MTI – formerly of the U.S. Navy)

Fink, John (Honeywell)

Flicker, Juergen (Mannheim Institute of Technology)

Foucher, Bruno (Aerospatiale)

Gerrish, Curt (Rochester Electronics)

Govind, Anant (LSI)

Gulliver, Mike (Smiths Industries)

Halle, Hank (Boeing)

Häss, Simon (Audi)

Heggli, William (White Microelectronics)

Hemens-Davis Chantal (Nortel)

Huang, Zhenya (Motorola)

Kelkar, Nikhil (National Semiconductor)

Kendall, Roxy (Celestica)

Kraft, Sandy (GIDEP)

Ladakos, Joanna (Aspect Development)

Liebman, Fely (AMD)

Lillard, Dale (Lansdale Semiconductors)

Love, Paul (Lucas Aerospace)

Mayfield, Dennis (Boeing)

McCullen, Larry (General Motors)

Newman, Robert (Lucas Aerospace)

Porter, Zell (Boeing)

Ramakrishnan, Arun (LSI Logic)

Ranade, Yogi (National Semiconductor)

Rao, Saijee (Aspect Development)

Richter, Jason (General Dynamics)

Schwach, Clifton (Rockwell Collins)

Searls, Damion (Intel)

Sherwood, Everett (Motorola)

Tomczykowski, Walter (ARINC)

Torri, Tom (Delphi-Delco)

Wyler, John (Smiths Industries)

An Additional Acknowledgment

The CALCE Electronic Products Systems Center has been asked to assess and review a wide variety of documents and standards used by different companies for parts selection and management. Many of these documents are confidential and were not used in preparing this guidebook. Nevertheless, these documents did provide an opportunity for comparing and strengthening our ideas and methodology. A list of the reviewed documents follows.

Advanced Micro Devices, *Specification 00-001: Quality Manual*, Sunnyvale, CA, January 18, 2001.

AlliedSignal Inc. [now Honeywell], *Electronic Component Management Plan*, Tucson, AZ, June 1997.

American Institute for Aeronautics and Astronautics, *Recommended Practice for Parts Management, R-100-1996*, Reston, VA, 1996.

Automotive Electronics Council, *CDF-AEC-Q100 Stress Test Qualification for Automotive Grade Integrated Circuits*, Revision E, Detroit, January 31, 2001.

Automotive Electronics Council, *CDF-AEC-Q200 Stress Tests Qualification for Automotive Grade Discrete Semiconductors*, Revision A, May 15, 1996.

Automotive Electronics Council, *CDF-AEC-Q200 Stress Test Qualification for Automotive Grade Passive Components*, Revision A, Detroit, August 18, 2001.

Bellcore, *GR-2840-CORE: Generic Requirements for Environmental Stressing Applied to Telecommunications Products*, Issue 1, Piscataway, NJ, June 1995.

Boeing Inc., *D6-55583: Electronic Component Management Program*, Revision A, Seattle, WA, February 1997.

British Standards Institution, *CECC 00804: 1996 Guidance Document: Interpretation of "EN ISO 9000: 1994," Reliability Aspects for Electronic Components*, London, February 1996.

Eldec Corp., *Electronic Component Management System*, Lynnwood, WA, March 21, 1997.

Electronic Industries Association, *EIA-599: National Electronic Process Certification Standard*, Arlington, VA, August 1992.

Hughes Aircraft [now part of Boeing], *Guidelines for Use of Plastic Encapsulated Microcircuits (PEMs) in Naval Avionics Equipment*, Indianapolis, IN, September 29, 1997.

JEDEC Standard, *JESD34: Failure Mechanism Driven Reliability Qualification of Silicon Devices*, Electronic Industries Association, Arlington, VA, March 1993.

Lockheed Martin Control Systems, *Requirements for Suppliers of Bought-Outside-Complete (BOC) Hardware Using Plastic Encapsulated Microcircuits (PEMs)*, Revision DR, Bethesda, MD, August 19, 1996.

Motorola, Inc., *Parts Management Plan for Iridium™ Comm Module*, Revision XA, Chandler, AZ, June 15, 1992.

Motorola, Inc., *BR1202/D: Motorola Corporate Quality Systems Review Guidelines*, Revision 4, Austin, TX, July 1996.

Motorola, Inc., *BR518/D: Reliability and Quality Handbook*, Revision 5, Austin, TX, 1992.

Rose, G. L., N. Viramani and J. S. Kadesch, *Plastic Encapsulated Microcircuit (PEM) Guidelines for Screening and Qualification for Space Applications*, NASA, Greenbelt, MD, January 1997.

Smiths Industries, *Electronic Component Management Plan*, Cheltenham, UK, February 1, 1997.

STACK International, *Specification 0001 Issue 12.2 at Notice 2: General Requirements for Integrated Circuits*, Herts, UK, November 19, 1999.

Sun Microsystems, *950-2484-01: ASIC Component Engineering Requirements (CER)*, Revision 3.3, Santa Clara, CA, June 25, 1996.

Texas Instruments, *TI-29814: Quality System*, Dallas, TX, October 1992.

United States Department of Defense, *MIL-STD-965: Parts Control Program*, Columbus, OH, September 26, 1996.

Editor

Michael Pecht, Ph.D., has a B.S. in Acoustics, an M.S. in Electrical Engineering, and an M.S. and a Ph.D. in Engineering Mechanics from the University of Wisconsin at Madison. He is a Professional Engineer, an IEEE Fellow, an ASME Fellow, and a Westinghouse Fellow. He has received the 3M Research Award, the IEEE Undergraduate Teaching Award, and the IMAPS William D. Ashman Memorial Achievement Award for his contributions. He has written 16 books on electronic products development. He served as chief editor of the *IEEE Transactions on Reliability* for eight years and on the advisory board of IEEE Spectrum. He is the founder and Director of the CALCE Electronic Products and Systems Center at the University of Maryland and a Chair Professor. He is chief editor of *Microelectronics Reliability* and an associate editor of *IEEE Transactions on Components and Packaging Technology*. He has consulted for over 50 major international electronics companies, providing expertise in strategic planning, design, testing, intellectual property, and risk assessment of electronic products and systems.

Authors

Ray Biagini is a partner in the Washington, D.C., office of McKenna and Cuneo, LLP. He is a member of the firm's Management Committee. Mr. Biagini's practice focuses on product liability litigation, and he has substantial trial and litigation experience. As a leader of the firm's Tort Defense Practice Group, he has defended manufacturers, system integrators, and component providers involving weapons systems, and products made for civilian government agencies, as well as consumer products. He also lectures and writes articles in the area of products liability/toxic torts and conducts seminars on minimizing manufacturers' product liability risks.

Paul Casey is a Research Assistant at the CALCE Electronic Products and Systems Center, University of Maryland, where he is currently analyzing the legal aspects of lead-free assembly with emphasis on lead-free patents, infringement concerns, and managing lead-free intellectual property for risk reduction. He is also conducting research on potential uses of nanotubes as mechanical and chemical sensors. He has had two internships at the Ionizing Radiation Division of the NIST Physics Laboratory in Gaithersburg, MD, and his interest in health physics continues with his involvement in GenTag, where he is contributing to the development of threat surveillance technologies.

Diganta Das received the B.Tech. degree (with honors) from the Indian Institute of Technology, India, and the Ph.D. degree in mechanical engineering from the University of Maryland. He is a Research Associate at the CALCE Electronic Products and Systems Center, University of Maryland. His primary research interests are environmental and operational ratings of electronic parts, uprating, obsolescence prediction and management, technology trends in electronic parts, and their effects on parts selection and management methodologies. He has published in the areas of electronic part uprating and operational environments of electronic parts, organized international conferences and workshops, and worked in international standards developments. He also provides services to scholarly journals and magazines.

Daniel N. Donahoe graduated with a B.S. in general engineering, and an M.S. in mechanical engineering, both from the University of Illinois at Urbana-Champaign, and an MBA from the University of Santa Clara. He has held a variety of engineering positions at Lockheed Missiles and Space Corporation, Motorola, Government Electronics Group, Ford Aerospace, Teledyne, CME, Compaq Computer Corporation, and Iomega Corporation. These firms cover most of the scope of the electronics industry. He holds seven U.S. patents. He served as an Associate Editor of *IEEE Transactions on Components and Packaging Technologies* for three years, and is a Senior Member of IEEE. He is a Registered Professional Engineer in California and Arizona. He is currently an Assistant Research Scientist at the University of Maryland, CALCE EPSC.

David Erhart received the B.S. degree in chemical engineering from the Illinois Institute of Technology, Chicago, and the Ph.D. degree in chemical engineering from the University of California, Berkeley. He is the Director of Reliability Engineering at ON Semiconductor, Phoenix, AZ, where he is responsible for developing and implementing global reliability strategies and methods. Prior to joining ON Semiconductor, he held various reliability engineering positions at Motorola Semiconductor Products Sector, Phoenix.

Yuki Fukuda is a Research Assistant at the CALCE Electronic Products and Systems Center, University of Maryland. She received a B.S. degree in mathematics from Chiba

University, Japan, and an M.S. degree in mathematics and computer science from Ochanomizu University, Tokyo, Japan. She is a member of ASME, IMAPS, JIEP, and SMTA.

David Humphrey received the M.S. degree in electrical engineering from Youngstown State University, Youngstown, OH. He has worked in the automotive industry in advanced design, electronics quality, and reliability. Currently, he is with Honeywell International, Inc., Tucson, AZ, in reliability and quality of commercial and military avionics and electronics parts selection and management. He served as the Chairman of the Industrial Advisory Board of the CALCE Electronic Products and Systems Consortium, University of Maryland.

Margaret Jackson earned the M.S. degree in mechanical engineering from the University of Maryland. She is an engineer at Northrop Grumman's reliability analysis laboratory, developing the process for commercial parts insertion.

Anant Mathur obtained his B.S. degree in mechanical engineering from the MS University of Baroda, India. He completed his M.S. degree in mechanical engineering at the University of Maryland. He worked as a Research Assistant at the CALCE Electronic Products and Systems Center. His interests lie in electronic parts selection and management. He is currently with Hughes Network Systems in Maryland.

Steven Murray received the B.S. degree in mechanical engineering from the University of Wisconsin, Madison and the M.S. degree in mechanical engineering from the University of Maryland. He worked as a Research Assistant at the CALCE Electronic Products and Systems Center. Prior to joining the Center, he worked on the design of cellular phones at Motorola. He is currently with Northrop Grumman.

Neeraj Pendse received the B.E. degree in electrical engineering from Visvesvaraya Regional College of Engineering, Nagpur, India and the M.S. degree in mechanical engineering from the University of Maryland. He was a Research Assistant at the CALCE Electronic Products and Systems Center, University of Maryland. He is with the National Semiconductor Corporation, Santa Clara, CA, where he works on evaluation of electrical performance of electronic packages and interconnect design.

Peter Sandborn is an Associate Professor in the CALCE Electronic Products and Systems Center at the University of Maryland. His interests include technology trade-off analysis for electronic packaging, system life cycle and risk economics, hardware/software codesign, and design tool development. Prior to joining the University of Maryland, he was a founder and Chief Technical Officer of Savantage, Inc., a technical contributor at Nu Thena Systems, and a Senior Member of Technical Staff at the Microelectronics and Computer Technology Corporation. He received the B.S. degree in engineering physics from the University of Colorado and the M.S. degree in electrical science and the Ph.D. degree in electrical engineering, both from the University of Michigan. He is the author of over 50 technical publications and one book on multichip module design. He is an Associate Editor of *IEEE Transactions on Components, Packaging, and Manufacturing Technology –– Part C: Manufacturing.*

Rajeev Solomon obtained his Ph.D. degree from the Department of Mechanical Engineering at the University of Maryland. He completed his M.S. degree in mechanical engineering from Villanova University in Pennsylvania and his B.S. degree in power plant engineering from Jadavpur University, India. His interests lie in electronic parts selection and management, plastic encapsulated microcircuits, and thermal modeling of electronic sub-systems. He works for Nortel Networks.

Laurie Sullivan has a B.A. degree in communications and an M.A. degree in creative writing. She has worked in the electronics industry for more than 16 years. She is currently a journalist for EBN, an industry trade publication, covering electronic components distribution and supply chain business trends.

Toby Syrus received the B.A. degree in manufacturing engineering technology from Midwestern State University, Wichita Falls, TX, and the M.S. degree in mechanical engineering from the University of Maryland. His research interests include fabrication and assembly of electronic devices and quality and reliability of electronic devices and products. He is a member of IMAPS and the Society of Manufacturing Engineers. He is currently with the United States Peace Corp.

Sanjay Tiku received his B.S. degree in mechanical engineering from the Regional Engineering College, Trichy, India. He worked at the Research Center of Tata Engineering in India before accepting the post of Lecturer in Mechanical Engineering at Government College of Engineering and Technology, Jammu, India. There he taught undergraduate courses in the area of mechanical system design and manufacturing. He is currently a Ph.D. candidate in mechanical engineering at the University of Maryland. His research interests include quality and reliability of electronic products and electronic parts selection and management. He is a member of the IEEE and IMAPS.

Ramgopal Uppalapati received the B.S. degree from the Indian Institute of Technology, Chennai (formerly Madras), India, and the M.S. degree in mechanical engineering from the University of Maryland, College Park. He is currently with Intel Corporation, Dupont, WA, as Structural and Thermal Engineer. His research interests include quality and reliability assessment of electronic packaging.

Niranjan Vijayaragavan is a graduate research assistant at the CALCE Electronic Products and Systems Center, University of Maryland. He holds a B.E. degree in mechanical engineering from Regional Engineering College, Trichy, India and an M.S. degree in mechanical engineering from the University of Maryland. He is now with AMD. His research interests include electronic hardware and packaging design, reliability assessment and testing, and failure analysis of electronics. He is a member of the IMAPS.

Chris Wilkinson MIEE holds a B.Sc. degree in electronics engineering from the University of London. He is a Senior Research Scientist with CALCE Electronic Products and Systems Center at the University of Maryland and Program Manager for the PASES program. His research interests include parts management, uprating, and parts obsolescence management. He has published papers in the areas of reliability assessment, parts obsolescence management, and parts uprating. He was previously with Smiths Industries, Cheltenham, UK, for 20 years, latterly as a Principal Engineer in the Corporate Research Department.

Jingsong Xie holds a B.S. degree in engineering mechanics from Tsinghua University, China, an M.S. degree in naval architecture and ocean engineering from Tokyo University, Japan, and a Ph.D. degree in mechanical engineering from the University of Maryland. He is currently a component reliability engineer at Microsoft and was previously a visiting assistant professor at the CALCE Electronic Products and Systems Center at the University of Maryland. He has written over 20 technical publications and is a member of IEEE, ASME, and IMAPS.

Liyu Yang received his B.S. degree from Xian Jiaotong University in China and M.S. degrees from Nanyang Technological University in Singapore and Xian Jiaotong University in China. He is a senior component/package development engineer at Intel Corporation. His research interests include fabrication and assembly as well as quality and reliability of microelectronics products. Before joining Intel, he worked as a Research Assistant at the CALCE Electronic Products and Systems Center and as a Package Development Engineer at Advanced Micro Devices in Singapore.

Acronyms

AC	Alternating Current
AC	Autoclave
ADC	Analog to Digital Converter
AEC	Automotive Electronics Council
AEGIS	Advanced Electronic Guidance and Instrumentation System
AME	Advanced Microcircuits Emulation
AMR	Absolute Maximum Ratings
ANSI	American National Standards Institute
AOQ	Average Outgoing Quality
APEX	Acquisition Planning and Execution
ASE Group	Advanced Semiconductor Engineering Inc.
ASICs	Application Specific Integrated Circuits
ASQC	American Society for Quality Control
ASTM	American Society of Testing and Materials
AWACS	Airborne Warning and Control System
AWG	Avionics Working Group
BAT	Brilliant Anti-Armor Submunition
BGA	Ball Grid Arrays
BiCMOS	Bipolar Complementary Metal Oxide Semiconductor
BIT	Binary Digit
BITE	Built-in Test Equipment
BOM	Bill of Materials
BSI	British Standards Institution
CALCE EPSC	Computer-Aided Life Cycle Engineering Electronic Products and Systems Center
CAST	Computer-Aided Software Testing
CCA	Circuit Card Assembly
CDF-AEC	Chrysler-Delco-Ford Automotive Electronics Council
CDR	Critical Design Review
CER	Component Engineering Requirements
CFM	Cubic Feet per Minute
CM	Contract Manufacturers
CMOS	Complementary Metal Oxide Semiconductor
CMP	Chemical Mechanical Polishing
COEX	Core/Executive Interface
COTS	Commercial-off-the-Shelf
CPLD	Complex Programmable Logic Devices
CRT	Cathode Ray Tubes
CSP	Chip Scale Packaging
CTE	Coefficient of Thermal Expansion
DAC	Digital to Analog Converter
DC	Direct Current
DESC	Defense Electronics Supply Center
DFARs	Defense Federal Acquisition Regulations
DFM	Design for Manufacturing

DIP	Dual Inline Package
DISC	Defense Industrial Supply Center
DLA	Defense Logistics Agency
DMEA	Defense Microelectronic Activity
DMS	Diminishing Manufacturing Sources
DMSMS	Diminishing Manufacturing Sources and Material Shortages
DoD	Department of Defense
DoE	Department of Energy
DPA	Defense Production Act
DRAM	Dynamic Random Access Memory
DSCC	Defense Supply Center, Columbus
DSP	Digital Signal Processing
DTAM	Distribution Total Available Market
DTL	Diode Transistor Logic
ECL	Emitter Coupled Logic
ECP	Engineering Change Proposals
EDAC	Error Detection and Correction
EDO	Extended Data Out (RAM)
EE	Electrical Engineering
EEC	European Economic Community
EEPROM	Electrically Erasable Programmable Read-Only Memory
EFSOT	Environmentally Friendly Soldering Technology
EIA	Electronic Industry Association
EIAJ	Electronic Industries Association of Japan
ELFNET	European Lead-Free Network
EMF	Electromotive Force
EMI	Electromagnetic Interference
EMS	Electronic Manufacturing Services
EOL	End-of-Life
ERAI	Electronic Resellers Association International
ESD	Electrostatic Discharge
EU	European Union
FAA	Federal Aviation Administration
FAE	Field Application Engineer
FCC	Federal Communication Commission
FET	Field Effect Transistor
FFF	Form, Fit and Functionality
FFOP	Failure-Free Operating Period
FIFO	First In, First Out
FIT	Failures in Time
FPGA	Field Programmable Gate Arrays
FRU	Field Replacement Units
GATT	General Agreement on Tariffs and Trade
GEIA	Government Electronics and Information Technology Association
GEM	Generalized Emulation of Microcircuits
GIDEP	Government Industry Data Exchange Program
GIFAS/SPER	Groupement des Industries Francaise Aeronutiques et Spatials (France's aerospace manufacturer's association)/Sybdicat des Industries de Materiels

	Professional d'Electronique et de Radiocommunication
GPC	Government Procurement Committee
GSA	General Services Administration
GSI	General Semiconductor
HALT	Highly Accelerated Life Testing
HAST	Highly Accelerated Stress Test
HCMOS	High-Speed CMOS
HDPUG	High Density Packaging User Group
HP	Hewlett-Packard
HTOL	High-Temperature Operating Life
HTS	High-Temperature Storage
IACC	International Anti-Counterfeiting Coalition
IC	Integrated Circuit
IDEA	Independent Distributors of Electronics Association
IDT	Integrated Device Technology
IEC	International Electrotechnical Commission
IEEE	Institute of Electrical and Electronics Engineers
IMEC	Interuniversity Micro-Electronics Center
I/O	Input/Output
IP	Intellectual Property
IPC	Institute for Printed Circuits
IRC	International Rectifier Corporation
ISI	International Semiconductor, Inc
ISO	International Standards Organization
IT	Information Technology
ITRI	International Tin Research Institute
JEDEC	Joint Electron Device Engineering Council
JEIDA	Japan Electronic Industry Development Association
JEITA	Japan Electronics and Information Technology Association
JESD	JEDEC Standards Division
JIEP	Japan Institute of Electronics Packaging
JIT	Just in Time
LAR	Lot Acceptance Rate
LCEP	Life Cycle Environment Profile
LOT	Life-of-Type
LRU	Line Replaceable Units
LSI	Large-Scale Integration
LSL	Lower Specification Limit
LTOL	Low Temperature Operating Life
MAIS	Major Automated Information System
MDAP	Major Defense Acquisition Program
MEP	Major Electronics Procurement
MII	China's Ministry of Information
MIL-HDBK	Military Handbook
MIL-SPEC	Military Specification
MIL-STD	Military Standard
MITI	Ministry of International Trade and Industry
MLRS	Multiple Launch Rocket System
MMU	Memory Module Unit
MOSFET	Metal Oxide Semiconductor Field Effect Transistor

MPU	Microprocessor Unit
MTBF	Mean Time Between Failures
MTC	Management to Cost
NAECON	National Aerospace and Electronics Conference
NAVSUP	Naval Supply Systems Command
NCMS	National Center for Manufacturing Sciences
N/C	Not Calculated
NDIA	National Defense Industrial Association
NEC	Nippon Electric Company
NEDA	National Electronic Distributors Association
NEDO	New Energy Development Organization
NEMI	National Electronics Manufacturing Initiative
NMOS	N Channel Metal Oxide Semiconductor
NOVRAM	Non-Volatile Random Access Memory
N/P	Not Provided
NPI	New Product Introductions
NPV	Net Present Value
NRE	Nonrecurring Engineering
OEL	Oki Electronics Limited
OEM	Original Equipment Manufacturers
ONR	Office of Naval Research
PBB	Polybrominated Biphenyls
PBDE	Polybrominated Diphenyl Ether
PBT	Persistent Bioaccumulative and Toxic Substances
PC	Personal Computer
PCB	Printed Circuit Board
PCI	Peripheral Component Interconnect
PCN	Product Change Notification
PCT	Patent Cooperation Treaty
PDN	Product Discontinuance Notice
PDR	Preliminary Design Review
PEM	Plastic Encapsulated Microcircuits
PICMET	Portland International Conference on Management of Engineering and Technology
PLC	Product Life Cycle
PLCC	Plastic Leaded Chip Carriers
PLD	Programmable Logic Devices
PLM	Product Lifecycle Management
PMOS	P Channel Metal Oxide Semiconductor
PoF	Physics-of-Failure
POURS	Point of Use Replenishment System
PPM	Parts Per Million
PQFP	Plastic Quad Flat Pack
PVC	Polyvinyl Chloride
PWB	Printed Wiring Board
PWM	Pulse Width Modulation
PZT	Lead Zirconate Titanate
QML	Qualified Manufacturers List
QPL	Qualified Parts List
QS 9000	Quality System (developed by Big 3 – GM, Ford, Chrysler)

QSA	Qualification Site Approval
RACE	Random Access Computer Equipment
RAM	Random Access Memory
RCA	Radio Corporation of America
RF	Radio Frequency
RH	Relative Humidity
RoHS	Reduction of Hazardous Substances
RTD	Resistive Temperature Detectors
RTL	Resistor Transistor Logic
SAE	Society of Automotive Engineers
SAMPE	Society for Advanced Materials and Process Engineering
SBE	Single Bit Error
SBIR	Small Business Innovation Research
SD	Standard Deviation
SDDV	Stress Driven Diffusion Voiding
SDRAM	Synchronous Dynamic Random Access Memory
SDU	Shop Discardable Units
SGRAM	Static Graphics Random Access Memory
SHARC	System Hardware Availability and Reliability Calculator
SIA	Semiconductor Industry Association
SIMM	Single Inline Memory Module
SIP	Single Inline Package
SMART	Self-Monitoring Analysis Reporting Technology
SMD	Standard Microcircuit Drawings
SMT	Surface Mount
SOA	Safe Operating Area
SOAR	Survivability, Operability, Availability and Recoverability
SOIC	Small Outline Integrated Circuit
SOJ	Small Outline J-Leaded
SOP	Small Outline Package
SPC	Statistical Process Control
SRAM	Static Random Access Memory
TC	Temperature Control
TDDB	Time Dependent Dielectric Breakdown
TH	Through-hole
THB	Temperature Humidity Bias
TI	Texas Instruments
TO	Transistor Outline
TQFP	Thin Quad Flat Pack
TRIPS	Trade-Related Aspects of Intellectual Property Rights
TS	Thermal Shock
TSI	Threshold Single-to-Independence Ratio
TTL	Transistor Transistor Logic
UDR	Urgent Data Requests
UL	Underwriters Laboratories
U.S.C.	United States Code
U.S.C.A.	United States Code Annotated
USL	Upper Specification Limit
UTMC	United Technologies Microelectronics Center
UV	Ultraviolet (radiation)

VHDL   Hardware Description Language
VHSIC   Very-High-Speed Integrated Circuit
VLSI   Very-Large-Scale Integration
VME   Virtual Machine Environment
VRAM   Video Random Access Memory
WEEE   Waste in Electrical and Electronic Equipment
WIPO   World Intellectual Property Organization
WTO   World Trade Organization
XC   Cross-Connect

Chapter 1

Motivation for a Parts Selection and Management Program

Sanjay Tiku and Michael Pecht

This chapter presents an overview of technology trends, market challenges, and the evolving electronic systems supply chain and then presents the objectives of a parts selection and management program suited to function in a rapidly changing world.

1.1 Technology advances

The growing demands for product performance, reliability, versatility, and miniaturization at a competitive price have imposed major challenges upon electronics companies. New device packaging and manufacturing technologies have overcome some challenges, but sometimes the new technologies have been disruptive.[1] For example, when Intel developed the 100 MHz CPU, changes were required in the design, manufacturing, and testing of memory modules and motherboards, affecting members of the supply chain who provided those supporting products and services [1]. Managing the selection and use of new technologies is fundamental to the parts selection and management program.

1.1.1 Device technology trends

The pressure to improve integrated circuit technology is great, as each new generation of electronic systems aims for smaller size, lower weight, lower cost, and higher speed than the previous one. Bipolar processes were dominant in the 1960s and 1970s, and offered high-speed operation. CMOS became dominant in the 1980s by offering higher device density and lower power dissipation. Today, CMOS also offers a high-speed performance, and new technologies such as GaAs show great promise for next-generation high-speed circuits. The most dominant commercially available GaAs technologies are the field effect transistor (FET) logic. Requirement for low power consumption and reduction in risk of failures caused by high supply voltages have been driving the transition towards low-voltage technology. Over the years, the supply voltage values have transitioned from 5V to sub-1V values. Table 1.1 shows device technology trends. Figure 1.1 shows the transistor count trends.

Another field of rapid technological advance is in the area of application specific integrated circuits (ASICs). Since 1990, the overall ASICs market has expanded dramatically, and many competing technologies are now available in the market. The worldwide ASICs market is projected to grow to more than $42 billion in 2005 [2].

Programmable logic devices (PLDs) can be considered ASICs which can be programmed for a specific application, thus offering the advantage of a custom product. The use of programmable logic can lead to lower development costs and shorter design times as compared with a full custom ASIC. In fact, PLDs like field programmable gate

---

[1] A technology is disruptive if it does not support the current manufacturing practices of a company, if it disrupts its current capability defined by an established market, or if the service or physical products generated by utilizing the technology are not valued by the existing customers [9].

Table 1.1: Device technology trends [3]

Technology generations	2001	2005	2011
Memory			
Generation @ production ramp	512M	2G	16G
Bits/cm^2 @ sample/introduction	490M	1.63G	9.94G
Logic (high-volume, cost-performance MPU)			
Logic transistors/cm^2	13M	44M	269M
ASIC			
Transistors/cm^2	40M	133M	811M
Functions/chip			
DRAM bits/chip	2.15G	8.59G	68.7G
Chip size (mm^2)			
DRAM introduction	445	790	1580
MPU introduction	340	408	536
ASIC production	800	800	800
Number of chip I/Os			
Chip-to-package, high-performance	2400	4000	7300
Chip-to-package, cost-performance	1195	1970	3585
Number of package pins/balls			
ASIC, high-performance	2007	3158	6234
MPU/controller, cost-performance	912	1384	2589
Chip frequency (MHz)			
On-chip, across-chip clock, high-performance	1454	2000	3000
Power supply voltage (V)			
Minimum logic voltage	1.2–1.5	0.9–1.2	0.5–0.6
Power (W)			
High performance with heat sink	115	160	175

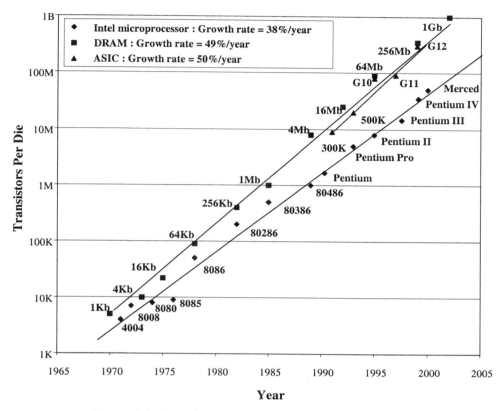

Figure 1.1: Transistor count trends with time [13], [16]

arrays (FPGAs) were so popular that approximately one-half of all chip design starts used FPGAs during the mid-1990s [12]. On the other hand, the difficulty of electrical testing of PLDs makes their validation testing harder, since qualification and acceptance tests must be conducted after the part is purchased and programmed.

1.1.2 Packaging and assembly trends

Packaging and assembly needs are driven as much by market application requirements as by developments in device technology. Package trends are driven by:

- *communication*: high speed and miniaturization. Examples are routers, Internet appliances, cellular phones, and pagers.
- *computers*: high speed, increasing functionality, improved cost/performance, and portability, leading to denser interconnections [11]. Examples are desktop and laptop computers. Servers are important for thermal management and reliability advances.
- *consumer electronics*: lighter weight, lower costs, and improved performance. Examples are digital cameras and camcorders.

While the number of I/O, is expected to increase to more than 3000 by the year 2005 on some advanced ASICs [2], package size is continuously decreasing, which indicates an increase in package density. For many applications, area array surface mount packages, such as chip scale packaging (CSP) and flip chip technology, provide promising solutions. However, new test methods and assessment criteria will be in order for the new package designs.

An electronic assembly or system contains multiple parts interacting with one another. Consequently, part selection for different applications is affected by manufacturability, assembly compatibility, routing compatibility, and test and rework acceptability. Verification at the front end of the design process helps avoid postdesign assembly conflicts that can be costly to fix and can affect the marketing schedule for the product. For example, the transition to lead-free parts and soldering has impacted assembly compatibility. In particular, the higher reflow temperatures for lead-free alternatives are giving rise to new assembly requirements for parts and printed wiring boards.

1.2   Market challenges

Prior to the 1980s, the military was a major influence on developments in the electronics industry. Throughout the 1980s and 1990s, the balance of influence within the electronics industry shifted decisively towards the consumer electronics and information technology product markets. Figure 1.2 illustrates industry market changes since 1984. Most of the electronic part manufacturers concentrate on the lucrative volume-driven electronics market (e.g., computers, consumer products, and portable communication equipment). With increasing global market demand for these products, part manufacturers are developing new higher-performance, smaller, lighter, and less costly parts at an unprecedented rate. At the same time, customers are experiencing, and expecting, a decrease in the time between new product offerings.

1.3   Supply chain trends

The era of vertically integrated electronics companies that made everything from chips to software to final systems is over. Technological advances and market dynamics have rendered uneconomical the concept of achieving market dominance through vertical integration. Companies adopting horizontal solutions, such as Microsoft in software and Intel in semiconductors, have demonstrated that they are the more likely candidates for success [20]. Companies have been adopting a new approach to doing business and tend to spin off their noncore businesses, allowing each spinoff to concentrate on its own core competencies. Spinoff companies are a valuable link between research institutes, industry, and the market.[2] Even when vertically integrated companies do stay together (e.g., Motorola or IBM), their units are given the independence to make business decisions, resulting in benefits similar to those available to spinoffs.

Due to the trend towards horizontal integration, most companies in commercial, computer, and telecommunication industries have witnessed a growth in the number of sources from which to select parts. However, companies in the avionics, aerospace, and military sectors are faced with a reduction of sources. These changes have made the task of parts selection and management more challenging.

---

[2] The 1995 breakup of AT&T into three publicly traded global companies is a case in point. AT&T provides communication services; Lucent Technologies provides communications products, semiconductors, and testing services; NCR Corporation provides computer business services.

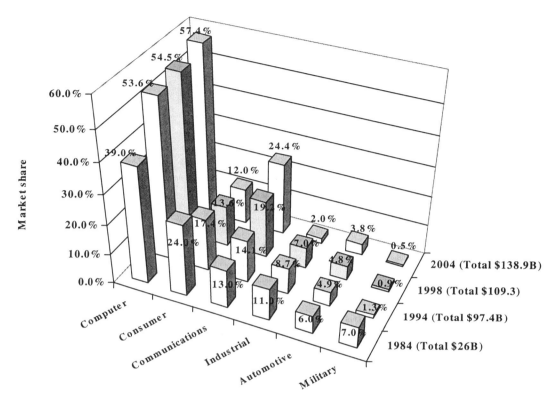

Figure 1.2: Worldwide trends in the electronics parts marketplace [5], [16]

Electronic products and systems are most often the result of the integration of parts and subsystems that are manufactured and assembled by different companies. As a group, these companies, specializing in their respective competencies, yet interacting with one another, can best be described as a "supply chain." Table 1.2 presents the definitions of various participants in an electronics industry supply chain. Figure 1.3 depicts a simplified example of a supply chain for avionics systems. The parts that comprise an electronic product have their own supply chains. When these supply chains are taken into consideration, their cross-linking results in a complex "supply web," which represents the interaction between all the different members contributing to the final product.

It is commonly thought that the participants at the higher levels of a supply chain are able to dictate requirements down the chain. In reality, relationships between members of the supply chain are bidirectional; therefore, the policies and practices of one member have consequences both up and down the supply chain. In addition to being affected by the participants in a supply chain, each member is also subjected to outside influences through the actions of private and public standard organizations.

These outside organizations can affect the modes and costs of operation for companies in the supply chain and may themselves be influenced by government policies and by public opinion and perceptions. Other forces exist in the form of consumer advocate groups, trade journals, and user groups. Such organizations may lack authorization for legal enforcement but can effectively influence the buying habits of users and the decisions of regulatory agencies.

Table 1.2: Definitions of participants in the electronics industry supply chain

Entity	Definitions
Part manufacturers	Producers of electronic parts (e.g., logic devices, microprocessors, capacitors, and resistors).
Authorized distributors (franchisees)	Warehouse operators and retailers of electronic parts who perform storage, handling, order processing, shipping, and other services. Franchisees sign selling and marketing contracts with part manufacturers for distribution of goods and services identified by franchisers' trademarks.
Independent distributors	Aftermarket sources of parts that make a one-time purchase of parts without continued commercial relationship with the manufacturer.
Part brokers	Scouting agencies for hard-to-find replacement parts and components.
Contract manufacturers	Assemble parts on printed circuit boards for the equipment manufacturers. They may also buy parts.
Equipment manufacturers	Procurers of electronic parts who assemble products or systems, e.g., line replaceable units (LRUs) for airplanes, field replacement units (FRUs) for telecommunication companies, hard disks for computer companies.
System designers	Companies or entities that design products, subsystems, and systems. They may outsource parts procurement and circuit card assembly to contract manufacturers. Can be the same as equipment suppliers.
Customers	Developers and integrators of final products, e.g., aircraft manufacturers, computer manufacturers, and cell phone manufacturers.
Users	Users of the final product like airlines, data processing firms, engineering companies, and the public.
Regulatory agencies	Authorities who determine the specifications for final products, including military standard agencies, the Federal Aviation Administration (FAA), large utilities, companies, and major buyers of office equipment, such as the General Services Administration (GSA).

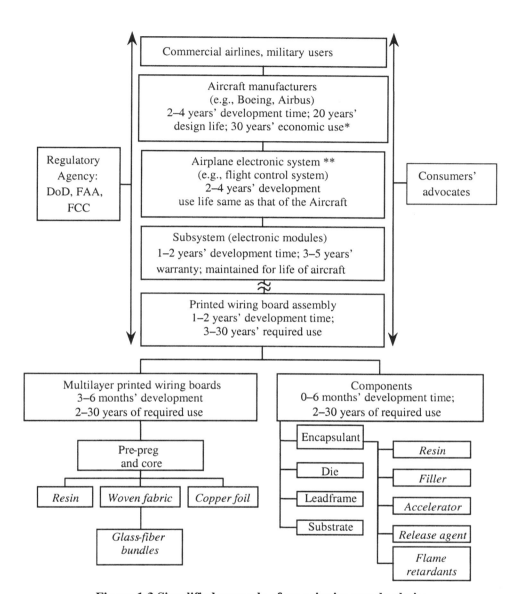

Figure 1.3 Simplified example of an avionics supply chain

---

* Can be used longer with proper maintenance.
** The aircraft manufacturers are involved in the design and assembly.

The industry has often been considered to be number one in value added to the U.S. economy [19]. The explosive growth has resulted in a supply chain power balance shift reducing the equipment (product) manufacturers' influence. Today, the name of the component used in the computer may be as important in a purchase as the brand name of the computer. Thus, decisions made by dominant part manufacturers affect the pace of technology change in the industry. The emergence of electronic parts distributors and contract manufacturers as major players in the electronics industry supply chain has also had a profound effect on the electronics supply chain. These companies have not only changed the business situation, but have also influenced technology development and insertion. Electronics part distributors sell 40% to 70% of all semiconductor products [4]. Distributors are among the part manufacturers' largest customers. Franchisees or authorized distributors build close relationships with the part manufacturers and provide added services, such as warranty and return responsibilities, programming, kitting, lead finishing, solderability evaluation, and marking. Part manufacturers believe that through these value-added services and close customer contacts, distributors can actually help create more demand for their products. Many distributors also serve part manufacturers by handling their increasingly frequent part discontinuations.

Contract manufacturers perform printed circuit board (PCB) assembly work for equipment suppliers. These manufacturers also play a major role in the electronics industry and perform up to 70% of the semiconductor component purchases [15]. Contract manufacturers are also becoming centers for technology development and integration as they assume responsibility for design, manufacturing, and testing work. Contract manufacturers and distributors are also joining hands and creating direct business and technical service relationships [4].

1.4   Availability and life cycle mismatch issues

The time required to transition silicon from the bare wafer stage to a functional chip is about 18 to 20 weeks. Thus, part manufacturers must predict market demands for part types 18 to 20 weeks into the future. The resultant predictions, if in error, can lead to cycles of overproduction or scarcity of part types. Depending on the equipment manufacturer's program schedule, these availability cycles may influence the time-to-profit.

The availability of an electronic part is a measure of the ease with which the part can be procured. Availability is assessed by determining the amount of inventory at hand, the number of parts required for units in production and those forecasted, the economic order quantity for the parts, the lead time between placing an order for the parts and receiving the parts, production schedules and deadlines, and part discontinuance plans. However, minimizing the availability problem by designing products for which parts are always available brings a new challenge – obsolescence.

Newer, smaller, faster, and lighter parts are constantly introduced into the market in response to consumer demand for product upgrades and peer competition. The part manufacturer's decision on which product lines should be continued is based on market economics, both current and anticipated. Conversely, when demand is greater than supply, as it is for newer parts, the decision is simple–nonstrategic products with the lowest unit margin are dropped.

The billions of parts that are produced each year often have shorter life cycles than the equipment manufacturer's products, and equipment manufacturers often have difficulty servicing their products with replacement parts after the parts have been discontinued.

1.5   Standards reorganization

In the late 1950s, the U.S. Department of Defense (DoD) began creating standards and specifications for all aspects of electronic parts design, manufacture, test, acceptance, and use. The purpose of these documents was to narrow the uncertainty or judgment area with respect to part quality, environment, interoperability, and documentation [14]. Without standardization, the government believed that each military contractor might develop its own methods, making part and product evaluation and comparison against competitive designs difficult. Not only were these standards used by the U.S. military suppliers (as required), but the standards were also voluntarily adopted by much of the commercial industry.

In November 1969, the military issued the first "General Specification for Microcircuits," MIL-M-38510.[3] The manufacturing lines for these devices were certified to MIL-STD-976,[4] "Certification Requirements for Microcircuits." In November 1974, a "List of Standard Microcircuits," MIL-STD-1562,[5] was published, followed by MIL-STD-965,[6] "Parts Control Program," in April 1977. Electronic parts selection and management was governed by MIL-STD-454,[7] "Standard General Requirement for Electronic Equipment."

By the late 1970s, semiconductor manufacturers were less willing to meet the special demands of the DoD. According to Braub and McDonald [10], "In the fifties and sixties, the military was a main market for semiconductor products and often the only one for new and expensive components. New firms often found the military market a particular boon. That situation has now changed and the *industrial, computer and consumer* markets are either actually or potentially of far more importance."

The Braub and McDonald statement was accurate. By the mid-1980s the computer, industrial, and consumer sectors dominated the semiconductor industry, accounting for over 90% of the market, and trends pointed towards continued growth. By the late 1980s, these new leaders started relying less on government specifications and more on their own internal standards and on international standards like ISO 9000 (introduced in 1987) as a means and a necessity to stay competitive both nationally and internationally.

---

[3] MIL-M-38510 defined the general requirements for microcircuits and the quality and reliability requirements for the acquisition of military microcircuits. Because qualification to MIL-M-38510 was a long and expensive process, often requiring several years to complete, most of the parts available to the specification were no longer leading-edge technology (United States Department of Defense, *MIL-M-38510: General Specification for Microcircuits*, Revision J, Columbus, OH, November 15, 1991).

[4] MIL-STD-976 established requirements for documented process control, specified plans for device screening and preconditioning, and specified criteria for lot acceptance testing (United States Department of Defense, *MIL-STD-976: Certification Requirements for Microcircuits*, Columbus, OH, November 15, 1994).

[5] MIL-STD-1562 contained the requirements established by the DoD for the selection of microcircuits used in the design, manufacture, and support of military equipment. It provided lists of parts and their sources, which could not be tailored without explicit approval from DESC (now DSCC). It lagged behind the available technology so much that even before canceling the document, the U.S. military exempted all new devices from MIL-STD-1562 requirements (United States Department of Defense, *MIL-STD-1562: List of Standard Microcircuits*, Columbus, OH, December 15, 1995).

[6] MIL-STD-965 was part of the contracting activity that required contractors to all military suppliers to establish a program parts selection list. This standard led to the use of old technologies and caused the obsolescence problem (United States Department of Defense, *MIL-STD-965: Parts Control Program*, Columbus, OH, September 26, 1996).

[7] MIL-STD-454 (introduced in 1965) controlled the design and development of all military electronic equipment. It provided an order of preference for parts for all military systems. The standard hindered introduction of advanced technologies, especially plastic packaging into military systems, and generally increased the costs by requiring custom military parts (United States Department of Defense, *MIL-STD-454: Standard General Requirement for Electronic Equipment*, Columbus, OH, May 4, 1995).

By the early 1990s, the military's electronic parts selection and management standards were identified as non-value added and cancelled. Their main drawbacks included "rigid" task overheads, lack of synchronization with new technologies, and higher life cycle costs. As a result, on June 29, 1994, the U.S. Military Specifications and Standards Reform initiative decreed the adoption of performance-based specifications[9] and industry standards for acquiring and modifying weapon systems, leading to the cancellation of many military specifications and standards [18]. The directive stated: "Performance specifications shall be used when purchasing new systems, major modifications, upgrades to current systems, and nondevelopmental and commercial items, for programs in any acquisition category. If it is not practical to use a performance specification, a non-government standard shall be used. The use of military specifications and standards is authorized as a last resort, with an appropriate waiver." Various U.S. DoD actions and various legislative actions for reorganizing federal procurement procedures followed this DoD directive. The "Perry Directive," as it came to be known, opened the door for change in the military, avionics, space, and traditional "hi-rel" industries but had little impact on the commercial world, as many commercial companies had already stopped using military specifications.[10] The French and British Ministries of Defense followed the U.S. example in 1995 [7], [8]. French and other European equipment suppliers, customers, and users worked to set new validation or qualification standards. Participants in this effort include professional bodies like GIFAS/SPER[11] and major part manufacturers and equipment suppliers like Aerospatiale Matra, Thomson CSF, SNECMA, Sagem, and Alcatel.

Standards reorganization further reduced the U.S. military standards structure and resulted in the elimination of many military specifications for electronic parts and systems. For example, MIL-HDBK-217, "Reliability Prediction of Electronic Equipment," once the standard for military reliability prediction, became obsolete due to the absence of government plans for its revision. Subsequently, the U.S. Air Force redirected the mission of the Rome Air Development Center (the preparing activity of MIL-HDBK-217) away from reliability. Similarly, the Qualified Manufacturer Lists (QML) program of the U.S. military faced a lack of leading technologies, lack of manufacturer support, issues associated with quality, reliability, and obsolescence, and the cost of maintaining the QML program [17]. Such sweeping changes have affected the operations not only of the U.S. government, but also those of private organizations that have become dependent on these standards in their purchasing and management practices.

The rethinking of the role and scope of standards in selecting, testing, and using microelectronics was not limited to defense electronics. Automotive, telecommunications, aerospace, computer, and consumer electronics industries have been and are still working to modify their acquisition standards to make them cost-effective and applicable to the latest technology. For example, the Nippon Telecom, British Telecom, and SAE reliability prediction methods are no longer in effect due to difficulties in modifying their databases to reflect modern technological changes.

Throughout the 1990s, commercial industries developed their own standards, such as ISO 9000, EIA 599, R-100-1996 from the American Institute for Aeronautics and

---

[9] Performance-based specifications are those that set forth the characteristics that a product must have without indicating how the supplier should design or manufacture these characteristics into the product.

[10] NASA and some other government agencies, not associated with the DoD, have been exceptions and have continued with the mil-spec paradigm.

[11] GIFAS stands for Groupement des Industries Françaises des secteurs Aeronautique et Spatial (Group of French Industries in Aeronautic and Space Sectors); SPER stands for Syndicat des Industries de Matériels Professionnel d' Electronique et de Radiocommunication (Syndicate of Industries in Professional Materials for Electronics and Radiocommunication).

Astronautics,[11] Q100 from the Chrysler-Delco-Ford Automotive Electronics Council (CDF-AEC)[12] Q-100, STACK 0001,[13] and IEC/PAS 62 239. However, a growing consensus indicates that these standards, while offering an improvement over the military standards through their flexibility and relevance to new technologies, often do not go far enough to control processes and ensure product quality and reliability. Adoption and certification by an external auditing agency does not ensure that a company produces high-quality parts; it only ensures that a quality system is practiced and audited.

Electronic equipment manufacturers and systems integrators, who buy parts for use in products, are responsible for ensuring the reliability of the parts throughout their life cycle in their product application. They are responsible for ensuring that parts selected for their applications can meet the electrical functionality needs under their application conditions. However, each part manufacturer is free to choose its own quality and qualification test types, conditions, sample size, frequency, and internal acceptability limits; there are no standards for these. Manufacturers guarantee the electrical performance parameters and functionality only under certain specific environmental and operational conditions. If the application conditions for a part do not match the manufacturer-specified conditions or the system is subjected to variabilities, it is incumbent upon the equipment manufacturers to ensure that those variations will not affect the part's performance. For the same reason, equipment manufacturers must differentiate between potential distributors. To remain competitive, equipment manufacturers must further ensure that their integrity assessment techniques will add value to their products at an affordable cost. Equipment manufacturers must also assess and ensure the reliability of parts in the intended applications, and parts management programs must effectively address all these concerns.

One example of a specific need for performance assessment standards is in the category of temperature rating of parts. Electronic parts are categorized according to operating temperature ranges specified by the manufacturer. These ranges nominally indicate the temperatures over which the manufacturer guarantees the electrical parameters specified in a data sheet. The most common part temperature ranges are given in Table 1.3. Many part manufacturers have discontinued producing extended temperature range parts, have reduced their offering of those parts, or have stopped introducing new technologies in extended temperature ranges altogether.[14] Equipment manufacturers who require extended temperature range components face a unique challenge in procuring those parts, because there are no standards to select parts for extreme temperature applications [6].

Similar issues also exist for applications whose voltage, radiation, or electromagnetic compatibility requirements are different from the ratings of the parts. For those factors, where there are no standards for use, it is also necessary to develop and implement risk mitigation techniques within the parts selection and management process.

---

[11] American Institute for Aeronautics and Astronautics, *Recommended Practice for Parts Management, R-100-1996*, Reston, VA, 1996.
[12] Automotive Electronics Council, *CDF–AEC–Q100 Stress Test Qualification for Automotive-grade Integrated Circuits*, Revision E, Detroit, January 31, 2001.
[13] STACK International, *Specification 0001 Issue 12.2 at Notice 2: General Requirements for Integrated Circuits*, Herts, UK, November 19, 1999.
[14] For example, in 1994 the TI Boundary Scan Logic Databook had 46 new logic functions for bus transceiver applications. All these were offered in the industrial temperature range only [4].

Table 1.3: Operating temperature ranges for common part categories

Part category	Temperature range
Commercial	0 to 70°C
Industrial and automotive grade 3	−40 to 85°C
Automotive grade 2	−40 to 105°C
Automotive grade 1	−40 to 125°C
Military	−55 to 125°C

1.6   Objectives of a parts selection and management program

Due to changes in technology trends, the evolution of complex supply-chain interactions and new market challenges, shifts in consumer demand, and continuing standards reorganization, a cost-effective and efficient parts selection and management process is needed. The goal is to provide an "eyes-on, hands-off" approach to parts selection and management that enables companies to:

- employ risk assessment and mitigation techniques to address technology insertion
- organize and conduct fact-finding processes to select parts with improved quality, integrity, application-specific reliability, and cost effectiveness
- make an informed companywide decision about parts selection and management based upon company resources, policies, culture, and goals and customer demands
- understand and evaluate the local environment the part sees within a product's life cycle and thereby choose the most appropriate technique to fit the part to its intended environmental requirements
- maximize product support by preparing for and meeting the challenge of parts' obsolescence during product life
- improve supply-chain interactions and communications with regulatory agencies in order to minimize time-to-profit.

A flexible parts selection and management process guides a company through the fact-finding mission needed to make informed decisions. The final decision is a company-specific function, where company resources, policies, philosophies, goals and customer demands govern trade-offs.

1.7   References

[1]   Lammers, D. and R. Wilson, "Intel Drags the PC Industry to 100-MHz Terrain," *EE Times*, <http://pubs.cmpnet.com/eet/news/98/1003news/intel.html>, accessed April 10, 1998.

[2]   McClean, B., B. Matas, T. Yancy, and R. D. Skinner, *The McClean Report*, IC Insights, Scottsdale, AZ, 2001.

[3]   Semiconductor Industry Association, *National Technology Roadmap*, San Jose, CA, 1997 and 1999.

[4]   Bruner, D., "Distribution's Role Redefined by Outsourcing, Value-Added Work," *Electronics News*, <http://www.sumnet.com/enews/DT98/DTStory2.html>, accessed January 5, 1998.

[5]   Karls, J., H. Dickens, and L. Shon-Roy, *Status 1998, a Report on the Integrated Circuit Industry*, ICE, Scottsdale, AZ, 1998.

[6]   Barrett, J., "The Use of Non-Military Discrete Semiconductors and Microcircuits in AEGIS," Diminishing Manufacturing Sources and Material Shortages (DMSMS) Conference, San Antonio, TX, 1997.

[7]   "Directive CONZE," Ministry of Defense, France, July 31, 1995.

[8]   "UK Defence Standards 59-36," Ministry of Defence, UK, Issue 5, UK Defence Procurement Bulletin, April 1995.

[9]   Bower, J. L. and C. M. Christensen, "Disruptive Technologies: Catching the Wave," *Harvard Business Review*, pp. 43–53, January–February 1995.

[10]   Braub, E. and S. MacDonald, *History and Impact of Semiconductor Electronics*, Cambridge University Press, Cambridge, 1977.

[11]   Daniels, R. G., "A Participant's Perspective," *IEEE Micro*, Vol. 16, No. 6, pp. 21–31, 1996.

[12]   Harper, C. and H. C Jones, *Active Electronic Component Handbook*, McGraw-Hill, New York, NY, 1996.

[13]   Jayant, M., ed., Microprocessors Division of the *Electronic News Magazine*, e-mail communication, September 8, 2000.

[14]   Kanter, H. and R. Atta, *Integrating Defense into the Civilian Technology and Industrial Base*, Office of the Assistant Secretary of Defense for Production and Logistics, Washington, DC, February 1993.

[15]   Liptak, R. and J. Gembicki, "Time to Technology and the Contact Manufacturer," *SMT International – Asia Pacific Edition*, pp. 30–34, September–October 1997.

[16]   McClean, B., B. Matas, N. Yancey, and T. Yancey, *The McClean Report*, IC Insights, Scottsdale, AZ, 1999.

[17]   Pecht, M., J. Fink, E. Hakim, and J. Wyler, "An Assessment of the Qualified Manufacturer List (QML)," *IEEE Aerospace and Electronic Systems Magazine*, Vol. 12, No. 7, pp. 39–43, July 1997.

[18]   Perry, W., "Specifications and Standards – A New Way of Doing Business," Internal Memorandum, U.S. Department of Defense, June 29, 1994.

[19]   Semiconductor Industry Association, *Turbocharging the U.S. Economy: A Report*, San Jose, CA, March 19, 1998.

[20]   Yoffie, D., "CHESS and Competing in the Age of Digital Convergence," *Competing in the Age of Digital Convergence*, Harvard Business School Press, Boston, MA, 1997.

Chapter 2

Methodology for Parts Selection and Management

Margaret Jackson, Peter Sandborn, and Michael Pecht

This chapter presents an overview of a risk-informed methodology for electronic parts selection and management which addresses both application-independent processes (part availability, part cost, part manufacturer quality, distributor quality, and part quality and integrity) and application-specific processes (determination of the local environment, part performance, part reliability, assembly, and life cycle obsolescence). Following the discussion of parts selection activities, risk management activities are addressed.

2.1   Responsibilities and composition of the parts management team

A parts selection and management methodology helps a company make risk-informed decisions when purchasing and using electronic parts. The parts selection and management process is usually not carried out by a single individual, but rather by a multidisciplinary team. The parts management team, as a whole, is responsible for:

- allotting financial and physical resources to develop the necessary infrastructure
- establishing the infrastructure to support parts selection and management
- assigning parts selection and management responsibilities to groups within the company
- establishing communication channels within and outside the company
- managing information flow within the team and to departments outside the team
- identifying process and assessment criteria acceptability levels, both application-specific and application-independent, for each step
- applying the parts selection and management methodology to the candidate part
- identifying potential supplier intervention procedures, authorizing such action when required, and ensuring the associated effectiveness
- monitoring periodically and making improvements to the methodology continuously.

The parts selection and management team includes experts from diverse fields[1] including the following:

Marketing conveys the customer's expectations for the product.

Systems engineering identifies the product-level macro requirements and their constraints within the system.

Product engineering defines and ensures the overall functionality and performance requirements of the final product.

Design engineering identifies components based on the electrical functionality of a part and its mechanical design requirements, and considers parts based on new technologies as suitable candidates for technology insertion.

---

[1] The concept of a parts management team has been in existence since the early days of the electronics industry; however, the composition of the team has evolved as the industry has changed [1].

Quality assurance defines acceptability criteria for the purchases made by the company. *Purchasing* identifies purchasing policies that may inhibit the parts selection and management team's directives, and acts as a catalyst for making the necessary changes to the policies.

Reliability engineering provides inputs for assessing application-specific reliability[2] and acceptability of integrity tests and test results.

Component engineering defines part and product life cycle mismatch criteria and part availability and life cycle mismatch solutions.

Manufacturing engineering addresses manufacturing and assembly requirements of the candidate part.

Testing assesses the feasibility, cost, and schedule implications of testing incoming parts, if necessary.

Legal department identifies those legal considerations that must be addressed with respect to the implementation of the parts selection and management team's directives.

2.2   The overall parts selection and management methodology

An effective parts selection and management methodology provides an "eyes-on, hands-off" approach to parts selection and management, which enables a company to:

- employ risk assessment and mitigation techniques to address technology insertion
- organize and conduct fact-finding processes to select parts with improved quality,[3] integrity,[4] application-specific reliability, and cost effectiveness
- make an informed companywide decision about parts selection and management based upon company resources, policies, culture, and goals and customer demands
- understand and evaluate the local environment the part sees within a product's life cycle, and thereby choose the most appropriate technique to fit the part to its intended environmental requirements
- maximize product supportability by preparing for and meeting the challenge of parts' obsolescence during product life
- improve supply-chain interactions and communications with regulatory agencies in order to minimize time-to-profit.

An overview of the parts selection and management methodology is given in Figure 2.1. Unacceptable results from any one step in the process are sufficient to reject the part as a candidate for inclusion in a product. However, acceptable results from any one or all steps in the process are not sufficient to cause the part to be approved. Even if the results from every step in the parts selection methodology are acceptable, the entire set of results must be considered with respect to the product requirements for the final part acceptance. The decision to use a part is a company-specific function that requires a trade-off governed by company resources, policies, quality culture, and goals and customer demands. Once a part is accepted, risk management, which includes supply chain management, manufacturing and assembly feedback, manufacturer warranties management, and field failure and root-cause analysis, is initiated.

---

[2] Application-specific reliability refers to a part's ability to perform as intended, without failure and within specified performance limits, for a specified time, in its life cycle environment.
[3] Quality is defined as a measure of a part's ability to meet the workmanship criteria of the manufacturer.
[4] Integrity is defined as a measure of the appropriateness of the tests conducted by the manufacturer and the part's ability to survive those tests.

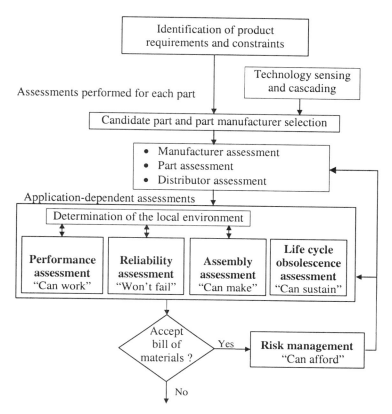

Figure 2.1: Parts selection and management methodology

2.3   Product requirements and constraints identification

A product's requirements and constraints are defined in terms of customer demands and the equipment supplier's core competencies, culture, and goals. The results of capturing product requirements and constraints allow the design engineers to choose parts that conform to product-specific and company objectives. If the product is for direct sale to end users, marketing usually takes the lead in defining the product's requirements and constraints through interaction with the customer's marketplace (to find customer inputs, standards, and requirements), past sales experience, market studies, technology roadmaps and trends, examination of current product sales figures, availability of supply chains, corporate objectives and strategies, and analysis of the competition. If the product is a subsystem that fits within a larger product, the requirements and constraints are determined by the product into which the subsystem fits.

The requirements and constraints definition process begins with the identification of an initial set of requirements and constraints defined by either the marketing activity of a specific customer or the product into which the subsystem fits. The initial requirements are formulated into a requirements document, where they are prioritized. The requirements should satisfy engineering and business objectives, such as maintaining market share and customer loyalty, improving products for differentiation, opening new markets, satisfying the needs of customers, and demonstrating experience with new technologies and methodologies. Thus, the requirements document should be approved by the team. Once the requirements are approved, engineering prepares a preliminary specification indicating the exact set of requirements that are practical to implement. Disconnects between the requirements document and the preliminary specification become the topic of trade-off analyses (usually cost/performance trade-offs), and if, after analyses and negotiation, all the requirements cannot be implemented, the requirements document may be modified. When the requirements document and preliminary specification are agreed upon, a final specification is prepared and the design begins.

2.4   Technology sensing and cascading

The purpose of technology sensing and cascading is to investigate technology, market, and product development trends and apply the knowledge gained to products in a manner consistent with the company's technology insertion initiatives and business plans. Company success often requires establishing corporate technology roadmaps that ensure that the company remains competitive.

Technology and market awareness is gained by inviting technology experts to discuss technology trends, obtaining industry-generated roadmaps, or studying product information provided by manufacturers. For example, a company may decide to include an advanced part within its products that is not yet in general use, anticipating the part's widespread use in the future and attempting to establish an infrastructure to support the part's use prior to its necessity. Knowledge of the future direction of emerging technologies often leads to unique opportunities for product differentiation.

2.5   Candidate part and part manufacturer selection

A candidate part is one that conforms to the functional, electrical, and mechanical requirements of the product, considering the equipment supplier's technology direction and development. In addition to technical conformance, a candidate part must be available at a reasonable cost. The cost of the part is assessed relative to the product's budget during candidate part selection. Availability of an electronic part is a measure of the ease with which the part can be procured. Availability is assessed by determining the amount of inventory at hand, the number of parts required for units in production and those forecasted, the economic order quantity for the parts, the lead time between placing an order for the parts and receiving the parts, production schedules and deadlines, and part discontinuance plans. If the search for parts yields no results, the product design must be modified.

2.6   Manufacturer, part, and distributor assessments

Manufacturer assessment is an evaluation of the part manufacturer's ability to produce parts with consistent quality. Part assessment is the determination of the candidate part's quality and integrity. Distributor assessment is the evaluation of the distributor's ability to provide parts without affecting initial quality and integrity and to provide required value-added services. The equipment manufacturer's parts selection and management team defines the minimum acceptability criteria for these assessments based on the equipment manufacturer's requirements. If the manufacturer, part, and distributor satisfy the minimum acceptable criteria, the candidate part then moves on to the application-dependent assessments.

If a manufacturer, part, and distributor are not found which satisfy the minimum criteria for acceptance, then the candidate part should be rejected (see Figure 2.2). However, at times it may be necessary to use a rejected candidate part, especially if no acceptable alternative part is available. In this case, the equipment suppliers may use intervention techniques in order to mitigate potential risks associated with using an unacceptable candidate part [2]. Here, the intervention action items are identified and their cost and schedule implications are analyzed through the "risk management" process step, as shown previously in Figure 2.1.

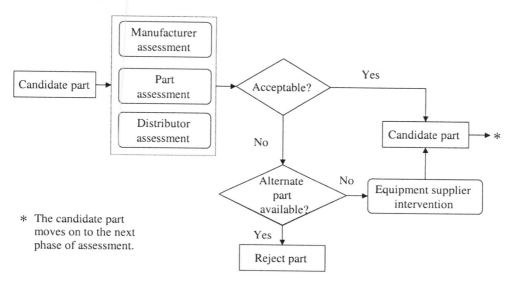

Figure 2.2: Manufacturer, part, and distributor assessment process

2.7   Determination of the local environment

The local environment is the environment in the vicinity of the part, which often varies from the environment the overall product experiences. The local environment of the part can impact part design and development decisions, qualification and specification processes, parts selection and management, quality assurance, safety, warranty, support commitments, and regulatory conformance.

The part life cycle environment includes all part assembly, storage, handling, and scenarios for the use of the part, as well as the expected severity and duration of the environment. Specific load conditions may include steady-state temperature, temperature ranges, cycles and gradients, humidity levels, pressure loads and gradients, vibrational or shock loads and transfer functions, chemical exposure, acoustic levels, sand, dust, electromagnetic radiation levels, and stresses caused by power, current, and voltage. These conditions, either individually or in combination, may influence the performance and reliability of the product.

The application life profile conditions include the application length, the number of applications in the expected life of the part, the part utilization or nonutilization profile (for example, storage, testing, and transportation), the deployment operations, and the maintenance concept or plan. This information is used to group usage platforms (whether the part will be installed in a car, boat, airplane, satellite, or underground), to develop duty cycles (on-off cycles, storage cycles, transportation cycles, modes of operation, and repair cycles), to determine design criteria, to develop screens and test guidelines, and to develop support requirements to sustain reliability and maintainability objectives.

External operational conditions include the anticipated environment(s) and the associated stresses that the part will be required to survive. These conditions may be determined through experimentation or through the use of numerical simulation techniques. Experiments are performed by creating environmental monitoring systems, consisting of sensors placed near the part, which are capable of monitoring the loads that the part experiences. Numerical simulation techniques combine material property, geometry, and product architecture information with the environmental data to determine the life cycle environment based on external stresses.

Internal operational conditions are associated with part-generated stresses, such as power consumption and dissipation, internal radiation, and release or outgassing of potential contaminants. If the part is connected to other parts or subsystems in a system, the stresses associated with the interfaces (i.e., external power consumption, voltage transients, voltage spikes, electronic noise, and heat dissipation) may also be needed.

2.8   Performance assessment

The goal of performance assessment is to evaluate the part's ability to meet the functional and electrical performance requirements of the product. There is often a minimum and a maximum stress limit beyond which the part will not function properly, or at which the increased complexity required would not offer an advantage in cost effectiveness. Part manufacturers' ratings or users' procurement ratings are generally used to determine these limiting values. Equipment manufacturers who integrate such parts into their products need to adapt their design so that the parts do not experience conditions beyond their absolute maximum ratings, even under the worst possible operating conditions (e.g., supply voltage variations, load variations and signal variations) [3].

It is the responsibility of the parts selection and management team to establish that the electrical, mechanical, and functional performance of the part is suitable for the operating conditions of the particular product. The manufacturer-specified environmental ranges are important to consider as a starting point, but the actual local environment defines the conditions in which the parts will truly operate within. Parts that satisfy the electrical requirements and meet the needs of the local environment should be identified, or it may be possible to modify the local operating environment to meet the needs of selected parts. The usage profile of the application may also be modified to ensure that the part is operated only when the environment is within the manufacturer-specified temperature range. For example, with sufficient warm-up time, the part can be brought to a temperature within the manufacturer-specified operating temperature range before the part's function is required. Another method in which part performance over the temperature environment may be ensured is local thermal management. The part may be moved to a different location with the system, cooling equipment may be added to the system, heat sinks may be installed, and other passive or active cooling (or if necessary, heating) arrangements may be made. Thermal management options may add associated costs from added EMI shielding, size and weight constraints, and additional maintenance requirements.

The use of parts outside their manufacturer-specified temperature limits requires a process called "uprating,"[5] which is defined as a process to assess the capability of a part to meet the functionality and performance requirements of the application in which the part is to be used outside the manufacturer's specifications. Uprating for operating temperatures outside specifications is called "thermal uprating." Uprating may be considered as an option to mitigate environmental mismatch, only when no other feasible alternative can be found.

There are three methods of thermal uprating: parameter conformance, parameter recharacterization, and stress balancing. In parameter conformance, the part is tested to assess if its electrical parameters satisfy the manufacturer-specified parameter limits over the target temperature range.[6] In parameter recharacterization, part functionality is assessed, and part parameters are characterized over the required target temperature range. The result may be the development of new parameter limits. In stress balancing, at least one of the part's electrical parameters is kept below its maximum allowable limit to reduce heat generation, thereby allowing operation at a higher ambient temperature than that specified by the part manufacturer.

2.9   Reliability assessment

Reliability assessment provides information about the ability of a part to meet the required performance specifications in its life cycle environment for a specified period of time. Reliability assessment is conducted through the use of integrity test data, virtual qualification results, or accelerated test results (see Figure 2.3).

Integrity monitoring tests are conducted by the part manufacturer to assess part/process changes and the ongoing material or process changes specific to the part. Integrity test data (often available from the part manufacturer) are examined in light of the application life cycle stresses and applicable failure mechanisms and models. If the magnitude and duration of the application life cycle loads are less severe than those of the integrity tests, and if the test sample size and results are acceptable, then the part's reliability is acceptable. However, if the magnitude and duration of the application life cycle loads are not as severe as those of the integrity tests, then integrity test data cannot be used to validate part reliability in the application, and virtual qualification must be considered.

Virtual qualification is a physics-of-failure-based simulation methodology that is used to identify and rank the dominant failure mechanisms associated with the part design and the applied life cycle loads, conduct design trade-offs to build in reliability, and plan warranties and maintenance activities based on actual life cycle loads. Physics-of-failure can also be used to determine acceleration factors to aid in life estimation and accelerated test development.

If virtual qualification proves insufficient to validate part reliability, accelerated testing must be performed. Once the appropriate test procedures, conditions, and sample sizes are determined, accelerated testing can be conducted by the part manufacturer or the equipment manufacturer to provide members with information on how to access third-party test facilities. Accelerated testing results are used to predict the life of a product in its field application by computing an acceleration factor that correlates the accelerated test conditions and the actual field conditions. Whether integrity test data, virtual qualification results, accelerated test results, or a combination thereof are used, each applicable failure mechanism to which the part is susceptible must be addressed.

---

[5] The term "uprating" was coined by Michael Pecht to distinguish it from "upscreening," which is a term used to describe the practice of attempting to create a part equivalent to a higher quality by additional screening of a part (e.g., screening a JANTXV part to JANS requirements).

[6] "Target temperature range" refers to the operating temperature range of the part in its required application.

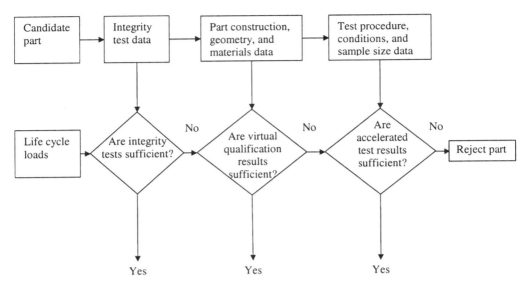

Figure 2.3: Application-specific reliability assessment process

If part reliability is not ensured through the reliability assessment process, the equipment manufacturer must consider an alternate part or product redesign. Redesign options include thermal management techniques, vibration damping, and modifying assembly parameters. If product design changes are made, part reliability must be reassessed.

2.10   Assembly assessment

During part selection, several issues associated with the assembly of the product must be considered. A part may be unacceptable from an assembly viewpoint if (1) it is incompatible with the assembly equipment or process; (2) it is impossible or impractical to wire the part into the system (routing compatibility); or (3) it cannot be acceptably tested or reworked.

2.10.1   Assembly compatibility

Parts must conform to a range of manufacturability constraints associated with their assembly into products. Assembly compatibility addresses whether a system that contains the part can be manufactured (assembled). There are three categories of assembly constraints that must be considered when designing a product:

- *Assembly process compatibility* involves comparing the part's size, shape, and mounting method to the processes that will be used to assemble the boards containing the part.
- *Proximity checking* involves checking the location of the component relative to other parts assembled on the board and the edge of the board and evaluating the orientation (rotation) of the part.

- *Artwork verification* involves checking the board layout for the correct orientation and location of fiducials (alignment marks), alignment holes, and other structures necessary to facilitate assembly.

2.10.2 Routing compatibility

Routing compatibility pertains to the layout and routing of the board. If the selection of a particular part causes significant layout or routing problems within the board, the part may be rejected. Rejection of a part is usually based on its use of routing resources within the board.

Two routing issues must be considered: how much board area is required to wire the part to the rest of the system and how many layers of the board are required to "escape route" the part.

A limiting requirement for parts with an area array format connection to the board (i.e., a flip chip or ball grid array package) is escape routing. If a part's I/Os are in an area array format (as opposed to a peripheral format), the part cannot be wired into the system until all of its I/Os are routed out from under the part.

A component is virtually always "routable" given a sufficient number of board layers. If the rest of the parts on the board are known, routing estimation techniques can be used to determine the effective routing limited footprint of a part under the constraints posed by the board design rules (lines, spaces, via/hole capture pad diameter) and layer count. If a candidate part exceeds the fraction of board wiring resources budgeted to it based on board growth and cost constraints, it may be rejected.

2.10.3 Test and rework acceptability

Test and rework acceptability assesses whether the candidate part can be adequately and economically tested and reworked during assembly. The cost of testing the part (to a specified quality level) prior to assembly and the cost of replacing the part if it needs to be repaired after assembly must be considered.

The cost of testing a part is related to the level of testing performed by the part manufacturer, whether the part is in a package or bare format, the function that the part performs, the number of gates or bits contained by the part, and the test equipment. If the part does not come from the manufacturer fully tested (e.g., a bare die), then test costs may need to be assessed. Test costs include the cost of creating the test patterns (or obtaining them from the manufacturer) and the cost of applying the test to the part. Predicting testing costs is of little value unless the corresponding test coverage (fraction of defects detected by the test) is also predicted.

Another key assembly-related cost is the cost of replacing a part that has been identified as defective during the assembly process. The cost of removing a defective part is a function of how the part is mounted to the board, the size of the part, and its proximity to other parts.

2.11 Life cycle mismatch assessment

To remain competitive, part manufacturers often introduce new parts and discontinue older parts [5]. In general, electronic products go through six phases during their life cycle: design, manufacturing, growth, maturity, decline, and discontinuation. A life cycle mismatch occurs between a product and its constituent parts if parts are not available to support the product throughout its life cycle. When factors such as lead time, risk of part obsolescence, or estimation of the product market are ignored or improperly judged during the design phase, the consequences can be costly. The net outcome can be a financial loss for the company.

A successful life cycle mismatch assessment process is one that prevents the selection of all possible parts that are already obsolete or soon to be discontinued. The process depicted in Figure 2.4 is a strategy whereby the risk associated with a life cycle mismatch between a product and its parts is reduced [5]. Part selection depends on the degree of mismatch and the decision to adopt an obsolescence management strategy (for example, redesign, lifetime buy, buy from aftermarket sources, or part substitution). The strategy is intended to mitigate obsolescence risks associated with using the part at some future point in time in the life cycle of the product. If the equipment manufacturer finds the life cycle mismatch between part and product unacceptable, the part is unsuitable and should be rejected.

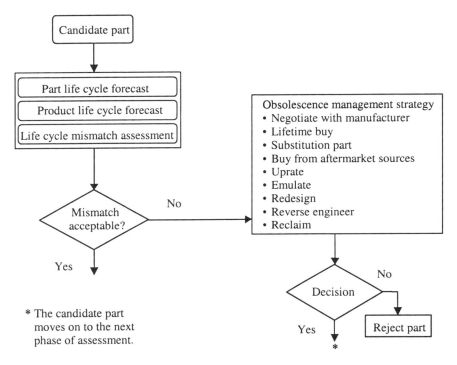

Figure 2.4: Life cycle mismatch assessment

2.12   Risk management

After a part is accepted, resources must be applied to managing the life cycle of the part, including supply chain management, obsolescence assessment, manufacturing and assembly feedback, manufacturer warranties management, and field failure and root-cause analysis. The risk management process is characterized by using the risks identified in the parts selection process to determine the resources needed to support a part throughout its application life cycle. The key metric used to determine whether risks should be managed or not is resources, which include time, data, opportunity, and money. The process of evaluating the risks is seen in Figure 2.5.

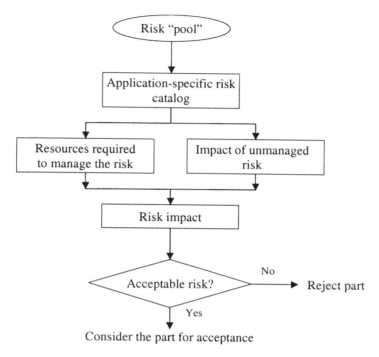

Figure 2.5: Risk assessment process

The risks associated with including a part in the product fall into two categories:

- *managed risks* – risks that the product development team chooses to proactively manage by creating a management plan and performing a prescribed regimen of monitoring the part's field performance, part manufacturer, and manufacturability.
- *unmanaged risks* – risks that the product development team chooses not to proactively manage, often because the probability of a risk materializing is considered too low to merit management activities or the cost of management exceeds the cost of ignoring the risk.

Feedback regarding the part's assembly performance, field performance, and sales history may be essential to ascertain the validity of the predictions made during the part selection process. If feedback suggests changes in selection criteria, the changes must be incorporated into the part selection process. Prospective parts must then be judged based on the altered part selection criteria. Part monitoring is also needed to make changes in parts that are already in use. For example, field data might indicate that a change in operating conditions is required for the part to perform satisfactorily.

2.13   References

[1]   Lee, L. K. and F. A. Paul, "A New Profession, Component Part Engineering," *IRE Transactions Component Parts*, Vol. PGCP-1, pp. 1–2, March 1954.

[2]   Schwach, C. A., A. Mathur, R. Solomon, P. Sandborn and M. Pecht, "Equipment Supplier Intervention Techniques," *Future EMS International*, No. 3, pp. 97–100, 2000.

[3]   International Electrotechnical Commission, *IEC 60134: Rating Systems for Electronic Tubes and Valves and Analogous Semiconductor Devices* (last reviewed in July 1994 by the IEC Technical Committee 39 on Semiconductors), Geneva, Switzerland, 1961.

[4]   Wright, M., D. Humphrey and P. McCluskey, "Uprating Electronic Components for Use Outside Their Temperature Specification Limits," *IEEE Transactions on Components, Packaging, and Manufacturing Technology*, Part A, Vol. 20, No. 2, pp. 252–256, 1997.

[5]   Pecht, M., P. Sandborn, R. Solomon, D. Das and C. Wilkinson, *Life Cycle Forecasting, Mitigation, Assessment and Obsolescence Strategies*, CALCE EPSC Press, College Park, MD, 2002.

[6]   Lee, L. K. and F. A. Paul, "A New Profession, Component Part Engineering," *IRE Transactions Components Parts*, Vol. PGCP-1, pp. 1–2, March 1954.

[7]   Lewis, E. E., *Introduction to Reliability Engineering*, John Wiley & Sons, New York, NY, 1996.

[8]   Sage, A. P. and W. B. Rouse, *Handbook of Systems Engineering and Management*, John Wiley & Sons, New York, NY, 1999.

Chapter 3

Product Requirements, Constraints, and Specifications

Peter Sandborn and Chris Wilkinson

An identified need that can be satisfied through known technological and engineering expertise fundamentally triggers the conception and creation of a product. The need may be expressed in explicit terms by a corporate entity or may just be a nebulous expression of expectations, even wish lists by individuals. The translation of these expectations into practical requirements for the product is the first task during any product development process.

The concept team establishes a generalized set of requirements within the context of product effectiveness attributes, which for electronic products may include parameters such as speed, computations per second, and accuracy, as well as constraints on size, weight, reliability, schedule, and cost. In addition to these primary requirements, some aesthetic considerations and other market-differentiating features may also drive the product design, e.g., the use of lead-free materials. Often, contractual requirements, company policy, and intellectual property constraints also dictate the product type, such as the use of an in-house technology or the exclusion of a certain type of technology.

3.1 Product requirements definition and realization constraints

The product development process involves understanding why, defining what, and then determining how. The why and what constitute product definition, and the how constitutes the product realization. The why and what are directly derived from the needs and constraints of the market. Determining how requires technical expertise and engineering ingenuity. The final success in marketing any product is based on the ability to deliver products that meet current market requirements in terms of size, weight, shape, portability, functionality and performance, ergonomics, aesthetics, and cost [1], [2].

Defining a product is a process of transforming broad goals and vague conceptions into realizable, concrete requirements. The company's core competencies, culture, goals, and customers define the requirements and constraints for new product designs. Figure 3.1 shows that the product definition results from a combination of marketing- and business-driven product requirements, design/manufacturing constraints and other external influences. Marketing often takes the lead in determining the product's requirements and constraints. For subproducts that must be fabricated as part of a specific larger product (e.g., an ignition module that is attached to an automobile engine), the requirements and constraints are usually defined by the customer (the manufacturer of the product or system that the subproduct fits into) and the marketing function is less involved. In general, the product's actual definition is at the confluence of the marketing-driven product requirements, business considerations, design/manufacturing constraints, and external influences, as shown in Figure 3.2.

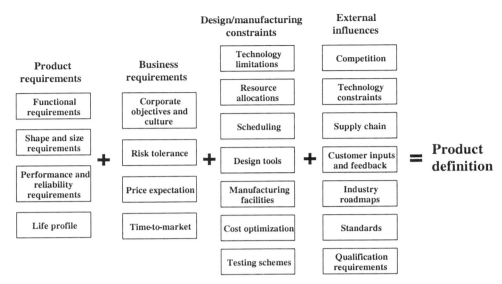

Figure 3.1: Product definition requirements

Figure 3.2 overviews the requirements and constraints process that results in a product realization. The process begins with the presentation of an initial set of requirements and constraints. The initial requirements are formulated into a preliminary requirements document, which must be approved by groups of people ranging from engineers to corporate management to customers (the actual people involved in the approval depends on the company and the product). Once the requirements are approved, engineering prepares a preliminary specification indicating the exact set of requirements that are practical to implement. Disconnects between the product definition and the preliminary specification are identified and rectified during a verification process to evolve a final specification for the product. If, after analyses and negotiation, all the requirements cannot be implemented, the requirements document may also be modified. When the requirements document and preliminary specification are agreed upon, a final specification is prepared and design begins.

Product designers make design decisions that balance all the different requirements, as per the final specification of the product, according to their best estimates. An optimization process follows the initial conceptual design, which involves trade-off analyses (usually cost-performance trade-offs) while simultaneously addressing the accessibility and the maintainability requirements for the product. Performance and cost of the product are closely tied to the knowledge and the risk tolerance of a product development team. The design may be adjusted to remove unnecessary costs or improved to correct performance deficiencies. Over the longer term, design iterations across product generations are frequently needed to improve the quality of the product while reducing the cost.

Testing is normally included in a product's requirements list. Qualification and customer assurance tests are often a part of product requirements. The supportability of the product may also be a requirement for products needing ease of maintenance and accessibility to spares, support equipment, and personnel to test and repair the product. A poor definition of requirements could lead to a design where, for example, the air conditioning compressor has to be removed to replace a spark plug or special tools are required to replace the oil in a car.

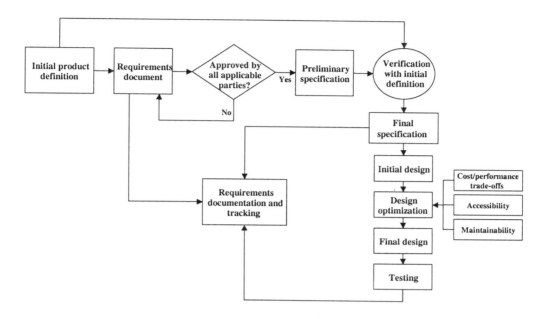

Figure 3.2: Requirements and constraints definition process

3.2   Who defines the requirements and constraints?

The IEEE Reliability Program Standard-1332 provides guidance for supplying products that satisfy the customer. The purpose of the standard is to ensure that the design and operability requirements do not conflict with the business goals. Although the standard is reliability oriented, it offers useful pointers for manufacturers in planning a program that suits the design philosophy, the product concept, and the available resources. The standard identifies three objectives:

- understanding the customer's requirements and product needs to generate a comprehensive design specification
- structuring a set of engineering activities to ensure that the resulting product satisfies defined requirements and needs
- developing activities to assure the customer that the requirements have been satisfied.

From the perspective of requirements and constraints, there are products that are targeted at a large number of customers, and single-customer products that are defined and manufactured for one or a few customers only. In both cases, a marketing group defines the requirements and constraints; however, the sources of information used by that marketing group and the business justifications for the products are different.

3.2.1   Multiple-customer products

Commercial applications are usually successful only when multiple customers exist. The objective of requirements and constraints definition is to maximize profit by appealing to the largest possible customer base. For commercial applications (multiple customer products), product requirements are often defined by technical marketing groups. To define the product requirements, these groups use information including:

- sales personnel from previous or similar product groups
- market focus groups
- market studies
- analysis of competition
- corporate objectives
- product roadmaps and trends

In addition to creating the requirements, marketing must also carefully consider the business reasons that justify the creation of the product. Business reasons that may be considered include:

Fill a perceived market need – A new product may be defined and produced because a compelling business case can be made that a market exists for the product that needs to be tapped.

Open new markets – A new product may be defined and produced (possibly at a financial loss) to facilitate the company's entry into a new market that the company considers to be strategic and believes that a market can be created for the product.

Remain competitive in a key market (maintaining market share) – A new product may be needed to remain competitive in a market that the company considers to be important. For example, in the cellular phone market, companies must continuously define and create more advanced phones (whether profitable or not) if they want to maintain market share. Skipping a generation of phones may cause loss of market share that can never be recovered.

Fill a need of specific strategic customers – Many companies tailor their business decisions to the needs of a few influential customers (or a key market segment). Product requirements may be defined solely to satisfy one customer's needs, either because it is felt that that customer is viewed as a leader in its marketplace or because that customer has a large enough demand for the product to justify the product's existence.

Improve maintainability/extensibility of an existing product – The cost of producing, supporting, or extending an existing product may be reduced to increase profit margins or competitiveness by redesigning or modifying the product.

3.2.2 Single-customer products

Many low-volume applications effectively have only a single customer who predefines the product's minimum requirements. Delivering products with these minimum requirements at the lowest life cycle cost and/or strategically exceeding the minimum requirements gain the customer's business. For single-customer products, the requirements and constraints are determined from concise customer definitions, analysis of competition, and technology roadmaps and trends.

The business reasons that justify the creation of the product include:

Maintain market share and customer confidence – Just as for commercial products, a new product may be needed to remain competitive in a market that the company considers important. A new product may also be important to retain the customer's confidence that the company is a long-term partner and a legitimate competitor in a strategic area.

Demonstrate experience with a new technology or methodology – In the single-customer market, customers are often influenced by prior experience with cutting-edge and niche technologies. Continued demonstration of this ability maintains customer confidence.

Reduce life cycle costs of an existing product – For single-customer products, the manufacturing cost of the product may not be as important as the costs of product sustainment (support, maintenance, upgrade, etc.). Product redesigns or modifications aimed at reducing the life cycle costs are often interesting to the customer.

3.2.3   Custom products

Some low-volume commercial products and "modular" single-customer products do not fit in either of the categories defined above. These are products that are designed with a minimum set of "generic" requirements and customized using each customer's specific requirements. Examples include supercomputers, military radios and navigation equipment (used across multiple Army, Navy, and Air Force platforms), and corporate Intranets. For these types of products, the guidelines defined above are relevant, with the additional requirement that the product be designed in a modular fashion. Reconfigurability with plug-and-play ability and upgradability may be the salient features of these modular designs.

3.3   Requirements and constraints definition risks

Two risks are prevalent in requirements and constraints definition:

Inclusion of irrelevant requirements – Inclusion of irrelevant requirements can involve unnecessary design and testing time and money. Irrelevant or erroneous requirements result from two sources: requirements created by persons who do not understand the constraints and opportunities implicit in the product definition and inclusion of requirements for historical reasons. The latter are requirements that get "cut and pasted" into requirements documents from a previous product. No one knows exactly why the requirement is included, but no one is brave enough to remove it, simply because no one wants to see the obvious (mistake) or to question the norm.

Omission of relevant requirements – Omission of critical requirements may cause the product not to be functional or may significantly reduce the effectiveness and the expected market size for the product.

3.4   The requirements document

The actual content of the requirements document is application specific; however, the requirements and constraints fall into the general categories shown in Table 3.1. In addition to defining values for the requirements listed in Table 3.1, the requirements document may also assign a priority to each requirement. When engineers read the requirements document and decide what subset of the requirements can be accommodated in the product, they need to know which requirements are the most important. Schedule and cost may not be included in the requirements document for some products if the requirements document is released to other (internal or external) organizations for bids. Table 3.2 provides three grades that can be used to prioritize requirements.

Table 3.1: Example of electronic product requirements and constraints

Requirement and/or constraint	Definition
Functionality	Describes what the product does and various features associated with its functionality.
Physical attributes Size Dimensions Weight Shape (aspect ratio)	Describes the physical size, shape, and weight of the final product, e.g., a critical constraint for electronics in mobile phones.
Performance Electrical performance Speed (delay and timing) Noise (switching, cross-talk) Power dissipation Cooling provisions	Describes the characteristics of operation of the product including electrical and thermal management performance attributes, e.g., the speed or clock frequency at which the product runs.
Environment Temperature Humidity Pressure Vibration Acceleration Flexure Electromagnetic interference (EMI)	Defines and constrains the physical environment and the internal and external loads under which the product must operate.
Reliability Useful life Acceptable redundancies Warranty periods	Defines the failure/maintenance free operating period required of the product.
Cost/quality Procurement costs Assembly costs Testing costs Final product quality	Defines the yield of the resulting product and the cost to field a tested product.
Qualification Cost Time Requirements	Defines how and under what conditions the product will be qualified.

Requirement and/or constraint	Definition
Schedule Time-to-market Product volumes	Defines when the first product needs to be delivered to customers and what volume needs to be delivered over the lifetime of the product.
Life cycle Maintainability Upgradability	Defines the ease with which the product can be maintained and upgraded during its fielded life.
End-of-life Disassembly costs Recycling Reuse	Defines what happens to the product after the customer is finished using it; also defines whether any end-of-life requirements exist, depending on legislation where the product is sold.

Table 3.2: Requirements grading

Grade	Definition
Must	The requirement is essential to the viability of the product.
Should	The requirement is not essential to the viability of the product, but it should be implemented either because it adds great value or because it is "low-hanging fruit" that can be easily implemented.
Could	The requirement is not essential to the viability of the product, and its development could be delayed either because the requirement is too costly to implement or because it adds only marginal value to the product.

3.5   Approving requirements and constraints – buy-in

A single group cannot realistically define product requirements that an entire corporate infrastructure must follow. To make the requirements realistic and useful, different groups within a company must study and approve, or "buy in," to product requirements. The level of buy-in varies, depending on the specific product, the customers involved, and the corporate culture. An example is given in Table 3.3.

Table 3.3: Example of requirements buy-in for a multiple-customer product

Role	Example buy-in level
Marketing	Approval authority
Engineering manager	Approval authority
Product manager	Approval authority

Role	Example buy-in level
Development engineers	Consulted prior to approval
Reliability engineers	Consulted prior to approval
Customer	Consulted prior to approval
Application engineers	Consulted prior to approval
Quality assurance	Informed after approval
Documentation	Informed after approval
Corporate management	Informed after approval

3.6   Preliminary specification

Once a set of requirements has been completed, the product engineering function creates a response to the requirements in the form of a preliminary specification.   The preliminary specification contains:

The requirements that must be met – The fact that a requirement appears in a requirements document does not guarantee that it will be achieved in the final product. Technical marketing does not always understand what can be successfully engineered in a specified time window. Requirements grading defines the priorities during specification development.

The methods by which the requirements will be met – This requires an outline of the basic process used to meet the requirements.   The outline may consist of flowcharts, block diagrams, manufacturing processes, and possible parts lists.

The schedule for meeting the requirements – The design, prototype, procurement, and manufacturing schedules for the product are identified. Proposed schedules for contract manufacturers may also be included.

An identification of those who will perform the work – The specific persons who will perform the work may be identified. In addition, groups within the company who perform specific functions necessary to fabricate the product, and all contract manufacturers from outside the company, should also be identified.

An identification of the potential risks in meeting the selected requirements – If any specific design, development, or manufacturing risks are known, they should be stated in this document. Possible second sources and backup plans should also be identified.

The preliminary specification is usually delivered back to the authors of the requirements document, and an iterative refinement of requirements and specification occurs until all parties agree.

3.7   Requirements tracking

Once product requirements are defined and the design process begins, the process of continuously comparing the product's requirements to the actual product design begins. As the product's design becomes increasingly detailed, through selection of parts and choice of system implementation technologies, it becomes increasingly important to track the product's characteristics (size, weight, performance, functionality, reliability, and cost) in relation to the original product requirements. The rationale for making changes should be documented. The completeness with which requirements tracking is performed can significantly reduce future product redesign costs. Planned redesigns or design refreshes through technology monitoring and use of roadmaps ensure that a company is able to market new products or redesigned versions of old products in a timely, effective manner to retain their customer base and ensure continued profits.

3.7.1   Technology roadmapping and monitoring

Technology roadmapping is the act of discerning those technologies that are already in use as well as those emerging technologies that could be beneficial to customers and to the manufacturer. Technology roadmapping can also be performed on the lagging side of technology (i.e., to identify which technologies are becoming obsolete and when). The identified set of technologies already in use, technology trends and forecasts, and "disruptive" or discontinuous technologies (technology obsolescence) all serve as inputs to this process.

Technology monitoring is an ongoing effort targeted at providing data on critical developments to support risk management decisions. Having identified the technology base and potential disruptive technologies for monitoring, one can perform a self-assessment of the company's competency with a particular technology [3]. This assessment provides some of the information needed for the analysis activity of risk management by indicating strengths or weaknesses. Following this self-assessment, initial candidates for technology evaluation and selection can be determined. The level of a particular technology's maturation is a key factor in selecting initial technology candidates for further technology evaluation and selection. Monitoring is conducted in many different ways, including:

- active participation in standards bodies and committees
- attendance at conferences
- literature research
- communication with vendors
- tracking developments in universities, research labs, start-up companies, and mainstream vendors.

3.8   Summary

Product requirements are usually defined by technical marketing groups and then passed on to other portions of the company's infrastructure for review and approval. The product engineering function creates a response to the product requirements in the form of a preliminary specification that defines what requirements will be implemented, who will perform the work, how the work will be performed, and a schedule. Requirements are tracked to ensure that the product remains in compliance as it is developed and during its marketing.

3.9   References

[1]   Jackson, M., A. Mathur, M. Pecht and R. Kendall, "Part Manufacturer Assessment Process" *Quality and Reliability Engineering International*, Vol. 15, pp. 457–468, 1999.

[2]   Jackson, M., P. Sandborn, M. Pecht, C. Hemens-Davis and P. Audette, "A Risk-Informed Methodology for Parts Selection and Management," *Quality and Reliability Engineering International*, Vol. 15, pp. 261–271, September 1999.

[3]   Mosier, S. P., S. A. Guenterberg and R. R. Raphael,   "The Relationship of Technology Change Management to Risk Management," *Proceedings of the 2000 IEEE Engineering Management Society*, Albuquerque, NM, August 13–15, 2000.

Chapter 4

Using the Part Datasheet

Diganta Das and Michael Pecht

The part datasheet provides information to select a part and determine how to design a part into a product [1]. This chapter is a primer on part datasheets.

4.1 The contents of a datasheet

A large volume of part information is developed by the part manufacturer from conceptual design through production, but not all of that information is published. The data-sheet is a "snapshot" of the part information that a manufacturer chooses to divulge.[1] From the part manufacturer's point of view, the datasheet serves as marketing literature, a technical fact sheet, and a business document, which may include disclaimers and limitations on the use of the part.

The International Electrotechnical Commission (IEC) publication 747-1 [2] lists the information to be included in a part datasheet but mentions that it is not compulsory to include data for all the items on the list. The standard format for the presentation of published data as per IEC publication 747-1 includes:

- part type and category
- information on outlines, terminal identification and connections, case material, and lead finish
- electrical, thermal, and mechanical ratings
- electrical and thermal characteristics
- mechanical data
- environmental and/or reliability data
- graphical representation of characteristics

The information in part datasheets may be complemented by associated documents, such as application notes and design guidelines. Table 4.1 lists the common information available in part datasheets and associated documents. The U.S. military has templates called "standard microcircuit drawings" (SMDs), which list the contents of a datasheet for acceptance as a "military" part. These drawings or part descriptions were within the part specification regime[2] created by the U.S. military procurement agencies since the 1960s. In the late 1950s, the U.S. Department of Defense (DoD) began creating standards and specifications for all aspects of electronic parts design, manufacture, test, acceptance, and use. The purpose of these documents was to narrow the uncertainty or judgment area with

---

[1] Not all datasheets are public. Some large-volume buyers may purchase a part built for a specific application by a part manufacturer. The datasheet of this part may be an internal proprietary document.

[2] In November 1969, the military issued the first "General Specification for Microcircuits," MIL-M-38510 [3]. The manufacturing lines for these devices were certified to MIL-STD-976 [4], "Certification Requirements for Microcircuits." In November 1974, a "List of Standard Microcircuits," MIL-STD-1562, was published [5], followed by MIL-STD-965, "Parts Control Program," in April 1977 [6]. Electronic parts selection and management was governed by MIL-STD-454, "Standard General Requirement for Electronic Equipment" [7].

respect to part quality, environment, interoperability, and documentation [8]. Without standardization, the government believed that each military contractor might develop its own methods, making part and product evaluation and comparison against competitive designs difficult. Not only were these standards used by the U.S. military industry (as required), but they were also voluntarily adopted by much of the commercial industry and received scattered acceptance overseas.

Table 4.1: Commonly available information in part datasheets and associated documents

Information in part datasheets	Information in associated documents
Part status	Definitions of terminology used
Part functionality	Thermal characteristics
Ratings	Programming guides
Electrical specifications	Design tips
Packaging information	Handling and assembly guidelines

The SMDs were also developed by the companies interested in participating in the Qualified Parts List (QPL) and later Qualified Manufacturers List (QML) programs. These programs were administered by the Defense Electronics Supply Center (DESC).[3] With the simultaneous decline of the military market share of the semiconductor market and withdrawal of major semiconductor companies (e.g., Intel, AMD, Motorola) from these programs, the parts control regime of the U.S. military has been practically dismantled.

4.2   The status of the part and datasheet

The datasheet provides information on the status[4] of the part – that is, the stage of product development (e.g., design, qualification, production). The date printed on a datasheet indicates that the datasheet parameters are valid on that date. It is good practice to contact the sales and support office of the part manufacturer to confirm that the datasheet is up-to-date before initializing a design.[5]

Most manufacturers publish their datasheets on their websites. Availability of a datasheet on a website does not mean that the corresponding part is available. For example, Fairchild continued to list datasheets for National Semiconductor military grade logic parts, even though Fairchild was not supplying the parts (the 54HC244 part datasheet was "available" on the Fairchild website until February 1999, but Fairchild had not been supplying the part since 1997).

The Electronic Industry Association/Joint Electron Device Engineering Council (EIA/JEDEC) publication 103 [9] provides a suggested list of product status, classifications,

---

[3] This agency has converted to a Defense Supply Center, Columbus (DSCC). It still deals with the electronic parts issues for the U.S. military but has less of a regulatory role.

[4] Some manufacturers provide information on the life cycle status of the part in the datasheet or associated documents to aid in obsolescence analysis. This life cycle status should not be confused with the datasheet status.

[5] Some part distributors and "information vendors" provide parts data that, although convenient, can be outdated, remarkably incomplete, or just plain wrong [10].

and disclaimers associated with the product stages. There are usually datasheets associated with three major stages of product development (i.e., preproduction, preliminary, final).[6]

4.2.1   Preproduction datasheet

During the design phase, a datasheet may be released to make design engineers aware of upcoming parts. Some terms used to describe this phase are "product preview" (Intel, Motorola), "advance" (Xilinx, Micron), and "objective specification" (Philips). These datasheets are often used to elicit feedback from prospective customers. Preproduction datasheet information is estimated from simulation and/or extrapolation from other devices or families [11], or is the design goal or target specification. Part manufacturers do not recommend these datasheet values for use in the final product design. In the worst case, the products described in these datasheets may never reach production or even advance beyond a market survey [9].

4.2.2   Preliminary datasheet

The preliminary datasheet is used for initial production lots. EIA/JEDEC suggests that the product stage be divided into advance information and preliminary information [9]. Intel follows the EIA suggestions [12]. Most companies, however, have only one stage covering both advance information and preliminary information. At Motorola, the preliminary datasheet indicates whether the device is in sampling, preproduction, or initial production stages [13]. Philips releases a preliminary datasheet when the product is in the qualification stage. Some manufacturers, such as Xilinx, state that changes to a preliminary datasheet are not expected, but most manufacturers warn that changes are likely.

4.2.3   Final datasheet

Most manufacturers indicate that a part is in its production stage by removing the status markings from the part datasheet. Philips marks the final part datasheet as "product specification." The "final" status does not indicate that the specifications will not be changed. Some companies, such as Motorola, state that the local sales office will be informed before any changes are made. The part manufacturers issue process change notices when there are changes to the form, fit, or function of a part. To make sure that the information being used from a datasheet is current and up-to-date, the part selection and management team needs to follow the process change notices [14].

4.3   The part number

The part number generally provides information on the technology type, functionality, package type, and temperature range for use. Examples of part numbers from several manufacturers designation given in Figure 4.1 to Figure 4.4 show that basic functionality (e.g., gates, flip-flops, counters), technology (e.g., TTL, BiCMOS, HCMOS), rating and packaging information (e.g., DIP, TO, PQFP) can be obtained from the part numbers. Part numbers also can be used to identify the status of the part. For example, Motorola [15] uses the prefix MC on part numbers only when the parts are fully qualified; before the final qualification, the prefix XC or PC is used to identify the part as not fully qualified.

---

[6] The electrical specifications in the part datasheets can change with the part status levels. For example, AMD sets objective specifications for electrical parameters in the design and development stage. In the validation stage, samples are tested over temperature and voltage ranges. In the manufacturability assessment stage, an early release of the product occurs before the final release. In the early release stage, electrical datasheet parameters are again tested over temperature and voltage ranges. During the production life of a part, sample parts are tested periodically to check if the electrical parameters specified in datasheet continue to be valid [16].

Multiple part numbers are sometimes reported on a datasheet when the same functionality is offered in various options (e.g., different package types and operating temperature range ratings). For example, digital logic parts with the same functionality in different packages or SRAM parts with the same memory density and organization with different access speeds have different part numbers but are usually reported on the same datasheet.

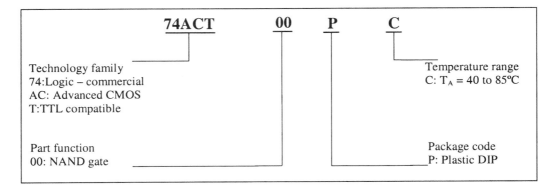

Figure 4.1: Logic part number format from Fairchild Semiconductor

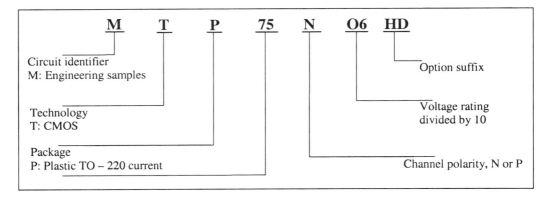

Figure 4.2: MOSFET part number format from ON Semiconductor

4.4   Ratings of an electronic part

Part datasheets provide two types of ratings: absolute maximum ratings and recommended operating conditions. The following observations can be made regarding these two ratings:

- Absolute maximum ratings are generally provided as a limit for the "reliable" use of parts.
- Recommended operating conditions are the conditions within which electrical functionality and specifications of a part are guaranteed.

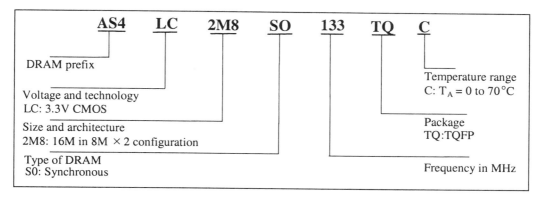

Figure 4.3: DRAM part number format from Alliance Semiconductor

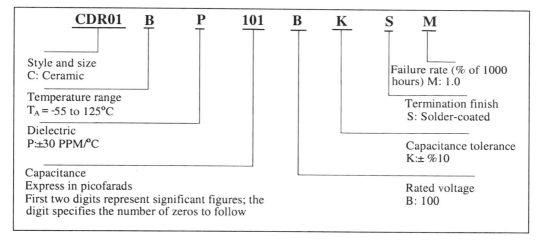

Figure 4.4: Capacitor part number format from Kemet

Intel [17] considers the difference between absolute and maximum ratings to be guidance to users as to how much variation from the recommended ratings can be tolerated without damage to the part. Motorola [18] states that, when operated between the recommended and absolute maximums, the part is not guaranteed to meet any electrical specifications on the datasheet, but that physical failure or adverse effects on reliability are not expected. Motorola notes that the recommended operating conditions are "guardbanded," that is, margins of safety are added to the absolute maximum ratings [18]. However, Motorola [19] also states that noise can push the environment beyond "destruct" limits when parts are operated near the absolute maximum ratings, necessitating the use of guardbands.

4.4.1   Absolute maximum ratings

The absolute maximum rating (AMR) section in the datasheet includes limits on operational and environmental parameters, including power, power derating, supply and input voltages, operating temperature, junction temperature, and storage temperature. The IEC [20] defines absolute maximum ratings as "limiting values of operating and environmental conditions applicable to any electronic device of a specific type as defined by its published data, which should not be exceeded under the worst possible conditions. These values are chosen by the device manufacturer to provide acceptable serviceability of the device, taking no responsibility for equipment variations, and the effects of changes in operating conditions due to variations in the characteristics of the device under consideration and all other electronic devices in the equipment. The equipment manufacturer should design so that, initially and throughout life, no absolute-maximum value for the intended service is exceeded with any device under the worst probable operating conditions with respect to supply voltage variation, equipment component variation, equipment control adjustment, load variations, signal variation, environmental conditions, and variation in characteristics of the device under consideration and of all other electronic devices in the equipment." In other words, the part manufacturers select the AMR values, and the companies that integrate electronic parts into products and systems are responsible for ensuring that the AMR conditions are not exceeded.

4.4.1.1   Effect of exposure to AMR conditions

Part manufacturers generally state that below the AMR but above the recommended conditions, the performance of the part is not guaranteed but the useful life of the part will not be affected – that is, there are no reliability concerns. Some manufacturers (e.g., Motorola) suggest that operating parameters within the recommended operating range are not guaranteed at or near the AMR, and there may be reliability concerns over the long term [21].[7] ZiLOG [22] states that "AMRs (Absolute Maximum Ratings) are given to allow our customers to understand at what point physical damage can occur to the device under stress. . . . Once the operating conditions exceed the AMR, damage may ensue." Philips clarifies their meaning of absolute maximum ratings: "The 'RATINGS' table (Limiting values in accordance with the Absolute Maximum System – IEC 134) lists the maximum limits to which the device can be subjected without damage. This doesn't imply that the device will function at these extreme conditions, only that, when these conditions are removed and the device operated within the recommended operating conditions, it will still be functional and its useful life won't have been shortened" [23].

4.4.1.2   Warning labels

Almost all datasheets contain some form of warning statement or disclaimer to discourage or prohibit use of the parts at or near AMRs. The most common wordings used in the warning labels regarding AMR are:

- Functional operation is not implied in these ranges.
- Stresses above these ratings can cause permanent damage to the parts.
- Exposure to these conditions for extended periods may affect reliability and reduce useful life.

4.4.2   Recommended operating conditions

Recommended operating conditions provided by part manufacturers include voltage, temperature ranges, input rise, and fall time. Part manufacturers guarantee the electrical parameters (typical, minimum, and maximum) of the parts only when they are used within the recommended operating conditions and standard circuit conditions. Philips notes that

---

[7] Some EIA/JEDEC documents refer to absolute maximum ratings as absolute maximum "continuous" ratings. In those documents, transient conditions under which these ratings may be exceeded are defined. For example, the JEDEC standard for description of low-voltage, TTL-compatible CMOS logic devices [24] states that "Under transient conditions these rating [AMR] may be exceeded as defined in this specification."

"The recommended operating conditions table [in the Philips datasheet] lists the operating ambient temperature and the conditions under which the limits in the 'DC characteristics' and 'AC characteristics' will be met" [23]. Philips also states that "The table (recommended operating conditions) should not be seen as a set of limits guaranteed by the manufacturer, but the conditions used to test the devices and guarantee that they will then meet the limits in the DC and AC characteristics table." ZiLOG [22] states, "[Recommended] operating conditions[8] are given so customers know the maximum and minimum conditions where normal performance is still available from the device. Once the normal operating conditions are exceeded, the performance of the device may suffer."

Some recommended operating conditions may not be explicitly marked as such but are listed within the electrical specifications. For example, input logic voltage levels are often listed only in the DC characteristics of the electrical specification section, but the input voltage levels are actually inputs to the part, within which the part will meet its output specifications [25]. The AC characteristics of microprocessors consist of output delays, input setup requirements, and input hold requirements [26]. The input setup and input hold requirements are practically recommended operating conditions, within which other parameters will meet specifications; however, these are included in the switching characteristics table of electrical specifications [1].

4.4.3   Factors used to determine ratings

Fairchild [27], [28] lists the following factors as crucial in determining the AMR and recommended operating conditions:

- *semiconductor fabricationrication process* – manufacturing processes and conditions that affect the temperature sensitivity of parameters
- *design rule limitations* – physical dimensions of device elements
- margins determined through electrical testing and procedures and methods used to set specifications from device test characterization data
- *competitors' second source advantages* – limits set to maintain parity with competitors' products.

Intel notes that simulation and testing is used for ratings determination. Intel [17] states, "voltage and current ratings are determined through statistical calculations with fabricationrication process related factors such as metal line width. Additionally, computer simulations are performed, and environmental stress and testing on controlled lots characterize the device under evaluation."

There are some JEDEC standards (e.g., JEDEC-7A for HC/HCT CMOS [29], EIA/JESD36 for Low Voltage TTL-Compatible, 5 V-Tolerant CMOS Logic Devices [30], EIA/JEDEC Standard EIA/JESD64 for 2.5 V CMOS Logic Devices with 3.6 V CMOS Tolerant Inputs and Outputs [31]) that list AMRs and recommended operating conditions for specific technologies.[9] However, adherence to these standards is not mandatory and one should not assume that a part of a particular technology will have ratings given in the standards. For example, the JEDEC-7A standard sets the recommended operating temperature range for the 74 series parts at −40 to 85°C; however, the Fairchild 74HC244 parts' rating is −40 to 125°C [29].

---

[8] ZiLOG term for "recommended operating conditions" is "normal operating conditions."
[9] JEDEC used to develop and publish these standards for mature part families, which were produced by multiple manufacturers. It is observed that with the shortened product life cycle for electronic parts and with few parts being produced by multiple manufacturers, JEDEC and other standard bodies are no longer developing these types of standards.

4.5   Reliability information

Reliability information is generally not provided in the part datasheets. Many part manufacturers use the term "useful life" frequently in relation to AMRs, but do not specify the duration of the useful life. The commonly understood reliability metrics (such as failure rate, time to failure) are not included in the part datasheets. Any reliability metric in a datasheet is an exception rather than the norm.

Some datasheets, such as those for certain memory parts, provide a reliability guarantee where parameters like data retention, endurance, and store cycles are listed. Table 4.2 shows an example of a Xicor part with some reliability specifications. The Hynix HY29F080 Flash Memory datasheet provides different data retention times in years for different operating temperatures.

Table 4.2: Reliability specifications of a nonvolatile VRAM X25401 from Xicor within recommended operating conditions [49]

Parameter	Minimum guarantee	Units
Endurance	100,000	Data changes per bit
Store cycles	1,000,000	Store cycles
Data retention	100	Years

Some types of additional reliability information in the datasheets may not be complete or may even provide misleading information:

- The 1301-ROP 101 1087 Uen Ericsson part is rated for operation over a free air temperature range of −40 to 85°C and a junction temperature range of −40 to 125°C. The storage temperature range is −55 to 125°C. The datasheet states that the absolute maximum junction temperature of 105°C is for a 10-year life goal. *The datasheet does not specify any failure mechanism associated with this life goal.*
- Motorola 51R99459A01 has the supply voltage rated for a maximum range of −0.5 to 2.1 volts, and the new designs are recommended to be operated below 1.95 volts. The voltage rating and recommendations given in the datasheet relate to the projected constant failure rates (in FITs) by the manufacturer. At 2.2 volts operation, the part reliability rating is given as 10 FITs maximum over 10 years at a 50% duty cycle. At 1.95 volts operation, the part reliability rating is given as 10 FITs maximum over 10 years at a 100% duty cycle. The datasheet information gives the false impression that the part failure rate is constant.

4.6   Thermal characteristics

The thermal characteristics of a part determine its ability to dissipate the heat generated within it without increasing its temperature to a level at which the performance and reliability may be degraded. Besides the temperature ratings, a part's thermal characteristics are governed by power dissipation and thermal resistance.

4.6.1   Power dissipation

Some manufacturers provide a value for maximum power dissipation in the AMR section of the datasheet. This value is usually based on the heat dissipation capacity of the package.[10]   In other words, the power dissipation limit is the maximum power that the manufacturer estimates the package can dissipate without causing damage to the part or raising the junction temperature above the manufacturer's internal specification. There are usually safety margins between the actual power dissipation capacity of a part and the rating on the data sheet.   The power dissipation level is sometimes linked to the temperature of operation through a "derating" factor.

4.6.2   Junction temperature

SEMATECH defines junction temperature as "the temperature of the semiconductor junction in which the major part of the heat is generated. The measured junction temperature is indicative only of the temperature in the immediate vicinity of the element used to measure the temperature" [46].   In commonly accepted usage, junction temperature is the average temperature of the die within the package [38].

4.6.3   Case temperature

SEMATECH defines case temperature as "the temperature at a specified, accessible reference point on the package in which a semiconductor die is mounted" [46]. Intel measures case temperature at the center of the top surface of the package [12]. The point to note is that a part manufacturer may select a point on the part surface that is not the highest temperature point on the part.

4.6.4   Ambient temperature

SEMATECH defines ambient temperature as "the temperature of the specified, surrounding medium (such as air, nitrogen, or a liquid) that comes into contact with a semiconductor device being tested for thermal resistance" [46]. For a complete description, the location of the measurement of the ambient temperature at test and system setup must be specified.[11]

4.6.5   Storage temperature

SEMATECH defines the storage temperature as "the temperature limits to which the device may be subjected in an unpowered condition. No permanent impairment will occur (if used within the storage temperature range), but minor adjustments may be needed to restore performance to normal" [46]. The storage temperature limit, when provided, is generally listed in the AMRs. The common ranges of storage temperature ratings are −65°C or −55°C to 150°C.

4.6.6   Lead temperature

The lead temperature rating is the maximum allowable temperature at a specified position on the leads of a part during the soldering process. This rating is usually provided only for surface-mounted parts.[12] Lead temperature ratings are typically in the 260°C to

---

[10] Some manufacturers, such as Philips and Motorola, provide supplementary information on how to estimate power dissipation for some of their parts and part families.

[11] For example, Intel specifies the measurement point of ambient temperature 12 inches upstream of the package in laboratory conditions and upstream close to the system environment [12].

[12] Surface-mounted parts usually go through reflow soldering, where the whole package is exposed to radiation or convective heat. Lead temperature limit and exposure time together provides a safeguard that the package and the

300°C range, with a maximum exposure time of 10 seconds. The temperature limit and exposure time depend on the thermal inertia of the package under consideration. However, when the same part is available in different package types, there usually is no difference in lead temperature ratings between package types. For example, the Unitrode controller, UC1637, is available in SOIC and PLCC packages, and both package types have the same 300°C/10s lead temperature ratings. Given the type of damage that can occur during the reflow soldering process [41], the allowable preheating level and the temperature gradients are important to report.

Some companies include details about the soldering conditions in separate documents. For example, Intel provides preheat, preflow, and reflow times, temperatures, and ramp rates in its packaging data book [12].

4.6.7   Thermal resistance

For a semiconductor device, Method 1012.1 of MIL-STD-883E defines thermal resistance as "a measure of the ability of its carrier or package and mounting technique to provide for heat removal from the semiconductor junction" [48]. The thermal resistance is given by the temperature difference between two specified locations per unit of power dissipation and is measured in °C/W. Lower thermal resistance indicates that the package is capable of higher heat dissipation. The most commonly used thermal resistance values for electronic parts are junction-to-case thermal resistance and junction-to-ambient thermal resistance. Definitions given by SEMATECH [46] are:

- *Junction-to-ambient thermal resistance* (θ_{JA}) is the thermal resistance from the semiconductor junction to a specified point in the surrounding ambient atmosphere.

- *Junction-to-case thermal resistance* (θ_{JC}) is the thermal resistance from the semiconductor junction to a specified point on the exterior of the package.

Part manufacturers determine thermal resistance values for devices and package families primarily through experimentation and thermal simulation. Some data may be extrapolated to determine the thermal resistances of other members of the package family. Part manufacturers may follow the EIA/JESD Standard 51 and its supplements in determining thermal resistance values [36]. Many companies provide descriptions of their thermal resistance determination methodology; for example, AMD describes how it follows the EIA/JEDEC standard [27], and Intel describes its in-house method [12].

Some datasheets include the thermal resistance information in the AMR section. In such cases, the thermal resistance values are to be taken as a constraint placed on the user. The user must ensure that the part is mounted and operated so that the thermal resistance does not exceed the rated maximum thermal resistance. Some datasheets specify mounting conditions – for example, mounting torque[13] – necessary to achieve the right value of thermal resistance [44]. Some manufacturers also provide thermal resistance values from the junction to a specified point on the package besides the case (such as lead or heat sink mounting locations).

Thermal resistance data are generally valid for a particular test or simulation condition [34], because they depend on factors such as the thermal conductivity of the printed circuit board, the proximity and power dissipation of neighboring devices, airflow speed and airflow pattern through the system [32], coolant physical properties, and thermal radiation

---

circuitry are not exposed to damaging high temperature. For insertion mount components, which usually are wave soldered, the component bodies are not exposed to direct heat and this rating is not essential.

[13] The mounting torque of a screw-mounted device determines the quality of thermal contact between the part and the board, and hence impacts the heat flow from the part to the board.

properties of the surrounding surfaces. Relating the thermal resistance data to the actual operating conditions is the responsibility of the product manufacturer.

Some companies make thermal resistance data available by package type in separate handbooks, on websites, or via technical support. There are cases of discrepancy between the thermal resistance values provided in datasheets (or websites) and the values provided by technical services. For example, thermal resistance values provided at the Fairchild Semiconductor website varied considerably from the values provided by Fairchild Semiconductor technical services through electronic mail.

4.6.8   Norms and deviations in temperature ratings

Semiconductor part temperature ratings are often categorized by the free air ambient temperature into commercial (0 to 70°C), industrial (–40 to 85°C), and military (–55 to 125°C). In addition, there are automotive temperature grades of –40 to 105°C and –40 125°C. Unfortunately, there are many exceptions to these categories. For example, in Figure 4.1, the Fairchild part is called commercial but the rating is for –40 to 85°C, which is commonly understood as an industrial temperature rating. Some manufacturers that have exited the military market continue to produce parts that are rated for –55 to 125°C but call those parts "extended" temperature range parts. Motorola's analog and integrated circuits group call their –25 to 85°C parts "industrial," –40 to 85°C parts "automotive," –40 to 105°C parts "extended," and –40 to 125°C parts "extended automotive" [15].

The IEC [2] and the Electronics Industry Association (EIA) [35] provide recommended lists of temperatures for use in ratings. However, part manufacturers are not obligated to supply parts that correspond to the common temperature ranges, and many manufacturers deviate from them. Table 4.3 shows some parts that are offered in temperature ranges that are not commonly used. Some of the values in Table 4.3 do not correspond to the recommendations of the IEC [2] or the EIA [35]. Many of these deviations from norms are driven by market conditions or actual part limitations.

Many discrete semiconductors (e.g., Siemens BSP 318S small signal transistor), most power devices (e.g., Siliconix MOSFET Si2305DS), and some special devices (e.g., IBM PowerPC 750, Xilinx FPGA XC5200) are rated by junction temperature. Some companies use the term "virtual junction temperature"[14] to rate some parts. For example, the TI voltage regulator UA7806 is rated for a virtual junction temperature up to 125°C. TI states that the term "virtual" is used because the junction temperature of a part is not directly measured, and the rating is based on testing a standard package, which may not include the same circuitry as the part being rated.

Table 4.4 lists some parts with recommended operating conditions specified by case temperature. Most microprocessors are rated by case temperature. The case temperature of a part is usually higher than the ambient (usually air) temperature because the case is at an intermediate point in the path between the heat-generating "junction" and the heat "ambient." Considering this, some manufacturers specify case temperature ratings that are higher than the corresponding ambient temperature ratings. For example, for commercial and industrial grade parts, AMD specifies case temperature limits of 85°C and 100°C, respectively, as opposed to the traditional ambient temperature ratings of 70°C and 85°C.

---

[14] A virtual junction is defined as "the theoretical point or region in a simplified model of the thermal and electrical behavior of a semiconductor device at or in which all the power dissipation within which the device is assumed to occur." The corresponding temperature is called the "virtual junction temperature" or "internal equivalent temperature" [50].

Table 4.3: Examples of parts specified by ambient temperature

Part number	Part type	Company	Temperature range
Alpha 21164	Microprocessor	Compaq	0 to 50°C
VSP3000	DSP	Burr Brown	0 to 85°C
MSM6895/6896	Communication	Oki	−10 to 70°C
E0C63458	Microcomputer	Epson	−20 to 70°C
SH7750	Microcontroller	Hitachi	−20 to 75°C
M37640E8	Microcontroller	Mitsubishi	−20 to 85°C
M54133FP/GP	Driver	Mitsubishi	−20 to 85°C
73M233	Communication	TDK	−25 to 70°C
IMP808	Analog	IMP	−25 to 85°C
MX7524/MX7624	Analog	Maxim	−25 to 85°C
UC2638	Controller	TI (Unitrode)	−25 to 85°C
LH1694	Logic	Sharp	−30 to 85°C

Table 4.4: Examples of parts specified by case temperature

Part number	Part type	Company	Temperature range
UT63M14x	Transceiver	UTMC	−55 to 125°C
XQ4000X	FPGA	Xilinx	−55 to 125°C
UT1750AR	Microprocessor	UTMC	−55 to 125°C
IDT79R3041	Microcontroller	IDT	0 to 85°C −55 to 125°C
Pentium with voltage reduction	Microprocessor	Intel	PQFP 0 to 85°C TCP 0 to 95°C
Pentium 4[*]	Microprocessor	Intel	5 to 78°C
AM486DE2	Microprocessor	AMD	0 to 85°C
Cyrix M II[†]	Microprocessor	Cyrix	0 to 70°C
ST6x86	Microprocessor	SGS – Thompson	0 to 70°C
CY7C1338	RAM	Cypress	0 to 70°C[‡]

[*] There are several different case temperature ratings for different Pentium IV parts. For Pentium IV with identifiers SL5SZ, SL5UH, and SL685 the maximum case temperature ratings are 74, 76, and 71°C, respectively.

[†] Now obsolete.

[‡] The rating is for "instant on" case temperature, i.e., the case temperature at the moment a part is switched on before part thermal dissipation has any effect on the case temperature [39].

Calculations provided by Intel [38] and AMD [1] show that the ambient temperatures required to maintain the 85°C case temperature limit are lower than 70°C, the maximum commercial grade ambient temperature limit. For the 133-MHz AMD 486 microprocessor (case temperature rating 85°C), the maximum ambient temperature allowable is between 38.9°C (still air) and 62°C (1000 linear feet per minute air speed).

Manufacturers of subassemblies such as VME boards that are purchased by military and avionics manufacturers may also rate their products for operating temperature. The rugged RACE (a type of VME form factor single board computer) specifications given by the manufacturer, Mercury, are shown in Table 4.5. The operating temperature ranges for operation of these types of boards are narrowing; in previous versions, the RACE boards were rated for –55 to 125°C operation.

4.7   Electrical specifications

Some of the DC and AC parameters of parts are specified by the manufacturers to be within a certain range for a given set of operational and environmental conditions. The datasheets provide tables for a set of parameters that the manufacturers consider relevant. De facto industry standards on electrical parameters exist for some common mature parts. For the voltage, current, and power values, the IEC [2] and the EIA [35] provide some recommendations for values to be used in semiconductor part specifications.[15]

Table 4.5: Examples of temperature ratings at subassembly level

	Operating temperature specification (Case temperature)		Maximum inlet air temperature	System examples minimum air flow (CFM/slot)
	Minimum	Maximum		
6U VME form factor				
12 SHARC (40 MHz)	−25°C	85°C	55°C	14
4 PowerPC (200 MHz)	−25°C	95°C	55°C	12
9U VME form factor				
48 SHARC (40 MHz)	−25°C	85°C	55°C	42
16 PowerPC (200 MHz)	−25°C	95°C	55°C	36

4.7.1   Standardization and changes

The various design and manufacturing methods used by different manufacturers can result in different capabilities for similar parts. For some mature parts, examination of equivalent parts from different manufacturers often shows very similar, if not identical, electrical parameter limits and test conditions. One reason for such similarity is availability of standards for families of parts. For example, JEDEC standards exist for HC/HCT CMOS [29], low-voltage, TTL-compatible CMOS [24], low-voltage, TTL-compatible, 5V-tolerant CMOS [30] logic, 54/74ABT and BC TTL-compatible BiCMOS [37], and other parts.

---

[15] The IEC recommendations are based on the ISO R-10 series of numbers, and the EIA recommendations are based on the ANSI R-10 series of numbers.

These standards list the electrical parameter limits for different types of parts that fall under these categories. The purpose of these standards was to provide uniformity and ease of use by electronic product designers.

The part manufacturers can change the electrical specifications of parts for various reasons including die shrinks, mask changes, and process modifications. Die shrinks allow part manufacturers to get more die from the same size wafer and can result in realizing higher revenue from same investment. Die shrinks (scale changes) are common in the semiconductor industry as a path to higher profits for the part manufacturer. Die shrinks also allow manufacturers to provide the smaller and lighter parts being demanded by the portable and consumer electronic industries. Mask changes involve a layout alteration of the die with or without die shrink. Mask changing is an expensive process, and it is often associated with process upgrades or manufacturing location changes. Process modifications occur at fabrication plants on a regular basis to improve productivity and yield. Process upgrades can occur every week or more often in newer technologies.

There are two major reasons for datasheet electrical parameter changes. The first involves new products, where the manufacturer discovers something after going to high-volume production that was not identified in the original product characterization; these problems sometimes discovered in the testing and sometimes identified by customers in their applications. Second when manufacturers move or expand wafer fabrication production to include new or different wafer fabrication sites, there can be slight variations that cause distribution shifts requiring the manufacturers to modify the specification to allow them to maintain a consistent product across all possible sources.

Sometimes the reasoning given by the part manufacturer for changes in electrical specifications is not direct. For example, TI changed the specification of the part TLC5920 for one parameter and stated that it was done "to maintain a stable supply for this product" [47]. When contacted for an explanation TI stated [33] that "The change is being made to align the specification with the current product distribution." This was "an indirect way of saying that we cannot get an acceptable or consistent yield from our process versus the published specification." TI stated that they usually do some market/customer analysis to be sure that their position remains competitive.

4.7.2   Parts with multiple temperature ranges

When a datasheet describes a part with multiple temperature ranges, usually multiple electrical specification ranges are provided, with wider electrical specifications for the "relaxed" wider temperature ranges. There are separate electrical specifications for the parts with different temperature ranges of operation. Table 4.6 shows a typical example of such a part. There are many variations of ratings for multiple temperature grade parts.

For example, a Motorola MC74HC242 transceiver is rated for –55 to 125°C, but it has three different sets of electrical specifications for three intermediate temperature ranges (Table 4.7). For some parts, such as the Xicor NOVRAM X25401, all the temperature grades (0 to 70°C, – 40 to 85°C, and –55 to 125°C) have the same electrical specifications (Table 4.8). When parts are available for various voltages and access speeds, different electrical specifications are given, as applicable. For example, Table 4.9 shows that Hynix HY58163210 16MB SGRAM parts have different electrical specifications for different access speeds.

Table 4.6: Excerpts from electrical specifications for parts with multiple temperature ranges

Parameter	Test conditions	VCC	TA = 25°C			SN54HC244		SN74HC244		Unit
			Min	Typ	Max	Min	Max	Min	Max	
V_{OH}	VI = V_{IH} or IL	2	1.9	1.998		1.9		1.9		V
	IOH = 20 µA	4.5	4.4	4.499		4.4		4.4		
		6	5.9	5.999		5.9		5.9		
II	VI = V_{CC} or 0	6 V		±0.1	±100		±1000		±1000	nA

Table 4.7: Excerpts from electrical specification of Motorola MC74HC242 *

Symbol	Parameter	Condition	V_{CC} (V)	Guaranteed limit			Unit
				−55 to 25°C	<85°C	>125°C	
V_{IH}	Minimum high-level input voltage	V_{out} = 0.1 V or V_{CC} = 0.1V |I$_{out}$ = 20 µA |	2.0	1.50	1.50	1.50	V
			4.5	3.15	3.15	3.15	
			6.0	4.20	4.20	4.20	
V_{IL}	Minimum low-level input voltage	V_{out} = 0.1 V or V_{CC} = 0.1V |I$_{out}$ = 20 µA |	2.0	0.3	0.3	0.3	V
			4.5	0.9	0.9	0.9	
			6.0	1.2	1.2	1.2	

*The part is rated for −55 to 125°C operation, but the datasheet provides different electrical specifications for three temperature ranges.

Table 4.8: Excerpts from electrical specification of XICOR NOVRAM X25401*

Symbol	Parameter	Limits		Units	Test conditions
		Min	Max		
I_{CC1}	V_{CC} supply current (TTL inputs)		10	mA	SCK = 0.4 V/2.4 V levels @ 1 MHz SO = open, all other inputs = V_{IH}
I_{CC2}	V_{CC} supply current (during AUTOSTORE)		2	mA	All inputs = V_{IH}. CS = V_{IL} SO = open. V_{CC} = 4.3 V
I_{SB1}	V_{CC} stand by current (TTL inputs)		1	mA	SO = open, CS = V_{IL} All other inputs = V_{IH}

*The part is available in three temperature grades. The DC operating characteristics limits are the same for all three temperature grades.

**Table 4.9: Excerpts from electrical specifications for parts with multiple options:
Hynix HY58163210 16MB SGRAM**

Symbol	Parametric condition	Speed			Unit
		−7	−8	−10	
I_{DD1S}	Operating current (average power supply current) Bust length = 4 t_{RC} = min., t_{CK} = min at each operation one bank active $0V \leq V_{IN} \leq V_{DD}$ outputs open addresses are changed up to three times during t_{RC}	250	230	190	mA

4.7.3   Electrical data beyond the rated temperature limits

Some datasheets provide electrical parameter data (in the form of characterization curves) beyond the part's recommended operating temperature limits. For example, Philips' low-power quad op-amp A2902 is rated for −40 to 125°C, but the datasheet provides electrical characterization curves for some parameters down to −55°C. The MITEL logarithmic amplifier SL2524 [42] is rated for an industrial temperature grade of −40 to 85°C, but the datasheet has characterization curves for some parameters for −55, 25, and 125°C. Philips also provides specifications on some AC parameters up to 125°C for the HC/HCT family of parts, although these parts are not rated for 125°C operation [45].

4.8   Derating and safe operating area

Derating and a safe operating area may be provided for power semiconductors and passive devices. The "advertised" current rating shown on the datasheets of most power MOSFETs is usually above the practically usable level of continuous drain current. This is because the case temperature adopted by the industry, to which this rating applies, is 25°C; most power MOSFETs operate above this case temperature. Most power MOSFET devices have current ratings at 25°C and at higher temperatures; a derating curve of continuous drain current is provided [40].

The continuous current rating of power MOSFET is of little direct use to the designer, other than as a benchmark, because power transistors are normally operated in switch mode, with duty cycles considerably less than 100%, and what is really of interest is the current-carrying capability of the device under the actual "switched" operating conditions. The pulse current rating is therefore higher than the continuous current rating because the package can dissipate heat between pulses. This rating is dependent on the pulse width. For different pulse width, a safe operating area of current and voltage can be constructed [40].

4.9   Summary

The information provided on the part datasheet is useful only when it is understood clearly and used in the context of the product requirements in which the part is to be used. Despite attempts, no universal standards have been set for the contents of a datasheet across different manufacturers. As a result, there are significant variations in the content and the format of part datasheets, both among part manufacturers and from part to part from the same manufacturer. In many cases, supplementary information or clarifications are required in order to assess a part's applicability in a product. In view of the frequent upgrades in part designs and manufacturing processes, the potential for erroneous product design because of improper or out-of-date information is increasing.

Product manufacturers must pay attention to the following areas to avoid misinterpretation of datasheet information:

- all definitions used by the part manufacturer
- validity of the information on the datasheet for the specific parts and part lots being considered
- stipulations associated with the absolute maximum ratings and recommended operating conditions
- the test methods and conditions used to determine the environmental ratings
- the test methods and conditions to determine thermal resistances
- the conditions under which the power dissipation capability of the part is guaranteed.

4.10   References

[1]   Das, D., N. Pendse, M. Pecht, L. Condra and C. Wilkinson, "Deciphering the Deluge of Data – Understanding Electronic Part Data Sheets for Part Selection and Management," *IEEE Circuits and Devices Magazine*, Vol. 16, No. 5, pp. 26–34, September 2000.

[2]   IEC Standard 747-1, Semiconductor Devices – Discrete Devices and Integrated Circuits, Geneva, Switzerland, 1983.

[3]   United States Department of Defense, MIL-M-38510: *General Specification for Microcircuits*, Revision J, Columbus, OH, November 15, 1991.

[4]   United States Department of Defense, MIL-STD-976: *Certification Requirements for Microcircuits*, Columbus, OH, November 15, 1994.

[5]   United States Department of Defense, MIL-STD-1562: *List of Standard Microcircuits*, Columbus, OH, December 15, 1995.

[6]   United States Department of Defense, MIL-STD-965: *Parts Control Program*, Columbus, OH, September 26, 1996.

[7]   United States Department of Defense, MIL-STD-454: *Standard General Requirement for Electronic Equipment*, May 4, 1995.

[8]   Kanter, H. and R. Atta, *Integrating Defense into the Civilian Technology and Industrial Base*, Office of the Assistant Secretary of Defense for Production and Logistics, Arlington, VA, February 1993.

[9]   EIA/JEDEC Publication EIA/JEP 103-A, *Suggested Product – Documentation Classifications and Disclaimers*, Alexandria, VA, July 1996.

[10]   Spiegel, R., "Parts Database Accurate?," *Electronic News*, Vol. 48, No. 29, pp. 1–4, July 15, 2002.

[11]   Xilinx, Data Sheet of QPRO™ XQ4000XL Series QML High-Reliability Field Programmable Gate Arrays, San Jose, CA, May 1998.

[12]   Intel, *Packaging Data Book*, Denver, CO, January 1999.

[13]   Moore, N., "Part Rating," e-mail to Diganta Das, Phoenix, AZ, December 14, 1998.

[14] Murray, S., M. Boru and M. Pecht, "Tracking Semiconductor Part Changes Through the Part Supply Change," *IEEE Transactions on Components and Packaging Technologies*, Vol. 25, No. 2, pp. 230–238, June 2002.

[15] Motorola, *Master Selection Guide*, Chapter 4, Amplifiers and Comparators, Austin, TX, 1998.

[16] Lieberman, F., "How to Select Parts and What Users Need to Know . . . . A Manufacturer's Perspective," presented at the *CALCE Parts Selection and Management Workshop*, June 16, 1998.

[17] Debra, T. (Intel), "Internet Technical Support," e-mail to Diganta Das, December 4, 1998.

[18] Basler, B. (Motorola), "Case #482385, Part #UA7806CKC," e-mail to Diganta Das, March 7, 1999.

[19] Graham, G., "Question on Ratings and Design Rules," e-mail to Diganta Das, May 21, 1999.

[20] IEC Standard 60134, Ratings System for Electronic Tubes and Valves and Analogous Semiconductor Devices, Geneva, Switzerland, 1961 (last review date 1994).

[21] Pickei, S. (Motorola), "Motorola," e-mail to Diganta Das, Phoenix, AZ, December 17, 1998.

[22] Bongiorno, B. (ZiLOG), "Absolute Maximum Ratings," e-mail to Diganta Das, December 4, 1998.

[23] Philips, Family Specifications: HCMOS Family Characteristics, March 1988.

[24] EIA/JEDEC Standard EIA/JESD52, Standard for Description of Low-Voltage TTL-Compatible CMOS Logic Devices, Alexandria, VA, November 1995.

[25] Harris Semiconductor, Datasheet of CD54HC00, Palm Bay, FL, 1997.

[26] AMD, Data Sheet of AM486DE2, Sunnyvale, CA, April 1996.

[27] Davison, J. (Fairchild), "Absolute Maximum Ratings," e-mail to Diganta Das, December 1, 1998.

[28] Davison, J. (Fairchild), "Specification Determination," e-mail to Diganta Das, December 2, 1998.

[29] JEDEC, JEDEC Standard No. 7-A: Standard for Description of 54/74HCXXXX and 54/74HCTXXXX High Speed CMOS Devices, Alexandria, VA, August 1986.

[30] EIA/JEDEC Standard EIA/JESD36, Standard for Description of Low Voltage TTL-Compatible, 5 V-Tolerant CMOS Logic Devices, Alexandria, VA, June 1996.

[31] EIA/JEDEC Standard EIA/JESD64, Standard for Description of 2.5 V CMOS Logic Devices with 3.6 V CMOS Tolerant Inputs and Outputs, Alexandria, VA, February 1998.

[32] Biddle, R., "Commercial Plastic Microcircuits – A Total Solution for Military Applications?" December 1999, Texas Instruments Incorporated, <http://www.ti.com/sc/docs/military/cotspem/cots_pem.pdf>, accessed June 28, 2002.

[33] Dell, G. (Texas Instruments), "PCN#20020528002," e-mail to Diganta Das, June 26, 2002.

[34] Dutta, V., "Junction-to-Case Thermal Resistance – Still a Myth?" *Proceedings of the 4th IEEE SEMI-THERM Symposium*, pp. 8–11, 1988.

[35] EIA Standard RS-419-A, Standard List of Values to Be Used in Semiconductor Device Specifications and Registration Formats, Alexandria, VA, October 1980.

[36] EIA/JEDEC Standard EIA/JESD51, Methodology for the Thermal Measurement of Component Packages (Single Semiconductor Device), Alexandria, VA, December 1995.

[37] EIA/JEDEC Standard EIA/JESD55, Standard for the Description of Low-Voltage TTL-Compatible BiCMOS Logic Devices, Alexandria, VA, May 1996.

[38] Intel, Application Note AP-480 – Pentium® Processor Thermal Design Guidelines Revision 2.0, Santa Clara, CA, November 1995.

[39] Koduru, S. (Cypress), "Instant on Case Temperature," e-mail to Diganta Das, March 1, 1999.

[40] Locher, R., Introduction to Power MOSFETs and Their Applications: National Semiconductor Application Note 558, Santa Clara, CA, December 1988.

[41] McCluskey, P., R. Munamarty and M. Pecht, "Popcorning in PBGA Packages During IR Reflow Soldering," *Microelectronics International*, No. 42, pp. 20–23, January 1997.

[42] MITEL Semiconductor, Datasheet of SL2524 1.3 GHz Dual Wideband Logarithmic Amplifier, Wiltshire, UK, July 1995.

[43] Motorola Semiconductor, MC68332 *User's Manual*, Austin, TX, 1995.

[44] Philips Semiconductor, Transistor Safe Operating Area (SOAR), Power Bipolar Transistors, Eindhoven, Netherlands, November 24, 1998.

[45] Philips Semiconductor, 74HC/HCT/HCU User Guide, Eindhoven, Netherlands, March 1997.

[46] SEMATECH, "SEMATECH Official Dictionary Revision 5.0, Technology Transfer 91010441C-STD," <http://www.sematech.org/public/publications/dict/images /dictionary.pdf, 1995>, Austin, TX, accessed August 2002.

[47] Texas Instruments, Inc., "PCN#20020528002," Dallas, TX, May 29, 2002.

[48] United States Department of Defense, MIL-STD-883: *Test Method Standards – Microcircuits*, Washington DC, 1996.

[49] Xicor, Datasheet of SPI Serial Autostore NOVRAM X25401, Milpitas, CA, 1996.

[50] JEDEC, JEDEC Standard No. 99: Glossary of Microelectronic Terms, Definitions, and Symbols, Arlington, VA, June 1985.

Chapter 5

Candidate Parts Selection: Making the First Cut

Diganta Das and Peter Sandborn

Every candidate part that is subjected to the parts selection process discussed in this book is initially preselected by engineers for use in a product. This chapter discusses the basis upon which the initial selection of candidate parts is performed (see Figure 5.1).

5.1   Candidate part identification process

Several properties of the part may be considered before selecting a candidate part. From the design perspective, the part with the best performance and functional specifications is a key factor. From the parts management perspective, the most generic industry standard part with the best quality rating that is least likely to become obsolete is important. From the procurement perspective, lead time and the availability of multiple sources are important. From the business perspective, the cost of the part is important. All these aspects are considered before a candidate part is selected.

The design engineer bases the initial part selection decisions on a combination of functional and environmental requirements. Besides previous experience and professional preferences, designers often use available databases,[1] websites, and promotional materials and brochures from part manufacturers to search for parts that conform to the required electrical and mechanical specifications. In some cases, the bill of materials (BOM) for a previous product or a previous version of the product is the starting point for a new BOM.

5.2   Part databases

Many electronic product manufacturers have internal part databases to archive information gathered on parts so that engineers and procurement specialists do not have to "reinvent the wheel" every time they consider a part. Many of these databases have evolved into "preferred" or "approved" parts lists containing only parts that have been approved for use in the company's products. Ideally, the selection of a part from an approved parts list is a guarantee to the design engineer that the part has been qualified for use within the company and can be successfully procured. In order for the parts database to perform this role, several key elements must be present and managed in the database:

- A formal method should be used for tracking and linking product change notifications (PCNs) to the parts in the database [3] so that the PCNs received by different functional groups (e.g., purchasing, component engineering) within a company are input to the database [1]. When a new PCN is received, a flag should be raised in the database to alert future potential users of the part changes.
- Expiration dates should be assigned to each part to indicate when the "approved" status of a part expires. The information gathered and analyzed for making part

---

[1] There are several part databases available from part distributors and other information service providers [1]. However, these parts databases, although convenient, can be outdated, incomplete, or just plain wrong [2].

selection decisions is time specific. For example, the "reliability" test results that are used for part family assessment are periodically (e.g., quarterly, biannually) updated by part manufacturers [4].

- Criteria for part reassessment must be defined. Examples of possible criteria include a fixed period, expiration of the part status, issuance of certain types of part changes (e.g., change in package type, site of manufacture), or unexpected or unexplained failures.

- The part database should link to the latest part datasheet source on which the assessment was made. The part datasheet can change with the part life cycle [5], and there is no guarantee that the original datasheet will be available from outside sources.

- The part database should be linked to other company databases such as those on assembly yield, test yield, warranty, and other field return analyses so that detailed information can be accessed by the parts team when necessary.

- Parts that have been considered for use in the company's products but rejected for any reason should be documented, along with the reasons for their rejection.

- Part life cycle information should be placed in the database to prevent design engineers from choosing parts that are designated for discontinuance or that are already obsolete.

- If the part assessment resulted in qualification only for a limited environment and application, that should be clearly noted.

- The availability and accessibility of the part by multiple company sites and by contract assemblers should be in the database.

- Each part in the approved parts list should have a distinct part number (even if it is the same part from different manufacturers). Manufacturers are assessed [6], [7] separately from parts in the part selection and management process, and a part is not uniquely identified unless a manufacturer is specified.

- Besides the part family assessment information, additional test data, which may be obtained by the manufacturer, by an independent testing facility, or by the company, should be linked to the database.

A part on the "approved" list should not mean that no further assessment is necessary; it only means that a part on that list can be a candidate part for further assessment. The process of adding a part to the approved parts list must provide design engineering with feedback as quickly as possible. Different companies have different policies regarding the use of parts during the approval process; however, in most organizations, design engineering can proceed with product development (using the part in their designs) while waiting for component engineering approval. If component engineering returns a "reject" decision, significant cost and schedule issues can occur if the approval process is lengthy.

Based on benchmarks at numerous companies, three weeks is a reasonable amount of time to turn around 90% of part approval requests. More lengthy approval processes may force design engineers to continue the design assuming that the part will be approved. If the part is eventually rejected by component engineering, it may be impractical, due to the solidification of the rest of the design, to reject the part. Regardless of where a product is in the design phase, if component engineering rejects a part, component engineering and the design engineers must have a process to resolve disputed rejections in a timely manner (in less than a week in most cases).

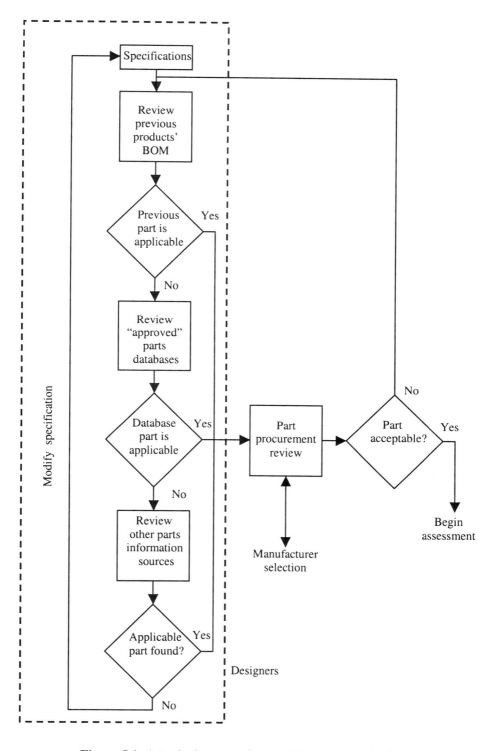

Figure 5.1: A typical process for candidate parts selection

Other types of part databases that may be useful but are not recommended include:

- The *demonstration database* contains parts for use in demonstrator products, i.e., those products designed and made for demonstration of a new technology or new system that is not intended for sale. Parts used in the demonstrator products are reviewed and signed off by purchasing, but other approvals are not generally necessary to put the part in the demonstration database.
- The *guaranteed parts database* contains all the information included on an approved parts list, but only for a subset of the most common parts used in the company's products. The additional attribute that sets this type of database apart from the approved parts list is that the parts in this database are guaranteed to be in stock at the company or at the contract assembler, so that part procurement (availability) problems are not an issue.

5.3   Part procurement

The procurement process for most companies in the electronics industry consists of several steps, including:

- selecting parts
- finding the right vendor (part manufacturer or distributor)
- negotiating the price
- getting approvals
- ordering the part
- receiving the part
- paying the invoice

The time, labor, and resources associated with procurement can be costly. Errors or delays in shipment add to procurement costs, slow down the design process, and leave design engineers waiting for parts they need to complete their jobs on schedule.

5.3.1   Part availability

Part availability is measured in terms of lead time and the volume of available parts and is usually addressed by the procurement process in the company. During the candidate part selection process, engineers provide procurement specialists with bills of materials. The procurement specialist performs several checks on each part in the bill of materials, including:

- It must be determined whether the part can be purchased from the company's existing qualified distributors and is manufactured by one of the company's qualified manufacturers. If not, then the procurement specialist must have engineering initiate a manufacturer and/or distributor assessment process.
- Depending on the quantity needed for the company's products and the short-term availability of the part, the procurement specialist may search for alternative sources of the specified part.
- If the part can be procured from an approved source, the cost of the part must be determined either from long-term strategic pricing agreements that the company already has in place, via catalog prices (for small quantities), or through quotes/negotiations with the supplier.
- The deliveries schedule must be determined, i.e., the number of parts the manufacturer/distributor can provide and when.

- The long-term availability of the part must be assessed in order to consider the necessity and viability of making a lifetime buy of parts with the initial purchase.
- Obsolescence of the part (i.e., whether it is already obsolete or soon to be obsolete) may be assessed by the procurement specialist.

5.3.2   Source of the part

To reduce the risks of obtaining low-quality or counterfeit parts (see Chapter 10) and to ensure support and the purchase, parts available only from brokers and aftermarket suppliers should be given low priority while deciding on the candidate parts. The risks associated with these part sources are discussed in other chapters of this book.

When the source of the part is not the part manufacturer, the obsolescence status of a part should be confirmed by the manufacturer. The distributor (or other part middlemen) may not have information regarding future part availability for the customers. A distributor may also inform the customer that a part is obsolete when it stops carrying the part, regardless of its actual availability from the manufacturer.

When the price and availability figures from one source of a part compared with those of other sources vary significantly, the reasons for the difference should be investigated before accepting that alternative. This difference may be a sign of possible problems with the part or the source of the part.

5.4   Summary

The candidate part selection process puts a part into the assessment flow. This step helps eliminate possible unpromising parts and balances between the need for functionality and the need to obtain costs and other practicality requirements for parts. It merges the input of functional groups besides electrical engineering into the part selection during the initial decision-making process and reduces the cycle time of information flow. Ordering, receiving, and paying occur only after the part is assessed and a positive decision is made regarding its use.

5.5   References

[1]   Roos, G. and M. Gold, "Software Tools Manage Selection of Components," *EE Times*, pp. 91–92, June 3, 2002.

[2]   Spiegel, R., "Parts Database Accurate?" *Electronic News*, Vol. 48, No. 29, pp. 1–4 July 15, 2002.

[3]   Murray, S., M. Boru and M. Pecht, "Tracking Semiconductor Part Changes Through the Part Supply Chain," *IEEE Transactions on Components and Packaging Technologies*, Vol. 25, No. 2, pp. 230–238, June 2002.

[4]   Syrus, T., M. Pecht and D. Humphrey, "Part Assessment Guidelines and Criteria for Parts Selection and Management," *IEEE Transactions on Electronics Packaging Manufacturing*, Vol. 24, No. 4, pp. 339–350, October 2001.

[5]   Das, D., N. Pendse, M. Pecht, L. Condra and C. Wilkinson, "Deciphering the Deluge of Data – Understanding Electronic Part Data Sheets for Part Selection and Management," *IEEE Circuits and Devices Magazine*, Vol. 16, No. 5, pp. 26–34, September 2000.

[6]   Syrus, T., U. Ramgopal and M. Pecht, "Manufacturer Assessment Procedure and Criteria for Parts Selection and Management," *IEEE Transactions on Electronics Packaging Manufacturing*, Vol. 24, No. 4, pp. 351–358, October 2001.

[7]   Jackson, M., A. Mathur, M. Pecht and R. Kendall, "Part Manufacturer Assessment Process," *Quality and Reliability Engineering International*, Vol. 15, pp. 457–468, 1999.

Chapter 6

Manufacturer Assessment Procedure and Criteria

Toby Syrus, Ramgopal Uppalapati, and Michael Pecht

Parts selection and management is a process designed to evaluate the risks inherent in the use of an electronic part (e.g., a resistor, diode, or integrated circuit) and then facilitate informed decisions regarding its selection and future management activities. One step in the process is the assessment of the part manufacturer, which involves comparing data acquired for the manufacturer with predetermined criteria to determine if the manufacturer's policies and procedures are consistent with producing quality and reliable parts. This chapter presents the manufacturer assessment process, along with the criteria developed for assessment. The process and criteria were developed by analyzing industry standards and existing company methods, consulting with industry and academic experts, and conducting a case study of 36 electronic part manufacturers. The results of the case study are also provided.

6.1   Manufacturer assessment process

The manufacturer assessment process is illustrated in Figure 6.1. The process begins with the selection of a candidate part. Documentation for the assessment procedure is then gathered from the manufacturer. Up to this point, the manufacturer to be assessed has not been formally identified – usually only an abbreviated company name. For example, the assessment team may know that the part is made by Vishay, but they do not know that the actual manufacturer is Vishay Dale at Yankton, SD Using this information, the actual manufacturer to be assessed is identified (see Section 6.2). The acquired data are then compared to predetermined criteria for a series of assessment categories, and an excellence level is assigned. If the manufacturer does not satisfy the criteria for a category or provides insufficient documentation, the lowest excellence level is given for the category. After an excellence level has been given for each assessment category, the manufacturer is given an assessment grade based on the results. If all criteria for all assessment categories are satisfied, the manufacturer becomes an acceptable manufacturer, and the candidate part continues on to the next step in the part selection and management process, which is part assessment. If the manufacturer is not acceptable, then the assessment team must decide if an acceptable alternative manufacturer is available. When no alternative is available, the assessment team may choose to pursue equipment manufacturer intervention techniques in order to mitigate the possible risks associated with using parts from an unacceptable manufacturer [28].

After a candidate part is chosen, the manufacturer of the part is identified. The manufacturer's quality policies are assessed with respect to the five assessment categories. If all criteria are satisfied, the candidate part continues on to the next part selection and management process. If not, the manufacturer, and consequently all parts produced by the manufacturer, are rejected (see Figure 6.1).

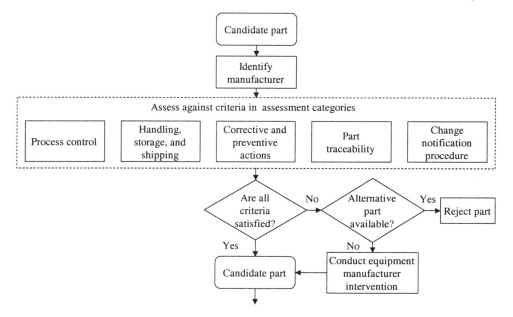

Figure 6.1: Manufacturing assessment process flowchart

6.2   Manufacturer identification

Manufacturer assessment is designed to evaluate a part manufacturer's ability to consistently produce quality parts. However, a common company name does not ensure the same quality system and policies at each of the manufacturer's different locations. A part can be fabricated and assembled at multiple locations around the world and subjected to different company quality policies. In fact, companies may have different quality certifications from site to site. Some examples are:

- Integrated Device Technology Inc. (IDT) part number IDT7006XS15PF is a static random access memory (SRAM) part that is fabricated at two different sites and assembled at another two sites [29]. It is unclear which site or how many sites need to be assessed.[1]
- Vishay Sprague at Sanford, Maine (USA), and Vishay Foil Resistors at Holon, Israel, are two locations of the company, Vishay Intertechnology Inc., but they have different quality system policies.
   - Vishay Sprague: According to the quality system policy manual, there is a policy of notifying customers of product or process changes which affect the customer's requirements [30].
   - Vishay Foil Resistors: According to the Holon quality manager, there is no documented policy of notifying customers of product or process changes [31].
- Fairchild Semiconductor Corporation has different certifications for each location, as shown in Table 6.1. Different certifications can cover different areas of a quality

---

[1] Using the criteria outlined below, only one site needs to be assessed. All IDT sites follow a common quality manual [32] and have ISO 9000 or equivalent certification. It is not important which site is assessed, because all sites have the same quality procedures and therefore will have identical results from manufacturer assessment.

system, and certification audits may examine different areas. This can leave uncertainties about the consistency of the audits carried out and the policy areas covered for each certification.

Table 6.1 shows Fairchild's manufacturing facilities, what quality certifications each have, and who the certifying body is.

Table 6.1: Fairchild Semiconductor Corporation certifications [33]

Site	Certificate type	Certifying body
Cebu, the Philippines	ISO-9001 and QS-9000	DNV
	Qualification of Transistors & Diodes for Delco Electronics	Delco Electronics
Kuala Lumpur, Malaysia	ISO-9002	UL
Penang, Malaysia	ISO-9001 and QS-9000	DNV
	Stack Level 2 Supplier Certification	Stack International
	AEC-A100 QSA Semiconductor Edition	AEC
Puchon, South Korea	ISO-9001 and QS-9000	BSI
South Portland, Maine	ISO-9001 and QS-9000	DNV
	Stack Level 1 Supplier Certification	Stack International
Sunnyvale, California	ISO-9001 and QS-9000	DNV
West Jordan, Utah	ISO-9001 and QS-9000	DNV
Wuxi, Jiangsu, China	ISO-9001 and QS-9000	TÜV Cert

Because a part may be manufactured at different sites within a company, and these sites may have different quality policies and certifications, the actual manufacturer of a candidate part must be identified prior to manufacturer assessment in order to ensure that the quality policies assessed are relevant to the candidate part. For assessment purposes, a single manufacturer encompasses all sites of a company that follow a common quality system policy manual and have (or do not have) certification (ISO 9000, ANSI/ASQC Q90, QS 9000, AS 9000, TL 9000, or any other acceptable certification) from an external auditor. Thus, a manufacturer is identified by a quality system policy manual and external audit (or lack of) certification. As a result, multiple manufacturing sites within a company may be identified as one manufacturer if the above criteria are satisfied, resulting in time savings for the assessment team.

6.3   Manufacturer assessment categories

After the part manufacturer has been identified, the assessment team must assess the manufacturer's policies and procedures. By studying industry methodologies and standards [7]–[10], [17]–[27] and conducting a case study of 36 manufacturers, a set of recommended assessment categories was formed: process control; handling, storage, and shipping controls;

corrective and preventive actions; product traceability; and change notification. These five categories contain the minimum set of criteria necessary to completely assess an electronic part manufacturer's ability to consistently produce quality parts. The criteria for each category will now be given.

6.3.1   Process control

Process control is defined as activities and techniques established by the manufacturer to regulate the manufacturing process in order to achieve and maintain product quality [34]. A major constituent of process control is statistical process control (SPC). In SPC, various methods are used to quantitatively measure and analyze a process. The objectives of SPC are to (1) improve the quality of the process output, (2) reduce process variability and achieve process stability, and (3) solve processing problems. Techniques used in SPC include control charts, histograms, Pareto charts, check sheets, defect concentration diagrams, scatter diagrams, and cause-and-effect diagrams [35].

An acceptable manufacturer must have defined and documented procedures in place for SPC. The manufacturer should also have a documented policy to train operators who perform process control. Implementation should be ensured through external audits and subsequent certification. The manufacturer should provide a copy of its current ISO 9000 or equivalent certificate. External audit certifications, such as ISO 9000, ensure that not only are the manufacturer's procedures documented, but they are also in use. The criteria for process control and all assessment categories can be found in Table 6.2.

Documentation for process control can often be found in a manufacturer's quality manual. Employee training and SPC are requirements under ISO 9001 sections 4.18 (Training) and 4.20 (Statistical techniques), respectively [36]. A quality manuals system policy set up in ISO 9001 format should contain sufficient documentation to satisfy all criteria for the process control category. However, not all manufacturer quality system policy manuals are set up parallel to ISO 9001; therefore, these manuals may or may not contain enough documentation to satisfy all criteria for process control. Manufacturers generally also have expanded sets of procedures separate from their quality system policy manuals, which may fulfill the criteria for process control if provided.

6.3.2   Handling, storage, and shipping controls

Handling, shipping, and storage controls are designed to prevent damage or deterioration of parts during the course of these processes. Packing and other post-production processes can have significant effects on part quality before the product reaches the customer. An acceptable manufacturer should have established procedures defined for handling, shipping, and storage. Quality system compliance should be ensured through external audits and ISO 9000 or equivalent verifiable certification. Handling, storage, and shipping controls are requirements covered in ISO 9001 section 4.15 (Handling, storage, packaging, preservation, and delivery) [36]; therefore, documentation may be found under this section in the manufacturer's quality manual.

6.3.3   Corrective and preventive action

Corrective and preventive actions involve identifying problems and potential problems, determining why the problem occurred, correcting the cause of the problem, and verifying that the changes worked. Once a problem has been identified, root cause analysis is used to determine the underlying cause of the problem. Preventive measures are taken to prevent the recurrence of the problem and to prevent the occurrence of potential problems. An acceptable manufacturer should have documented corrective and preventive action procedures, including root cause analysis, to correct nonconformities and prevent their causes. Quality system compliance should be ensured through external verifiable audits, such as ISO 9000. Corrective and preventive actions are requirements of ISO 9001, section 4.14 (Corrective and preventive action) [36].

6.3.4   Part traceability

Part traceability is the ability to trace a part and its constituent material to its lot, to the manufacturing processes it has gone through, and to its suppliers. The purpose of traceability is to trace a defective part to its origin and identify the location of other parts from the defective lot(s). Part traceability aids in root cause analysis and helps prevent the usage of additional parts from a defective lot. The equipment manufacturer should ensure that part traceability is documented and maintained within the manufacturer's processes and to its suppliers. Quality system compliance should be ensured through external verifiable audits such as ISO 9000. Part traceability is a requirement of ISO 9001, section 4.8 (Product identification and traceability) [36].

6.3.5   Change notification

Change notification involves notifying customers of changes in materials, processes, or the part that may affect the form, fit, or function of the part. The purpose is to inform the equipment manufacturer of any changes which may affect the equipment manufacturer's product quality and integrity or design process. An acceptable manufacturer should have documented change notification procedures to notify customers of changes in materials, processes, or the part before the changes are implemented. Quality system compliance should be ensured through external verifiable audits such as ISO 9000. Change notification is not a requirement of the ISO 9000 standards. However, in quality system policy manuals patterned after ISO 9001, change notification may be found in section 4.4.9 (Design changes) [36]. If this does not provide sufficient documentation, then a change notification procedure manual (if available) is necessary to satisfy the assessment criteria.

6.4   Results

The case study assessment results, using the recommended assessment criteria, for the 36 manufacturers can be seen in Table 6.3. The numbers in each assessment category represent the excellence level assigned to that part group. An excellence level of 0 indicates that the part did not satisfy the assessment criteria for the category. Twelve of the 36 manufacturers did not satisfy all the criteria for each assessment category and thus are not acceptable manufacturers.

The manufacturer assessment results for all 36 case study manufacturers are listed in. An excellence level of 1 is required for each assessment category for the manufacturer to be acceptable.

Eight case study manufacturers were not ISO 9000 or equivalent certified at the time of our assessment. In addition to these eight, one manufacturer, ELNA, stated that all facilities are QS9000 certified and some are ISO 9000 certified but did not provide copies of any certificates as proof. These nine manufacturers could not fully satisfy the criteria for any manufacturer assessment category because implementation of documented procedures could not be verified.

Two manufacturers, ITW Paktron and ELNA, did not provide a copy of their quality manual. For this reason, manufacturer identification could not be correctly completed. Also, documentation for the assessment categories is generally found in the quality manual;

therefore, these two manufacturers did not satisfy much of the manufacturer assessment criteria.

Thirty-five of the 36 manufacturers provided documentation of a policy to use statistical process control to monitor manufacturing processes, and 34 of 36 provided documentation of a policy to qualify and train operators performing process control activities. This information did not include the method used in training employees. Thirty-four manufacturers provided documentation of a policy to use handling, shipping, and storage control procedures to prevent damage or deterioration of the product. All 36 manufacturers provided a documented corrective and preventive actions policy to eliminate the causes of product nonconformities. Thirty-four manufacturers provided documentation of their policy to maintain part traceability within manufacturing processes and to their suppliers. Twenty-nine manufacturers provided documentation of their process to notify customers regarding product or process changes before the changes are implemented.

6.5   Summary

A methodology has been presented for the quality assessment of an electronic part manufacturer. The methodology and criteria were developed by studying numerous industry programs and standards and through a case study of 36 manufacturers. Manufacturer assessment begins with the collection of information. From this information, the manufacturer to be assessed is identified, and then the manufacturer's quality policies are evaluated by comparing them against the assessment criteria in Table 6.2. In order to be an acceptable manufacturer, all criteria for each category shown in the table must be satisfied. Otherwise, the manufacturer is rejected. Parts from a rejected manufacturer should not be used unless special circumstances warrant, such as when no alternative part is available. In this situation, equipment supplier intervention techniques should be used. The interested reader is referred to the article by Schwach et al. for more information on equipment supplier intervention [28].

Table 6.3 represents a summary of manufacturer assessment and can be a valuable tool during the assessment process. All assessment categories, criteria, and the minimum excellence levels for each are listed there. The assessment categories were developed to contain the minimum criteria necessary to fully assess a manufacturer's ability to consistently produce quality parts. The methodology does not require the equipment supplier to perform an in-house audit of the manufacturer. Although no amount of assessment will eliminate all risk associated with the use of a part, an effective manufacturer assessment program like the one described here can reduce the risk.

Table 6.2: Summary of manufacturer assessment criteria and categories

Item number	Assessment category	Assessment criteria	
		Excellence level 0	Excellence level 1
1	Process control	Criteria for excellence level 1 are not satisfied.	Documented SPC procedures are in use to monitor manufacturing processes, and operators are qualified and trained. Quality system compliance is ensured through external verifiable audits.
2	Handling, storage, and shipping controls	Criteria for excellence level 1 are not satisfied.	Documented handling, storage, and shipping procedures are implemented to prevent damage or deterioration of the part. Quality system compliance is ensured through external verifiable audits.
3	Corrective and preventive action	Criteria for excellence level 1 are not satisfied.	Documented corrective and preventive action procedures are in place, including root cause analysis, to correct and prevent the causes of nonconformities. Quality system compliance is ensured through external verifiable audits.
4	Part traceability	Criteria for excellence level 1 are not satisfied.	Part traceability is documented and maintained within the manufacturer's processes and to its suppliers. Quality system compliance is ensured through external verifiable audits.
5	Change notification	Criteria for excellence level 1 are not satisfied.	Documented change notification procedures are in place to notify customers of changes in materials, processes, or the part before the changes are implemented. Quality system compliance is ensured through external verifiable audits.

Table 6.3: Manufacturer assessment case study results

Manufacturer	Assessment category excellence level				
	Process control	Handling, storage, and shipping controls	Corrective and preventive actions	Product traceability	Change notification
	(out of 1)	(out of 1)	(out of 1)	(out of 1)	(out of 1)
AMD	1	1	1	1	1
Analog Devices	1	1	1	1	1
Atmel Corporation	1	1	1	1	1
AVX	1	1	1	1	1
Burr-Brown	1	1	1	1	1
Cornell-Dubilier	1	1	1	1	0

Manufacturer	Assessment category excellence level				
	Process control	Handling, storage, and shipping controls	Corrective and preventive actions	Product traceability	Change notification
	(out of 1)	(out of 1)	(out of 1)	(out of 1)	(out of 1)
Dielectric Labs	1	1	1	1	1
Diodes Inc.	1	1	1	1	0
Ecliptek Corp.	0	0	0	0	0
ELNA Inc.	0	0	0	0	0
Elpac Electronics	0	0	0	0	0
Fairchild	1	1	1	1	1
General Semiconductors	1	1	1	1	1
IDT	1	1	1	1	1
Intersil	1	1	1	1	1
IRC	1	1	1	1	1
ITW Paktron	0	0	0	0	0
Kemet	1	1	1	1	1
Monitor Products	1	1	1	1	1
Motorola	1	1	1	1	1
MTC	1	1	1	1	1
Murata Electronics	0	0	0	0	0
Precision Devices Inc.	0	0	0	0	0
Quick Logic	1	1	1	1	1
RCD components	0	0	0	0	0
Sensycon	1	1	1	1	1
Tecate	0	0	0	0	0
Telcom	0	0	0	0	0
Temic	1	1	1	1	1
Texas Instruments	1	1	1	1	1
Vishay Dale	1	1	1	1	1
Vishay Foil Resistors	1	1	1	1	0
Vishay Siliconix	1	1	1	1	1
Vishay Sprague	1	1	1	1	1
Vishay Vitramon	1	1	1	1	1
Xicor	1	1	1	1	1
Xilinx	1	1	1	1	1

6.6 References

[1] Jackson, M., A. Mathur and M. Pecht, "Part Manufacturer Assessment Process," *Quality and Reliability Engineering International*, Vol. 15, pp. 457–468, 1999.

[2] Foucher, B., R. Kennedy, N. Kelkar, Y. Ranade, A. Govind, W. Blake, A. Mathur and R. Solomon, "Why a New Parts Selection and Management Program?," *IEEE Transactions on Components, Packaging, and Manufacturing Technology – Part A*, Vol. 21, No. 2, pp. 375–378, June 1998.

[3] Jackson, M., P. Sandborn, M. Pecht, C. Hemens-Davis and P. Audette, "A Risk-Informed Methodology for Parts Selection and Management," *Quality and Reliability Engineering International*, Vol. 15, pp. 261–271, September 1999.

[4] Perry, W., "Specifications and Standards – A New Way of Doing Business," Internal Memorandum, U.S. Department of Defense, June 29, 1994.

[5] Ministry of Defense, France, Directive CONZE, July 31, 1995.

[6] Ministry of Defense, UK, UK Defense Standards 59-36, Issue 5, UK Defense Procurement Bulletin, April 1995.

[7] International Organization for Standardization, *ISO 9001:2000, Quality Management System-Requirements*, Geneva, Switzerland, 2000.

[8] Automotive Electronics Council, *CDF–AEC–Q100 Stress Test Qualification for Automotive-grade Integrated Circuits*, Revision E, Detroit, MI, January 31, 2001.

[9] STACK International, *Specification 0001 Issue 12.2 at Notice 2: General Requirements for Integrated Circuits*, Herts, UK, November 19, 1999.

[10] American Institute for Aeronautics and Astronautics, *Recommended Practice for Parts Management, R-100-1996*, Reston, VA, 1996.

[11] IEC/PAS (International Electrotechnical Commission) 62239, Edition 1, 2001-04, "Electronic Component Management Plans," 2001 (also being developed as GEIA 4899), Geneva, Switzerland.

[12] Solomon, R., P. Sandborn and M. Pecht, "Electronic Part Life Cycle Concepts and Obsolescence Forecasting," *IEEE Transactions on Components and Packaging Technologies*, Vol. 23, No. 3, pp. 707–717, December 2000.

[13] Quality Semiconductor, Inc., *Quality and Reliability, Second Quarter 1996 Accumulated Test Results*, Santa Clara, CA, April 1996.

[14] Yoffie, D., *Competing in the Age of Digital Convergence*, Harvard Business School Press, Boston, MA, 1997.

[15] Semiconductor Industry Association, *Turbocharging the U.S. Economy: A Report*, San Jose, CA, March 19, 1998.

[16] Silicon Valley Survey, "The Changing Dream," *The Economist*, p. 19, March 29–April 4, 1997.

[17] AlliedSignal Inc. [now Honeywell], *Electronic Component Management Plan*, Tucson, AZ, June 1997.

[18] Boeing Inc., *Electronic Component Management Program*, Seattle, WA, February 1997.

[19] Eldec Corp., *Electronic Component Management System*, Lynnwood, WA, March 21, 1997.

[20] Hughes Aircraft [now part of Boeing], *Guidelines for Use of Plastic Encapsulated Microcircuits (PEMs) in Naval Avionics Equipment*, Indianapolis, IN, September 29, 1997.

[21] Motorola Inc., *Parts Management Plan for Iridium™ Comm Module*, Revision XA, Chandler, AZ, June 15, 1992.

[22] Northrop Grumman Corp., *Exciter/Receiver/Processor Parts Control Program Plan*, Revision A, Baltimore, MD, August 8, 1997.

[23] Rockwell Collins, Inc., *Air Transport Systems Electronic Component Management Program*, Cedar Rapids, IA, January 7, 1999.

[24] Smiths Industries, *Electronic Component Management Plan*, Cheltenham, UK, February 1, 1997.

[25] Sun Microsystems, *950-2484-01: ASIC Component Engineering Requirements (CER)*, Revision 3.3, Santa Clara, CA, June 25, 1996.

[26] Electronic Industries Association, *EIA-599-1992: National Electronic Process Certification Standard*, Arlington, VA, August 1992.

[27] Electronic Industries Association, *EIA-670: Quality System Assessment*, Arlington, VA, June 1997.

[28] Schwach, C. A., A. Mathur, R. Solomon, P. Sandborn and M. Pecht, "Equipment Supplier Intervention Techniques," *Future EMS International*, Technology Publishing Ltd., London, Issue 3, pp. 97–100, 2000.

[29] Integrated Device Technologies, Inc., *Quality and Reliability Monitors*, Santa Clara, CA, July 1997.

[30] Sprague, V., PQP Index, Revision 16, Sanford, ME, April 16, 1997.

[31] Gizunterman, A., and Israel, V. [Vishay Foil Resistors], letter to Rahul Mahajan, AlliedSignal, April 28, 1998.

[32] Integrated Device Technologies, Inc., *Quality Policy Manual*, Revision 20, Santa Clara, CA, September 1, 1999.

[33] Fairchild Semiconductor, "Fairchild Semiconductor Quality Certificates," South Portland, ME, <www.fairchildsemi.com/ company/quality.html>, accessed June 27, 2001.

[34] Fraser, R. E., *Process Measurement and Control*, Prentice Hall, Upper Saddle River, NJ, 2001.

[35] Montgomery, D., *Introduction to Statistical Quality Control*, 3rd ed., John Wiley & Sons, Inc., New York, NY, 1996.

[36] British Standards Institution, *CECC 00804: 1996 Guidance Document: Interpretation of "EN ISO 9000: 1994," Reliability Aspects for Electronic Components*, London, February 1996.

Chapter 7

Part Assessment Guidelines and Criteria

Toby Syrus, Michael Pecht, and David Humphrey

Parts selection and management is a process designed to evaluate the risks inherent in the use of an electronic part (e.g., a resistor, diode, integrated circuit, connector, etc.) and then facilitate informed decisions regarding its selection and future management activities. One step in the process is the evaluation of a part's quality, integrity, and assembability. This involves comparing data acquired for the part with predetermined criteria to determine if the part will function acceptably in an equipment manufacturer's product.

This chapter presents the part quality, integrity, and assembability assessment process. The process and criteria were developed by analyzing industry standards and existing company methods, consulting with industry and academic experts, and conducting various case studies including a case study of 113 electronic parts is presented in this chapter. The case study parts are listed in Table 7.1.

Table 7.1: The case study parts

Device type	Manufacturer	Part number	Device type	Manufacturer	Part number
Amplifier	Analog Devices	AD620AR	Microcontroller	Motorola SPS	MPC509L3CFT33
		OP284ES			MC68332ACFC16
		OP285GS	Microprocessor	MTC	MTC1750 (UP)
	Texas Instruments	TLE2027ID		Intersil	ID80C86
		TL072ID			ID82C84A
		TLC2272AMD			IP80C86
		TLE2062AMD	Multiplexor	Intersil	HI1-506-2
ADC / DAC	Analog Devices	AD7895AR-2		Vishay Siliconix	DG506AAK
		AD7943BR	Octal buffer	IDT	74FCT244TSO
	Burr-Brown	AD7943BR	Oscillator	Ecliptek Corp	EC1100HSETTS-64MHz
Analog switch	Vishay Siliconix	DG412DY		Monitor Products	4183-02
Capacitor	AVX	TPSD107K010R0100			4183-51
		TPSE227K010R0100			4192-27
		TPSD226K025R0200			4203-02
	Cornell-Dubilier	MLS272M075EB0A			SM99121BSA-2.048MHZ

Device type	Manufacturer	Part number	Device type	Manufacturer	Part number
		MLS152M060OK0A			PDI 1506
	Dielectric Labs	C11CF270G5TXL	PDI		PDI 2111
		C17CF331G6TXL			PDI 2231R
	ELNA Inc.	RVJ-50V101MH10-R	PWM controller	Telcom	TC28C43EOE
		C5B104F			TC28C46EOE
		C5L103F	Rectifier	Diodes Inc.	ES1D
	Elpac Electronics	MD5L103F			ES3D
		PM5A105K		Intersil	RURD620CCS
		Z5L103F		IRC	PCF-W0805R031010B
	ITW Paktron	106K100CS4G			PCF-W1206R031000B
		475K100CS4G		Sensycon	SMD2101
	Kemet	T491B106K010AS	Resistor	Vishay Dale	LVR-3 0.05 ohm 1%
		C1206C224K5RCX			CRA06S0803103J087
	Murata	GRM426X7R471K500			CRCW08051054F
	Tecate	MXZZ-50/101M8X10TR			CRCW12061021F
	Vishay Sprague	293D106X9035D2			CRCW12101431F
		VJ0805A100JXB		Vishay Foil Resistors	SMR3D07K150
		VJ1206Y223KXB	Ripple counter	Motorola SPS	MC74HC595AD
	Vishay Vitramon	VJ1210A103JXA		Diodes Inc.	SMCJ24CA
		VJ1812Y105KXA	Transient voltage suppressor	General Semiconductor	SMCJ10
		VJ1812Y564KXA			SMCJ17
		VJ1812Y684KXA			SMCJ51
		VJ1812Y824KXA	Transistor		2N6796
Comparator	Texas Instruments	TLC3702MD		Intersil	HUF75639S3S
		TLC374MD			RF1S640SM
		TLC372ID			RF1S9640SM
CPLD	Xilinx	XC9572-15PQ100I		Motorola SPS	MMBF0201NLT1
		XC95108-15PQ100I			MJD31
		XC9536-10PC44I			MJD32

Device type	Manufacturer	Part number	Device type	Manufacturer	Part number
Diode	Diodes Inc.	BZX84C12			MJD41
	Fairchild	1N4148			MJD42C
Flip-flop logic	Texas Instruments	SN54LS174J			MJD47
FPGA	Quick Logic	QL2009-OPQ208I			MJD50
		QL24X32B208PQFP			2N2222
Inductor	Vishay Dale	IMC-1210 10UH 10%			2N2907
Jumper wire	RCD Components	ZJ-2			2N3634
Memory	AMD	AM29F400AB-90SE			2N3700
	Atmel	AT25160N-10SI			2N4033
		AT28C010E-20JI		Vishay Siliconix	SST4117
	Xicor	X28C256-20JM			SSTJ211
		X28C010D1-25	Voltage reference	Analog Devices	AD780BR
		X28VC256-JM90			

7.1   Part assessment process

The part assessment process is shown in Figure 7.1. The process begins with the selection of a candidate part. Next, documentation is obtained, either directly from the manufacturer or indirectly from sources such as the Internet. If the information is explicitly for the candidate part, then the assessment team moves on to assessing this information against the assessment criteria. If the information is not for the candidate part but for similar parts, the candidate part must be put into a part group before the data can be compared to the assessment criteria. Based on the part's satisfying (or failing to satisfy) the criteria, an excellence level is given. If insufficient documentation is provided, the lowest excellence level is given. If all criteria are satisfied, the part is deemed acceptable, and one proceeds to the next step in the parts selection and management methodology, as seen in Figure 7.1.

If the part is not acceptable, then the assessment team must decide if an acceptable alternative is available. When no alternative is available, the team may choose to pursue intervention techniques in order to mitigate the possible risks associated with using an unacceptable part [24]. The interested reader is referred to Chapter 17 [25].

Part assessment results may not be valid for parts manufactured before or after the assessment if changes are made to the part. Therefore, parts that do not pass the assessment process may be deemed acceptable after passing a subsequent assessment. Likewise, parts that are deemed acceptable may need to be reassessed periodically to ensure their acceptability.

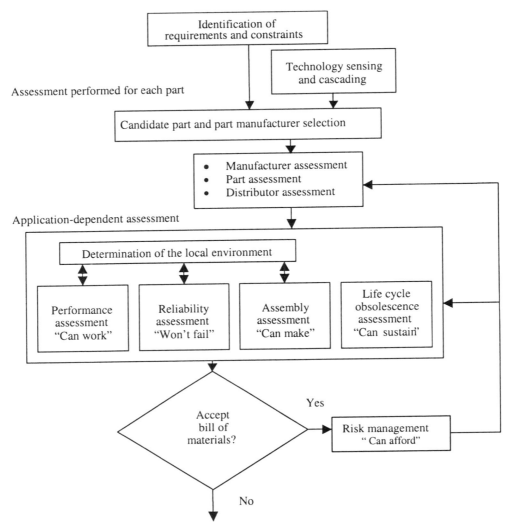

Figure 7.1: Part selection and management methodology

7.2   Part grouping

In general, manufacturers do not provide data for each individual part. Instead, data are provided for specific groups or families of parts. Manufacturers often classify parts into families based on fabrication technology (e.g., CMOS, bipolar), temperature rating (e.g., 0 to 70°C, −55 to 125°C), device type (e.g., diode, logic), and/or package type (e.g., BGA, DIP). Grouping parts into families facilitates the measurement of quality metrics (e.g., average outgoing quality, C_{pk}) and/or reporting of assembly guidelines. A manufacturer may thus inspect or test one part and apply the information to the entire family. Unfortunately, part families defined by different manufacturers are often different, even for similar devices and package types.

In some cases, quality and integrity data supplied by the manufacturer for a part family do not accurately portray all parts in that family. To remedy this, the assessment team should put the candidate part into a part group. A part group is defined as a group of parts that are from the same manufacturer and have the same manufacturing process flow, packaging flow, and assembly guidelines. The assessment team, not the manufacturer, identifies which parts belong to a particular part group. However, a part group and a part family may contain the same parts. Each part in a part group has a common set of criteria for quality data, integrity data, and assembly guidelines; therefore, the results from the assessment of two parts in one group will be the same, and only one part per group needs to be assessed.

When a manufacturer provides part family data for a candidate part, the assessment team must decide if the data are applicable. If the data is not applicable, then the team must request additional information. If applicable information is not available, then the part will receive the lowest excellence level for the particular category. Once applicable data is obtained, a part group is defined to contain all the parts for which the manufacturer-supplied information is applicable. Figure 7.2 depicts the process for part grouping.

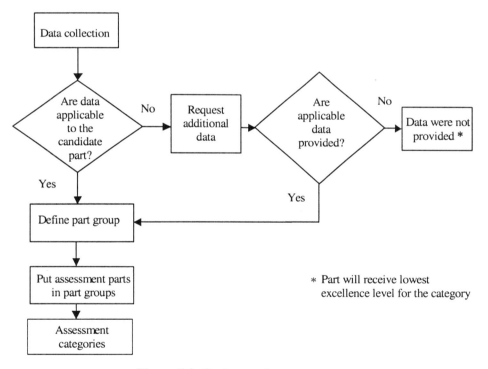

Figure 7.2: Part grouping process

After a candidate part is chosen, data are collected. If data cannot be found for the individual part but rather for a group of parts, then the part is put into a part group. Next, the part's quality and integrity are assessed with respect to the criteria in each assessment category. If all criteria are satisfied, the candidate part continues on to the next step in the part selection and management methodology. If the part is unacceptable, then the team must determine if an alternate is available. When an alternate is available, the unacceptable part is rejected. However, if no alternate is available, some form of equipment supplier intervention must be undertaken to make the part acceptable (see Chapter 11).

As an example of part grouping, take Analog Devices parts AD620AR (low-power instrumentation amplifier) and AD780BR (high-precision reference voltage). Both are made by the same fabrication technology (P3) and have the same package type (small outline, integrated circuit-surface mount package) [26]. Analog Devices provided the same documentation for both parts. Average outgoing quality values were based on fabrication technology, process capability indices (C_{pk}) were based on package type, integrity data were based on fabrication technology and packaging type, and assembly guidelines were given for surface mount packages. Based on this information, these two parts were put in a single part group.

Once the part group is identified, the equipment manufacturer should assess the part group with respect to the assessment categories, discussed in Section 7.3, in order to evaluate the acceptability of the part. From our 113 case study parts, 68 part groups were identified.

7.3   Part assessment categories

The three key elements of part assessment are quality, integrity, and assembability. Quality is evaluated through AOQ and process capability index (C_{pk}). Integrity is evaluated through integrity monitor test results. Assembability is evaluated through assembly guidelines provided by the part manufacturer (see Figure 7.3).

7.3.1   Average outgoing quality (AOQ)

AOQ is defined as the average nonconforming lot fraction from a series of lots [27]. It represents the total number of parts that are outside specification limits, as determined through sample tests conducted during the final quality control inspection. This number reflects the estimated number of defects received by the customer. AOQ is usually reported in parts per million (ppm) but may be reported in percent or parts per billion (ppb). Defective parts are identified through visual, mechanical, and electrical testing. Tests conducted for measuring AOQ may include visual and mechanical tests for marking permanence, package dimensions, coplanarity, solderability, bent leads, and hermeticity (if applicable). Electrical testing may include functional tests at room temperature, as well as at high and low temperatures [28]. AOQ tests are nondestructive and are conducted on a large number of devices in comparison with the sample size selected for integrity monitor tests.

AOQ reflects the effectiveness (or lack of it) of the manufacturer's quality management system. An effective quality management system will minimize the total number of nonconformities produced, as well as the number that reach shipping. AOQ is also a direct measure of part quality. High values represent a high defective count, implying poor quality management. Low values reflect high part quality. The parts management team should establish threshold AOQ requirements to determine part acceptability.

If all parts are tested prior to shipping, then theoretically the AOQ should always be zero because all nonconformities should be removed. However, due to the volume of parts produced, it is usually impractical to test all parts shipped. Instead, a sample is tested, and an estimation of the AOQ is calculated from it. AOQ can be calculated using different formulas, which could lead to problems when comparing AOQ values from different manufacturers. For example, Xilinx provided AOQ values using JEDEC standard JESD 16-A, which uses a lot acceptance rate (LAR) parameter to estimate ppm [27]. AOQ is calculated from JESD 16-A using the following formula [27]:

$$AOQ = P * LAR * 10^6 \qquad (7.1)$$

where P is fraction of nonconforming components found in all samples given by

$$P = \frac{D}{N} \qquad (7.2)$$

LAR is the lot acceptance rate given by [27]

$$LAR = \frac{AL}{TL} \qquad (7.3)$$

D is the number of defects in the sample size N, AL is the total number of accepted lots, and TL is the total number of lots tested.

Analog Devices calculates AOQ according to EIA-554 method B, which is shown in eqn. (7.4) [29]. The equation for EIA-554 is very similar to the one used by JEDEC.

$$AOQ = \frac{\sum (N_x * \dfrac{d_x}{n_x} * LAR)}{\sum N_x} * 10^6 \qquad (7.4)$$

when N_x is the total quantity in lot X, n_x is the sample quantity in lot X, d_x is the number of rejects in sample n_x, and LAR is the lot acceptance rate seen in eqn. (7.3).

IDT evaluates AOQ using the following formula:

$$AOQ = \frac{D}{N} * 10^6 \qquad (7.5)$$

Calculations using eqns. (7.1), (7.4), and (7.5) can give different AOQ values, even for the same sample size and number of defects. Equation (7.1) provides a lower AOQ value than eqn. (7.5) unless AL equals TL, in which case both equations produce the same value. To further exacerbate the problem eqns. (7.1) and (7.4) can provide different AOQ values for different sampling schemes. However, for large sample sizes, such as those used for calculating AOQ, eqns. (7.1), (7.4), and (7.5) give comparable results.

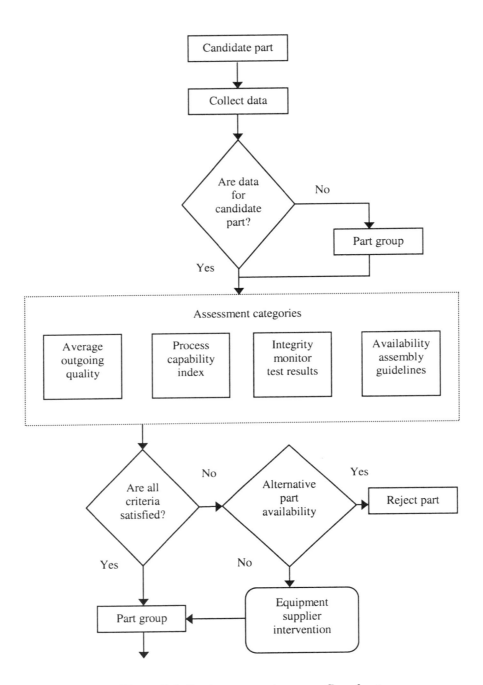

Figure 7.3: Part assessment process flowchart

In the case study, AOQ data were provided for 65 of the 68 part groups. The D and N values used in the calculation of AOQ were provided by 16 manufacturers. D and N could have been used by the assessment team to calculate AOQ with a common formula if all manufacturers provided this information. Thirteen manufacturers also provided the method used in their calculation of AOQ. Six manufacturers calculate AOQ according to EIA-554 [29], four follow JEDEC 16-A [27], Atmel uses the Hahn estimator [30], and two companies use eqn. 7.5. Manufacturers may provide functional, electric, visual/mechanical and hermetic AOQs. Therefore, manufacturers that follow this standard may provide up to four ppm numbers for a single part. EIA-554 is structured in a similar way [29]. Table 7.2 lists all manufacturers and shows which ones provided AOQ in terms of D and N and the method of AOQ calculation. An "X" indicates that the manufacturer did not provide AOQ values, and "NP" indicates that the AOQ calculation method was not given.

The equipment manufacturer should determine AOQ cutoff or accessibility limits depending on the level of risk involved. Cutoff limits should be defined by the equipment manufacturer to differentiate acceptable and unacceptable parts. Some factors to be considered include application, testability and diagnosability, production volume, and the cost and reworkability of the subassembly. These factors are only involved in determining AOQ cutoff limits and not in actual part assessment.

- *Application:* The part's application may influence the acceptable AOQ cutoff limit. Applications where the consequences of failure are severe, such as aerospace or life support, may necessitate high quality and hence a low AOQ cutoff limit. However, in an application where the consequences of failure are less severe, such as video games or electronic toys, a low AOQ value may be less important than other factors.

- *Testability and diagnosability:* If a part can be tested and functional faults diagnosed with low cost, then higher AOQ values may be allowed. This allows rework (when applicable) on a faulty subassembly or repair on a returned product to be initiated quickly. Parts that are difficult to test and diagnose may lead to high costs when attempting to identify the location and cause of the fault and potentially scrapping the entire subassembly.

- *Production volume:* Parts that are produced in high volumes may allow for higher AOQ values. If the equipment manufacturer discovers that parts are faulty or that additional parts are needed, then the volume of parts may make additional parts less expensive and more readily available compared to the same part produced in low volume. Low-volume parts may be more expensive to replace, due to per part costs, or may be unavailable at the desired time, therefore necessitating low AOQ values.

- *Cost and reworkability of the subassembly:* An inexpensive subassembly may allow for higher AOQ values. If a faulty part is found in the subassembly, it can be discarded with minimal financial loss to the equipment manufacturer. However, if a faulty part is found on an expensive subassembly, rework may be done to avoid the financial loss that would result from discarding faulty subassemblies. Rework leads to an investment of additional time, resources, and money (presumably less costly than discarding the entire subassembly); thus, easily reworkable subassemblies are preferred for parts with higher AOQ values. The chance of damaging other components and potentially scrapping the entire subassembly can also be reduced if rework can be accomplished with minimal effort.

Manufacturers may provide AOQ values separately for each test type. JESD 16-A [27] calls out four classes for reporting AOQ: functional, electric, visual/mechanical, and hermetic. Therefore, manufacturers that follow this standard may provide up to four ppm numbers for a single part. EIA-554 is also structured in a similar way [29].

Table 7.2: Manufacturers that provided D and N values and the AOQ calculation method

Manufacturer	D and N provided	AOQ calculation method	Manufacturer	D and N provided	AOQ calculation method
Advanced Micro Devices	Yes	NP	Motorola	Yes	JEDEC 16A
Analog Devices	Yes	EIA-554	MTC	No	NP
Atmel Corporation	Yes	Hahn	Murata Electronics	Yes	NP
AVX	Yes	EIA-554	Precision Devices, Inc.	No	NP
Burr-Brown	X	NP	Quick Logic	Yes	NP
Cornell-Dubilier	Yes	NP	RCD Components	Yes	NP
Dielectric Labs	X	NP	Sensycon	X	NP
Diodes Inc.	No	Eqn. 7.5	Tecate	No	NP
Ecliptek Corp.	No	JEDEC 16A	Telcom	No	NP
ELNA Inc.	No	NP	Monitor Products	No	NP
Elpac Electronics	No	NP	Texas Instruments	Yes	NP
Fairchild	No	NP	Vishay Dale	No	NP
General Semiconductors	Yes	NP	Vishay Foil Resistors	Yes	NP
Integrated Device Technology	Yes	Eqn. 7.5	Vishay Siliconix	No	JEDEC 16A
International Resistive Company	No	EIA-554	Vishay Sprague	Yes	EIA-554
Intersil	No	NP	Vishay Vitramon	No	EIA-554
ITW Paktron	No	NP	Xicor	Yes	NP
Kemet	Yes	EIA-554	Xilinx	No	JEDEC 16A

Figure 7.2 graphically shows the AOQ values reported for all 68 part groups separated by various device types. Here example cutoff limits for AOQ values are 2700, 460, and 63 ppm. Assuming a normal distribution of failures, these limits represent the number of parts

per million that are more than 3σ, 3.5σ, and 4σ away from the mean of the parameter. It is preferred that the manufacturer provide AOQ values for the past year. While only the current value may be used for assessment, past values may be cause for rejection if they show an unfavorable trend. If the trend indicates that the current value is an aberration or there is a significant upward trend, the assessment team may give the part a lower excellence level. Where no bar is visible in Figure 7.4, the reported AOQ was zero. Each bar represents the AOQ value for one of the part groups assessed. Part numbers have been left off to avoid congestion on the x-axis. Instead, part groups have been combined into six categories: integrated circuits, discrete semiconductors, inductors, capacitors, resistors, and oscillators. With the exception of inductors, where only one inductor was assessed, the minimum AOQ value for each category is zero. The maximum values, however, vary; they are all below 1000 ppm.

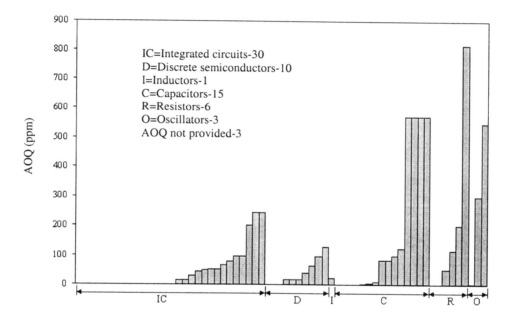

Figure 7.2: Graph of AOQ values for different device types

7.3.2   Process capability index

C_{pk} is a measure of process performance. The performance of a process is determined by how much margin exists between the design requirement characteristics and the actual value of these characteristics. Even though a process should reproduce its characteristics identically for each part, some variation occurs. This variation is usually measured in standard deviations (σ) from the mean, as shown in Figure 7.5.

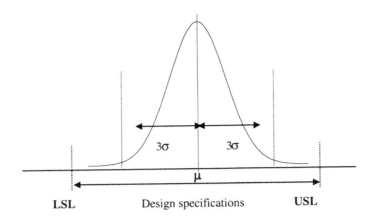

Figure 7.3: Distribution with 3σ process limits and upper and lower specification limits

Normal process variation is generally ± 3σ about the mean, which signifies that 2700 ppm fall outside the normal variation. Thus, to design a part virtually defect-free (zero ppm), specification limits for an individual process must be significantly more than ± 3σ away from the mean.

C_{pk} quantifies the variation in design width as compared with process width and is defined as

$$C_{pk} = \frac{\min((USL - \mu), \ (\mu - LSL))}{3\sigma} \qquad (7.6)$$

where *USL* stands for upper specification limit, *LSL* stands for lower specification limit, μ stands for process mean, and σ stands for process standard deviation [31].

An acceptable part should have C_{pk} values for all critical processes which are acceptable to the equipment manufacturer. Cutoff (acceptability) limits for C_{pk} should be determined by the equipment manufacturer, depending on the same factors as for AOQ (see Section 7.3.1). Some example cutoff limits are 1.00, 1.33, 1.67, and 2.00. These limits are based on 3σ, 4σ, 5σ, and 6σ from the mean of a normalized process distribution. The minimum acceptable cutoff limit should not be below 1.00. A C_{pk} of less than 1.00 indicates that the natural process spread is wider than the specification limits. Because most parts go through more than one process during manufacture, multiple C_{pk} values may be provided for each part. In this case, the lowest value, or worst case, is used for assessment.

Figure 7.4 shows the C_{pk} values reported by each manufacturer in the case study. Each bar represents the C_{pk} value the part groups assessed. Two bars have been truncated to allow greater resolution on the y-axis. Manufacturers did not provide process capability indices (C_{pk}) for 14 part groups. Of these, C_{pk} was not calculated for nine part groups and was proprietary for three part groups. No reason was given why C_{pk} was not provided for the remaining two part groups. Three part groups had a C_{pk} lower than 1.00. Two bars have been truncated to allow greater resolution on the y-axis.

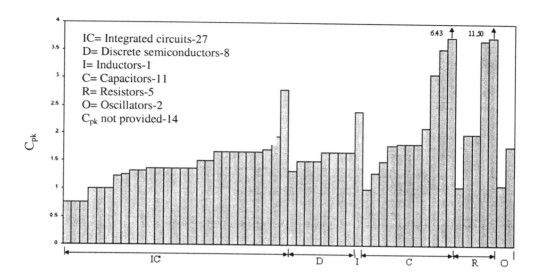

IC= Integrated circuits-27
D= Discrete semiconductors-8
I= Inductors-1
C= Capacitors-11
R= Resistors-5
O= Oscillators-2
C_{pk} not provided-14

Figure 7.4: Graph of C_{pk} values for different device types

7.3.3   Why qualification test results are not a part assessment category

Qualification tests are conducted prior to mass production and distribution of a part to baseline the effects of new materials, processes, and designs on production parts. Qualification tests are usually conducted only once. However, requalification tests may be conducted after major changes to a part or process. Examples of qualification tests include temperature cycling, highly accelerated stress test (HAST), high temperature operating life (HTOL), autoclave, thermal shock, bake, solderability, lead finish adhesion, and lead pull.

All manufacturers in the case study have a process for part qualification; however, qualification test results were provided for only 8 of the 68 part groups. Most manufacturers commented that qualification results might not be relevant if the part is mature or has undergone changes after initial qualification. Texas Instruments noted that "Availability of qualification test results will depend on how long it has been since qualification was performed. In the case of your listed [assessment] devices [for the case study], no data was found in our electronic database which includes 1994 to the present" [35].

Due to the unavailability of qualification tests and their possible irrelevance to a candidate part, qualification test results are not included as a category in the part assessment procedure. Instead, integrity monitor test results are used, which are more readily available and are more relevant.

7.3.4   Integrity monitor test results

Integrity monitor tests are those tests conducted by the part manufacturer to continuously monitor changes in the part, materials, or processes. Integrity monitor tests are also called "life tests," "continuous monitor tests," "environmental tests," and "reliability tests." However, integrity monitor tests are not actual reliability tests because they are application-independent and thus, do not measure reliability (see Chapter 14).

Integrity monitor tests are similar to qualification tests. They involve similar tests; however, fewer tests are usually conducted, and these tests are less severe in magnitude and duration than those used during qualification. Unlike qualification tests, integrity monitor tests are conducted periodically. In general, integrity monitor tests are destructive in nature. Sample sizes are relatively low compared to actual production levels or the sample sizes used during quality inspection.

Many different types of integrity monitor tests can be conducted. The test types, conditions, and duration vary from manufacturer to manufacturer and from part to part. The tests conducted are device and package specific. A test is selected to monitor failures intrinsic to a particular device and/or package. Due to the wide variety of parts, there is no set of integrity monitor tests that is applicable to all parts. Some of the more common integrity monitor tests are:

- low temperature operating life (LTOL)
- high temperature operating life (HTOL)
- high temperature storage (HTS)
- autoclave (AC)
- highly accelerated stress test (HAST)
- temperature humidity bias (THB)
- temperature cycle (TC)
- thermal shock (TS).

The results of integrity monitor tests are generally published in a written document, but some are also placed on the Internet. Results are most often published on a quarterly basis; however, some are published biannually. Most manufacturers provide integrity monitor test results in terms of the number of failures and the sample size tested.

For an acceptable part, the manufacturer should conduct integrity monitor tests based on acceptable industrial standards, such as JEDEC Standard 22 Series [36], MIL-STD 883 [37], STACK 0001 [13], or AEC-Q-100 [11], and provide the results to customers upon request. Results should include test types, test conditions, duration, sample size, and number of failures. Root cause analysis should be performed for all failures. If any valid failures are reported, parts manufactured during the same period as those that the data represents are not recommended for use. This is because the same kind of defect may exist in other nontested parts. Failures due to inspection and improper handling should not be considered valid failures.

In the case study, manufacturers provided integrity data for LTOL, HTOL, HTS, AC, HAST, THB, TC, and TS. Integrity monitor test results were not provided for 20 part groups from 14 manufacturers. Of these, one manufacturer, Elpac, stated that it did not conduct integrity monitor tests on the case study parts. Valid failures were found for six part groups from five different manufacturers. The results provided by each manufacturer can be seen in Table 7.3. The table includes the test types, conditions, number of failures per sample size and the results of any failure analysis conducted.

Table 7.3: Valid failures found during integrity monitor testing

Manufacturer and part group	Test type	Test conditions	Failures	Failure analysis
Analog Devices AD7xxx [38]	HTOL	MIL-STD-883, 125°C, 1000 hrs	3/250	Two silicon defects and one gate oxide
Analog Devices ADxxx [39]	HTOL	MIL-STD-883, 125°C, 1000 hrs	1/90	Marginal ICC shift
	THB	JEDEC-STD-22, 85°C, 85% R.H., 1000 hrs	1/150	Broken bond wires
	TS	MIL-STD-883, -65 to 150°C	0/45	Not applicable
Burr – Brown ADS7804 [40]	Life test	Not given	2/11	Wire shorted to lead frame and damaged wire bond
	HAST	130°C, 85% R.H., 33.5 psig, 100 hrs	1/12	Particle contamination
	TC	−65 to +150°C 500 cycles	1/15	Particle contamination
	AC	121°C, 100% R.H., 15 psig	0/3	Not applicable
Kemet T491 [41]	Life test	85°C, 2000 hrs	27/2215*	Failure analysis results were not provided
Motorola MC68332 [42]	Early dynamic operating life	125°C, 6.0 V, 48 hrs	1/171	CPU functional failure
	THB	85°C, 85% R.H., 1008 hrs	0/150	Not applicable
	TC	−55 to +150°C 1000 cycles	0/160	Not applicable
IDT 74FCT [43]	Dynamic high-temperature life	+5.5V, +150°C, 57500 total device hrs	1/115	Non-visual fabrication defect
	HAST	JEDEC-STD-22, Method A110; 5.5V, +130°C, 85% R.H., 33.3 psig, 240 hrs	0/45	Not applicable
	AC	5.5V, +130°C, 85% R.H., 33.3 psig, approx. +121°C, 100% R.H., 30 ±1 psig, 168 hrs	0/45	Not applicable
	TC	JEDEC-STD-22, Method A104; Temp. range: -65 to +150°C, 1000 cycles	0/45	Not applicable
* Results represent a 15-month moving interval.				

7.3.5   Assembly guidelines

Assembly guidelines are recommendations by the part manufacturer to prevent damage and deterioration of the part during the assembly process. Examples include a recommended reflow temperature profile, solder type, solder footprint on the PCB, cleaning agents, adhesives, moisture sensitivity, and ESD protection. Assembly guidelines could also include information regarding part compatibility with equipment or technologies. New solder materials, solder profiles, cleaning agents, and moisture and ESD standards continue to be developed; lead-free soldering is an area of increasing importance. As new technologies emerge, assembly guidelines become more important to ensure the quality and integrity of parts used by the equipment manufacturer.

For an acceptable part, the manufacturer should provide the necessary assembly guidelines to the equipment manufacturer to ensure that materials, processes, equipment, and additional parts are compatible with the candidate part. Parts for which guidelines acceptable to the assessment team are not provided are not recommended. Acceptable assembly guidelines were provided for 59 of the 68 part groups.

7.4   Case study results

The part assessment results for all 68 case study parts are listed in Table 7.4. Each of the 68 part groups, composed of 113 parts from 36 manufacturers, was assessed using the procedures described above. A minimum excellence level of 1 is required for each assessment category for the part group to be acceptable. Actual values reported by the manufacturer are given in parentheses for AOQ and C_{pk}. The numbers under each assessment category represent the excellence level assigned to that part group. An excellence level of 0 indicates that the part did not satisfy the assessment criteria for the category. "N/P" indicates that the metric is calculated and tracked but that the manufacturer did not provide the information. "N/C" indicates that the metric is not calculated and therefore could not be provided. Thirty-six of the 68 part groups were not recommended for use due to failure to satisfy all the criteria in the assessment categories.

Table 7.4: Part assessment results

Manufacturer	Part group	Assessment categories			
		AOQ (actual values in parentheses)	C_{pk} (actual values in parentheses)	Integrity monitor test results	Assembly guidelines
		(out of 3)	(out of 3)	(out of 1)	(out of 1)
Advanced Micro Devices	AM29F400	2 (200)	1 (1.23)	1	1
Analog Devices	AD7xxx	3 (0)	1 (1.00)	0	1
	ADxxx	3 (0)	1 (1.00)	0	1
	OP28x	3 (0)	1 (1.00)	1	1

Manufacturer	Part group	Assessment categories			
		AOQ (actual values in parentheses)	C_{pk} (actual values in parentheses)	Integrity monitor test results	Assembly guidelines
		(out of 3)	(out of 3)	(out of 1)	(out of 1)
Atmel Corporation	AT25	2 (80)	2 (1.50)	1	0
	AT28	3 (0)	2 (1.50)	1	0
AVX	TPS	3 (10)	3 (2.12)	0	1
Burr-Brown	ADS7804	0 (N/P)	0 (N/P)	0	1
Cornell-Dubilier	Flat Pack	2 (100)	2 (1.51)	0	1
Dielectric Labs	CxxCF	0 (N/P)	0 (N/P)	1	1
Diodes Inc.	BZX84C	2 (65)	2 (1.33)	0	1
	ESxD	3 (50)	2 (1.33)	0	1
	SMCJ	3 (40)	2 (1.50)	0	1
ELNA Inc.	EC1100HSETTS	2 (300)	1 (1.07)	1	1
Elpac Electronics	RVJ	3 (0.1)	3 (3.54)	0	1
	C Series	1 (575)	0 (N/C)	0	0
	MD Series	1 (575)	0 (N/C)	0	0
	PM Series	1 (575)	0 (N/C)	0	0
Ecliptek Corp.	Z Series	1 (575)	0 (N/C)	0	0
Fairchild	1N4148	2 (130)	2 (1.50)	1	1
General Semiconductors	SMCJ	3 (0)	1 (1.24)	1	1
Integrated Device Technology	74FCT	3 (30)	1 (1.25)	0	1
International Resistive Company	PFC	3 (52)	1 (1.04)	0	1
Intersil	8xC8x	3 (17)	0 (N/C)	1	1
	Diode	3 (0)	0 (N/C)	1	1
	HI1506	3 (17)	0 (N/C)	1	1
	Power MOSFET	3 (0)	0 (N/C)	1	1
ITW Paktron	CS4G	3 (3)	1 (1.01)	1	1
Kemet	C1206	3 (5)	1 (1.30)	0	1
	T491	3 (2.26)	3 (1.80)	0	1

Manufacturer	Part group	Assessment categories			
		AOQ (actual values in parentheses)	C_{pk} (actual values in parentheses)	Integrity monitor test results	Assembly guidelines
		(out of 3)	(out of 3)	(out of 1)	(out of 1)
Monitor Products	SM990	1 (550)	0 (N/P)	1	1
Motorola	2N	3 (18)	3 (1.67)	1	1
	MC68332	3 (54)	3 (1.67)	0	1
	MC74HC	3 (46)	3 (1.67)	1	1
	MMBF0201	3 (18)	3 (1.67)	1	1
	MPC509	3 (54)	3 (1.67)	1	1
	Mxx	3 (18)	3 (1.67)	1	1
MTC	MTCxxxx	2 (69)	3 (2.78)	0	1
Murata Electronics	GRM42-6X	3 (0)	3 (1.83)	1	1
Precision Devices Inc.	PDIxxxx	3 (4)	3 (1.78)	1	0
Quick Logic	QL2009 (pASIC1)	3 (0)	3 (1.67)	1	1
	QL24X32 (pASIC2)	3 (0)	2 (1.33)	1	1
RCD Components	ZJ-2	3 (2)	3 (2.00)	0	1
Sensycon	SMD2101	0 (N/P)	0 (N/P)	0	1
Tecate	MXZZ	2 (124)	3 (3.08)	1	0
Telcom	TC28C	3 (1)	0 (N/C)	0	1
Texas Instruments	SN54LS	3 (0)	2 (1.37)	0	1
	TL072	3 (0)	2 (1.37)	1	1
	TLC2272	3 (0)	2 (1.37)	1	1
	TLC37	3 (0)	2 (1.37)	1	1
	TLE2027	3 (0)	2 (1.37)	0	1
	TLE2062	3 (0)	2 (1.37)	1	1
Vishay Dale	CRA06S	1 (817)	3 (2.00)	1	1
	CRCW	2 (117)	3 (3.69)	0	1
	IMC	3 (23)	3 (2.41)	1	1
	LVR-3	3 (0)	3 (11.50)	1	1

Manufacturer	Part group	Assessment categories			
		AOQ (actual values in parentheses)	C_{pk} (actual values in parentheses)	Integrity monitor test results	Assembly guidelines
		(out of 3)	(out of 3)	(out of 1)	(out of 1)
Vishay Foil Resistors	SMRxD	2 (200)	0 (N/P)	0	0
Vishay Siliconix	DG412	2 (99)	3 (1.67)	1	1
	DG506	2 (99)	3 (1.67)	1	1
	JFET SOT-23	2 (99)	3 (1.67)	1	1
Vishay Sprague	293D	3 (0)	3 (6.43)	1	1
Vishay Vitramon	VJxxxxA	2 (85)	3 (1.82)	1	1
	VJxxxxY	2 (85)	3 (1.82)	1	1
Xicor	X28C010	3 (0)	0 (0.76)	1	1
	X28C256	3 (0)	0 (0.76)	1	1
	X28VC256	3 (0)	0 (0.76)	1	1
Xilinx	XC95xx-10PC	2 (245)	3 (1.71)	1	1
	XC95xx-15PQ	2 (245)	3 (1.95)	1	1

Three part groups from three manufacturers were not recommended because they failed to satisfy the criteria for AOQ. In all three cases, AOQ was not provided. Two manufacturers stated that this information was proprietary.

The criteria for C_{pk} were not satisfied for 17 part groups from nine manufacturers. C_{pk} was not provided for 14 of the part groups. Three manufacturers stated that C_{pk} is proprietary. C_{pk} was not calculated for nine part groups. Three part groups had a C_{pk} lower than 1.00.

In the case study, 26 part groups from 18 manufacturers did not satisfy the criteria for the integrity monitor test results category. Six part groups showed valid failures during their integrity monitor testing (see Table 7.3). Results were not provided for 16 part groups. Cornell Dubilier and Sensycon both stated that integrity monitor test results could be viewed on site. Elpac stated that integrity monitor tests are not conducted on their C Series, MD Series, PM Series, and Z Series part groups. Assembly guidelines were not provided for nine part groups from five manufacturers.

7.5   Summary

A methodology has been presented for the quality and integrity assessment of an electronic part. The methodology and criteria were developed from numerous industry programs and standards, in conjunction with a case study of 113 parts from 36 manufacturers. The part assessment procedure begins with the collection of information. If information is unavailable for the specific part, the part is put into a part group. Then the part's quality and integrity are evaluated by comparing them against the assessment criteria seen in Table 7.5. All assessment categories, criteria, and the minimum excellence levels for each are listed here. In order to be a recommended part, all criteria for each category shown in the table must be satisfied. If all criteria are not satisfied, then the part is rejected. A rejected part should not be used unless special circumstances warrant, such as when no alternative part is available. In this situation, equipment supplier intervention techniques should be employed [24].

Table 7.5: Summary of part assessment criteria and categories

Assessment category	Assessment criteria				Part assessment excellence level	Recommended minimum acceptable excellence level
	Excellence level 0	Excellence level 1	Excellence level 2	Excellence level 3		
AOQ (in ppm)	AOQ > 2700	$460 < AOQ \leq 2700$	$63 < AOQ \leq 460$	$AOQ \leq 63$		1
C_{pk}	$C_{pk} < 1.0$	$1.00 \leq C_{pk} < 1.33$	$1.33 \leq C_{pk} < 1.67$	$C_{pk} \geq 1.67$		1
Integrity monitor test results	Integrity test results are not provided or results show valid failures.	Integrity test results are provided and show no valid failures.	—	—		1
Assembly guidelines	Adequate assembly guidelines are not provided.	Necessary assembly guidelines are provided.	—	—		1

The assessment categories were developed to contain the minimum number of criteria necessary to assess fully a part's quality and integrity. The methodology does not require the equipment supplier to perform an in-house audit of the manufacturer. Although no amount of assessment will eliminate all risk associated with the use of a part, an effective part assessment program like the one described here can reduce the risk.

7.6   References

[1]   Jackson, M., A. Mathur and M. Pecht, "Part Manufacturer Assessment Process," *Quality and Reliability Engineering International*, Vol. 15, pp. 457–468, 1999.

[2]   Foucher, B., R. Kennedy, N. Kelkar, Y. Ranade, A. Govind, W. Blake, A. Mathur and R. Solomon, "Why a New Parts Selection and Management Program?" *IEEE Transactions on Components, Packaging, and Manufacturing Technology,* Part A, Vol. 21, No. 2, pp. 375–378, June 1998.

[3]   Kanter, H. and R. Atta, *Integrating Defense into the Civilian Technology and Industrial Base*, Institute for Defense Analyses, Alexandria, VA, February 1993.

[4]   United States Department of Defense, *MIL-M-38510: General Specification for Microcircuits*, Revision J, Columbus, OH, November 15, 1991.

[5]   United States Department of Defense, *MIL-STD-976: Certification Requirements for Microcircuits*, Columbus, OH, November 15, 1994.

[6]   United States Department of Defense, *MIL-STD-1562: List of Standard Microcircuits*, Columbus, OH, December 15, 1995.

[7]   United States Department of Defense, *MIL-STD-965: Parts Control Program*, Columbus, OH, September 26, 1996.

[8]   United States Department of Defense, *MIL-STD-454: Standard General Requirement for Electronic Equipment*, Columbus, OH, May 4, 1995.

[9]   British Standards Institution, *CECC 00804: 1996 Guidance Document: Interpretation of "EN ISO 9000: 1994," Reliability Aspects for Electronic Components*, London, February 1996.

[10]   Perry, W., "Specifications and Standards — A New Way of Doing Business," Internal Memorandum, U.S. Department of Defense, June 29, 1994.

[11]   American Institute for Aeronautics and Astronautics, *Recommended Practice for Parts Management, R-100-1996*, Reston, VA, 1996.

[12]   Automotive Electronics Council, *CDF–AEC–Q100 Stress Test Qualification for Automotive-grade Integrated Circuits*, Revision E, Detroit, MI, January 31, 2001.

[13]   STACK International, *Specification 0001 Issue 12.2 at Notice 2: General Requirements for Integrated Circuits*, Herts, UK, November 19, 1999.

[14]   AlliedSignal, Inc. [now Honeywell], *Electronic Component Management Plan*, Tucson, AZ, June 1997.

[15]   Boeing Inc., *Electronic Component Management Program*, Seattle, WA, February 1997.

[16]   Eldec Corp., *Electronic Component Management System*, Lynnwood, WA, March 21, 1997.

[17]   Hughes Aircraft [now part of Boeing], *Guidelines for Use of Plastic Encapsulated Microcircuits (PEMs) in Naval Avionics Equipment*, Indianapolis, IN, September 29, 1997.

[18]   Motorola, Inc., *Parts Management Plan for Iridium™ Comm Module*, Revision XA, Chandler, AZ, June 15, 1992.

[19]   Northrop Grumman Corporation, *Exciter/Receiver/Processor Parts Control Program Plan*, Revision A, Baltimore, MD, August 8, 1997.

[20]   Rockwell Collins, Inc., *Air Transport Systems Electronic Component Management Program*, Cedar Rapids, IA, January 7, 1999.

[21]   Smiths Industries, *Electronic Component Management Plan*, Cheltenham, UK, February 1, 1997.

[22] Sun Microsystems, *950-2484-01: ASIC Component Engineering Requirements (CER)*, Revision 3.3, Santa Clara, CA, June 25, 1996.

[23] Electronic Industries Association, *EIA-670: Quality System Assessment*, Arlington, VA, June 1997.

[24] Jackson, M., P. Sandborn, M. Pecht, C. Hemens-Davis and P. Audette, "A Risk-Informed Methodology for Parts Selection and Management," *Quality and Reliability Engineering International*, Vol. 15, pp. 261–271 , September 1999.

[25] Schwach, C. A., A. Mathur, R. Solomon, P. Sandborn and M. Pecht, "Equipment Supplier Intervention Techniques," *Future EMS International*, Technology Publishing Ltd., London, Issue 3, pp. 97–100, 2000.

[26] Analog Devices, *Product Cross Reference*, Norwood, MA, <www.analog.com/world/quality/read/xref1.html>, accessed July 13, 1999.

[27] JEDEC Standards, *Assessment of Average Outgoing Quality Levels in Parts Per Million (PPM), JESD16-A*, Electronic Industries Association, Arlington, VA, April 1995.

[28] Ackermann, C. S. and J. M. Fabia, "Monitoring Supplier Quality at PPM Levels," *IEEE Transactions on Semiconductor Manufacturing,* Vol. 6, No. 2, pp. 189–195, May 1993.

[29] Electronic Industries Association, *EIA-554: Method Selection for Assessment of Nonconforming Levels in Parts per Million (PPM)*," Arlington, VA, August 1996.

[30] Hahn, G. J., "Estimating the Percent Nonconforming in the Accepted Product After Zero Defect Sampling," *Journal of Quality Technology,* Vol. 18, No. 3, pp. 182–188, July 1986.

[31] Montgomery, D. C. and G. C. Runger, *Applied Statistics and Probability for Engineers*, 2nd ed., John Wiley & Sons, Inc., New York, NY, 1999.

[32] Kemet Electronics, *EIA-554 PPM-2 Performance: Ceramic Molded Axials*, Simpsonville, SC, July 29, 1999.

[33] Kemet Electronics, *EIA-554 PPM-3 Performance: Ceramic Molded Axials*, Simpsonville, SC, July 29, 1999.

[34] Zhang, N. F., G. A. Stenack and D. M. Wardrop, "Interval Estimation of Process Capability Index, C_{pk}," *Communications in Statistics – Theory and Methods,* Vol. 19, No. 12, pp. 4455–4470 , 1990.

[35] Biggerstaff, J., Texas Instruments, letter to A. Mathur, AlliedSignal, July 17, 1997.

[36] JEDEC Standards, *Test Methods, JEDEC Standard 22 Series*, Electronic Industries Association, Arlington, VA, April 2000.

[37] United States Department of Defense, *MIL-STD-883: Test Methods and Procedures for Microelectronics*, December 31, 1996.

[38] Analog Devices, *ADI-BiCMOS: Life-Test Data*, Norton, MA, <www.analog.com/world/quality/read/bicmos.html>, accessed July 13, 1999.

[39] Analog Devices, *ADI-Temperature Humidity Bias for SOIC Packages*, Norton, MA, <www.analog.com/world/ quality/read/thbr.html>, accessed May 1, 1999.

[40] Burr-Brown Corp., *Reliability Assurance Program,* Tucson, AZ, Quarter 2, 1999.

[41] Kemet Electronics, *Failure Rate Molded Chip-Tantalum*, Simpsonville, SC, July 1, 1998.

[42] Motorola Inc., *Reliability and Quality Monitor Report*, Austin, TX, Quarter 1, 1996.

[43] Integrated Device Technologies, Inc., *Quality and Reliability Monitors*, Santa Clara, CA, July 1997.

Chapter 8

Electronic Part Distribution and Distributor Assessment

Sanjay Tiku, Laurie Sullivan, Liyu Yang, and Michael Pecht

Distributors are the link that enables parts suppliers to reach deeper into the supply chain to serve thousands of original equipment manufacturers (OEMs) and electronic manufacturing service (EMS) providers. Warehousing, packaging, kitting, order tracking, credit terms, new product introductions (NPI), design-in services, and technical support are some of the more traditional offerings. Others include supply chain services such as forecasting, inventory management, and logistics, as well as product change and obsolescence notifications. Appendix A lists the top 25 franchised distributors. Appendix B lists product categories and services available from the major distributors.

Electronic distributors today are known as supply chain and inventory managers. The road, however, has been long and difficult. Electronic parts distributors started by supporting the specialized needs of amateur radio enthusiasts from the Courtland Street section of New York City (then known as Radio Row) in the early 1920s and evolved into the resistor, capacitor, and vacuum tube electrical retailers of the 1940s. In the 1950s, they began carrying connectors, and in the 1960s they added transistors and diodes, replacing vacuum tubes. It was not until the 1970s that integrated circuits emerged, primarily through Fairchild Semiconductor, Texas Instruments, and Motorola Semiconductor. In the late 1970s and 1980s computer systems and peripherals came on the scene.

From the mid-1980s, the distributors focused on just-in-time (JIT) delivery, engineering services, kitting, bar coding, assembly, and programming to meet specific customer requirements. By the mid-1990s, some North American distributors began to expand into Europe, Asia-Pacific, and Japan. Supply chain services such as forecasting and inventory management emerged in the late 1990s after distributors invested millions of dollars in sophisticated IT systems to streamline processes. A fee-based model emerged that separated the component price from the price of the service.

With the millennium emerged Web-based tools and supply chain services aimed at tracking, improving, and lowering transaction costs. These new services included technical information and specifications aimed at helping customers select and cross-check parts during product development cycles as well as managing inventory. For example, Avnet Inc., in Phoenix, Arizona, has taken its Point of Use Replenishment System (POURS) to the Web. The system operates from handheld Windows-based Pocket PC barcode scanners, and the supply chain tool's open architecture allows the user to connect wirelessly to Avnet and other suppliers to replenish stock.

8.1   Why part manufacturers use distributors

Distributors are the middlemen that specialize in assisting OEMs and EMS providers to procure a large mix of components from many parts manufacturers in smaller quantities (see Figure 8.1). Those OEMs and EMS providers whose devices are built in bulk and require a smaller mix of parts from a handful of part manufacturers normally procure parts direct from the part manufacturers to secure better pricing. Parts manufacturers also generally

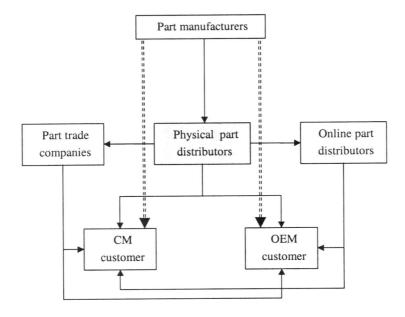

Figure 8.1: Electronics parts supply chain diagram

serve the OEMs and EMS providers that procure the highest quantities directly and rely on distributors to service the remainder. Distributors that aggregate customer procurement requests may be afforded more favorable pricing.

Part manufacturers and customers rely on the marketing and the inventory burden that distributors assume. Distributors also create demand, provide technical assistance, and extend payment terms to customers on behalf of the parts manufacturer. For example, companies like Altera and Xilinx that offer proprietary product lines, such as field programmable logic, will train the distributor's field application engineers to assist as their extended sales arm. Distributors also have begun to provide logistics support to companies that procure material directly from the parts manufacturers. Some parts manufacturers like Vitesse Semiconductor have publicly recognized the importance of reducing costs by tapping distribution rather than third-party logistics providers, UPS Supply Chain Services, or Exel, a technology and global freight management service.

Financial cutbacks and industry layoffs following the downturn in late 2000 forced part manufacturers to rely more on distribution for their sales and technical services. This enabled them to devote more working capital to research and concentrate on new product development.

The trend toward globalization has prompted a handful of distributors to follow parts manufacturers around the world to offer logistics services as EMS providers push manufacturing to low-cost regions like China and Eastern Europe. For example, Nu Horizons Electronics created Titan Logistics Corp., in Melville, New York, in 2000 to support parts manufacturers worldwide. The Melville company contracted with Vitesse Semiconductor in 2003 to support the integrated circuits (ICs) and optical modules parts manufacturers in North America and China.

8.2   Why customers buy from distributors

Quick access to parts, limited liability and the risk of noncancelable/nonreturnable or obsolete components, favorable credit terms, and the opportunity to maintain smaller inventories to improve cash-to-cash cycles are the top reasons for OEMs and EMS providers to procure material through distribution. Distributors that invested millions of dollars to upgrade global IT systems have become a critical link between parts manufacturers and their customers. These sophisticated IT systems enabled distributors to offer supply chain services such as forecasting, part change notifications, and information on when a part is nearing the end of its life. To focus on core competencies, customers are increasingly partnering with distributors like Arrow Electronics Inc., in Melville, New York, and Avnet, which can provide them with a variety of products and services.

EMS is the fastest-growing customer segment for distribution since OEMs began to divest their manufacturing facilities to eliminate fixed costs. Distributors continue to maintain relationships with top-tier OEMs and EMS providers. However, the "sweet spot" or the market segment from which distribution will garner its highest gross profit margins is the second-, third-, and fourth-tier OEMs and EMS providers that rely on distributors for logistics and materials management, as well as other value-added services such as chip programming, kitting and return privileges.

8.3   Types of electronic parts distributors

Electronic parts distributors are classified either as franchise broad-line or specialist, as well as independent or broker. "Franchise" refers to the product lines distributors are authorized to sell. Distributors classified as independents or brokers are not authorized by the parts manufacturer to sell the material. Independent distributors and brokers are not categorized as broad-line or specialists and are not restricted by product type, manufacturer, regions, or price.

8.3.1   Franchised distributors

Franchised distributors that support product lines have signed selling and marketing contracts with parts manufacturers. The agreement accompanies privileges such as better pricing, ship and debit to protect against price degradation, and stock rotations for slow-moving inventory [9]. However, the agreement also means that the distributor must make a pre-negotiated financial investment to warehouse and secure the product on its line card.[1] These well-defined contracts, also known as "terms" and "conditions," help to avoid disputes and reduce liability and risk concerns that may arise. Contracts are negotiated to determine specific stipulations in the agreement such as market price, credit terms, part qualities, and return policies.

Part manufacturers provide technical support and sales referrals as part of the franchise agreement. In return, distributors act as an extension of the part manufacturers' sales force, marketing to a broad customer base that the part manufacturers cannot serve economically. Franchised distributors are categorized as broad-line distributors and specialized distributors.

8.3.1.1 Broad-line distributors

Broad-line distributors carry numerous part manufacturers on their line cards. The products range from passives to semiconductors. Many times a distributor chooses to offer

---

[1] The term "line card" is used in the electronic distribution industry to mean the product offerings.

this type of line card to provide more products on the bill of material submitted in a request for quote by the procurement specialist at either the OEM or the EMS provider.

8.3.1.2 Specialized distributors

Specialized distributors focus on specific product lines such as passive or active parts. They typically have expertise in either interconnect, passive, and electromechanical parts or semiconductors and are experienced in specific technologies. TTI Inc., in Fort Worth, Texas, is an example of a distributor specializing in passive and interconnect parts. Memec, in San Diego, CA is an example of a distributor specializing in semiconductors. The company is particularly useful to parts manufacturers who believe that specialized distributors that focus on their product can serve the customer better. Sales, marketing, and technical employees at specialized distributors are generally more knowledgeable in specific areas. This expertise may give the customer a competitive advantage. There also are companies that stock several hundred thousand products distributed through paper- and Web-based catalogs, such as Digi-Key Corp., in Thief River Falls, MN and Newark Electronics, in Chicago, IL.

8.3.2   Independent distributors

Distributors not authorized by the parts manufacturer to sell their products are known as "independents" or "brokers." They can carry any supplier line and sell any product they wish. The distinction lies in whether the company stocks products or becomes the mediator that finds the seller of specific components for the buyer. While these distributors provide customers some valuable services, they are far from the conventional and the traditional bar-coding, kitting, tape and reel, device programming, or design-in services offered by franchised distributors.

Independent distributors were among the industry's earliest supply chain managers, evolving from working on kitchen tables or behind urban storefronts to become sophisticated, Web-savvy businessmen and businesswomen. They initially brokered radio and vacuum tubes and later turned their businesses into influential franchised distributors, e.g., Hamilton Avnet and Schweber Electronics, as well as All American Semiconductor Inc., in Miami, Florida [10]. Franchised distribution, in fact, emerged from the independent distributor business models.[2]

The value independent distributors such as Advanced MP Technology, America II Electronics, and Smith & Associates bring to the supply chain is the ability to locate hard-to-find parts in allocated markets. When material is plentiful, these distributors are the outlets for excess inventory for many OEMs and EMS providers. Condensed component life cycles have led to unplanned demand and shortages, making forecasting more complicated. When material is plentiful, they can manage both surpluses and obsolete parts for the customer. Additionally, independent distributors provide OEMs with access to spot market pricing to reduce component costs [1].

---

[2] "When the semiconductor industry emerged in the 1960s, the Vietnam War was in progress. The industry was unprepared to handle the surge in demand for products during that period" [10]. High-reliability parts for defense electronics applications were in great demand, and consumer spending was high. The same independent sources that had become an important sales arm for emerging players, such as Fairchild and Texas Instruments, attempted to keep up with the supply and demand but quickly realized that they did not possess the sales, marketing, and technical backup to fulfill industry needs. The concept of franchised distribution emerged, and most of these early players in distribution contracted with the part manufacturers to sell their devices within an agreed pricing structure and with limits on the number of products that could be sold to any one customer. In return, the component manufacturers supplied technical support. This was the fundamental concept of franchised distribution, which grew rapidly throughout the 1960s and became established as a potent force in the supply chain during the 1970s.

Following the industry downturn in late 2000, for example, OEMs and EMS providers with warehouses filled with excess inventory, created by grossly inflated end market demand forecasts, used independent distribution to dump billions of dollars of inventory in an effort both to redistribute the parts in the channel and to get the inventory off their books. These distributors can, however, make a profit and support customers simultaneously by procuring the parts from customers for pennies on the dollar, stocking the inventory until demand kicks in, and then selling the parts at the market price.

Lingering economic woes that plagued the industry in 2002 and 2003 increased independent distributors' role in assisting OEMs and EMS providers to secure more competitive component pricing. This increased their worth in the supply chain, enhanced their credibility, and earned them a place on the EMS provider's preferred parts sourcing list during a time when companies began to rationalize and reduce their supply base. In a pointed departure from historic sourcing practices, EMS providers in early 2002 began to give independent distributors preferred supplier positions on their approved-vendor lists as a way to bring order to chaotic procurement in past years [5]. This shift, in part, induced a more positive attitude by OEMs as well.

8.4   Distributor identification for assessment

Creating demand for a product or assessing the correct material to design into a customer's device is a critical process and requires the distributor to be technically savvy. Each company location, site, and facility are collectively considered one distributor for the purpose of assessment if they follow a common quality system policy manual, which undergoes a continuous external audit process such as ISO 9000, ANSI/ASQC Q90, QS 9000, AS 9000, TL 9000, or another equivalent certification process.

The distributor identification and assessment process must be cognizant of several factors:

- state of constant flux in the distribution market due to mergers and acquisitions
- additional services provided by the distributors where the line between distributors and other supply chain members is blurred
- global manufacturing and procurement activities
- Internet-based procurement.

8.4.1   Flux in the distribution market

The trend of acquisitions and mergers makes it difficult to determine the actual identification of a company. In some cases, North American distributors have been forced to divest their business due to the migration of manufacturing to China. For example, for Cleveland-based Pioneer-Standard Electronics Inc.'s Industrial Electronics Division, it came down either to divesting the business or making the financial commitment to expand globally and compete with the major players like Avnet or Arrow. In February 2003, management made the decision to sell the electronic components business to Arrow to concentrate on its computer business.

When someone else acquires a company, it may retain part or all of its previous identity and policies or may be completely absorbed. The company's quality policies and procedures may also change.

8.4.2   Additional services from distributors

Traditional distribution tasks include purchasing, stocking, order tracking, and shipping electronics parts. In addition, distributors offer other value-added services, e.g., kitting, warehousing, and device programming. These services often constitute a substantial percentage of the cost of acquisition and ownership of an electronic part.

Both EMS providers and distributors offer design services and on occasion compete for the same value-added business. Distributors that specialize in a specific technology niche like semiconductors or radio frequency and wireless offer standalone design services as another way to increase margins. However, many rely on the proceeds garnered from the silicon sale on manufacturing runs when the design goes into production. At the same time, their ability to influence part selection for OEMs and other users raises their standing among parts manufacturers. EMS providers, on the other hand, offer OEMs design services as a more efficient production technique enabled by a tighter coupling of design and manufacturing. Both parties (distributors and EMS providers) can strengthen ties with customers by offering those services [11].

The trend in the electronics industry is for both the distributors and the EMS providers to strengthen their OEM ties by involving them at the earliest stages of product development with Web-based design and supply chain tools. Major contract manufacturers are developing partnerships with enterprise design software companies to offer design services. Celestica is partnering with e4enet.com Inc. to provide a design share platform. Selectron is cooperating with ChipData Inc. for secure design transmission.

Distributors have also integrated software companies into their service offerings. For example, Arrow and Agile Software Corp. signed a partnership agreement in April 2003 aimed at providing purchasers with more detailed information on supplier part numbers. The Web-based solution combines Agile's Product Lifecycle Management (PLM) suite of tools with Ubiquidata, an electronic components database that the Melville, New York, distributor offers through its services group, Global Information Business. The combined Agile toolset and Ubiquidata database is expected to cut in half the time that procurement executives and design engineers spend researching information on components. Users also gain access to life cycle and cross-reference information on component supplier part numbers.

Both Arrow and Avnet began rolling out these types of services in 2000. Acceptance by OEMs and EMS providers for global support through Avnet Electronics Marketing Americas Supply Chain Services has increased in the past two years. Services offered through Avnet's Promiere supply chain services tools range from BOM Manager and Components Selector [7] to the POURS that supports KanBan, consignment inventory, and push/push inventory replenishment services. With more than 100 companies tapping POURS, Avnet took its server-based system to the Web, launching the tool in early 2003 through its Electronics Marketing Division geared to electronic components. The Web-based system gives users access to the identical system anywhere in the world.

8.4.3   International distribution

Globalization has opened new opportunities for the distribution industry. Arrow, Avnet, Future Electronics, Memec Group Holdings, and Richardson Electronics are among those with a significant presence in China through either acquisitions or organic growth. All American Semiconductor, Jaco Electronics, and Nu Horizons Electronics began to move into the region in 2002. With the opening of China's economy, the multinational distributors have captured a major market share in the distribution of electronic parts, with Arrow and Avnet emerging as the top two [2].

Estimates indicate that Avnet and Arrow had captured 32% of the global distribution total available market (DTAM) by 2001, rising to more than 50% in North America and Europe during 2003. Small to mid-size regional distributors actively seek mergers to stave off growing competition from multinational rivals. Some have tapped business practices learned from Western rivals and offer additional services such as logistics, demand creation, and product and design support.

The migration of manufacturing to China has made the Asian market a strategic focus for distributors. Avnet, one of the largest electronics distributors in the world, has relied on both acquisitions and partnerships in China to gain strength [4].

Avnet increased its presence from 3 to 23 offices throughout China when it acquired Sunrise Technology (2000 revenue − $230 million) in 2001 [4]. Sunrise is based in Hong Kong, with a warehouse facility in Pudong, New Area, and has support centers in 22 cities. Sunrise supported a 15-supplier line card that included Intel, Motorola, and National Semiconductor, and its customers included major state-owned telecom companies, making this a significant acquisition for Avnet. Post acquisition, the existing Sunrise management team remained with the company and took on the expanded responsibility of running Avnet operations in the region. Besides acquiring of Sunrise, Avnet has purchased shares in Shanghai Logistic and ChinaTronic Technology Ltd. In 2000, Avnet formed a partnership with China's Ministry of Information (MII) and Global Techmart Inc., a Chinese information and business development company, to launch a direct procurement portal and electronics part information website − ChinaECNet, which gives Avnet the ability to book orders from Chinese companies in Chinese currency [6].

The future of the Chinese distribution industry involves consolidation, tariff benefits, and local industry support. With thousands of small distributors in China that support one or two parts manufacturers, the market's fragmentation could present opportunities for North American distributors looking to make acquisitions in the region. Many small domestic distributors will disappear through mergers, acquisitions, takeovers, and loss of business. The tariff benefits available to the local distributors will erode with Chinese entry into the World Trade Organization (WTO). Local distributors will continue to support the industry by offering special terms (e.g., extended credit and small batch size orders). However, multinational companies will have their own space by providing support to foreign electronic companies and contract manufacturers [2], [4], [6].

Japan's distribution landscape is different from that in China. Japan's $43.9 billion (revenue year 2000) semiconductor market is largely out of bounds for foreign distributor participation. Most distributors in Japan are captive and owned by major semiconductor suppliers like Hitachi and Toshiba. Semiconductor distribution specialist Memec had expanded its presence in Japan through acquisitions of OEL and Okura Electronics in the late 1990s. Arrow made forays into this market by purchasing a minority stake (5.2%) in Marburn in 2001. Marburn is the largest (fiscal 2000 sales of $1.45 billion) noncaptive franchised distributor of electronic parts and supply chain services in Japan. Its lines include Samsung, Seiko, TI, and Xilinx. Many other major multinational distributors (e.g., Avnet) have decided not to try to gain a presence in Japan because, unlike China, electronics manufacturing in Japan is not a growth sector [8].

International issues to address in distributor assessment are:

- Which distributor to assess: The corporate conglomerate or the local incarnation? The one that takes the order or the one that services the order?
- How does one compare pricing and cost?
- What are the quality standards being used when buying in international markets?

8.4.4 Internet-based third-party organizations in electronic parts distribution

The Internet revolution made many typical part-trading activities available and more accessible. Internet transaction services have gained popularity, and customers are beginning to place more orders via the Web. Different types of third-party organizations are listed in Table 8.1. Table 8.2 lists the support and services customers require from distributors' websites to conduct business.

Table 8.1: Types of Internet-based third-party organizations in electronic parts distribution

Aggregators	Auctions	Exchange	Reverse auction/ request for quote
Questlink/ NetBuyChipCenter ISuppli.com SupplyView	USBid Virtual Chip Exchange	FastParts Converge	RiverOne FreeTradeZone HighTechMatrix netComponents

Table 8.2: Distributor support and services provided through the Internet

Types of distributor services used by customers	Types of distributor services provided on the Web	Top improvements areas in distributor websites
Immediate access to the newest parts/technology Technical information Timely new product information Device cross-referencing Contract engineering services Device programming FAE visits/assistance Contract manufacturing Worldwide product availability Reference design ASIC designs	Pricing information Technical information on parts Availability Accessing vendor data books Links to manufacturers Technical service information Company information Contact names, phone numbers Transactions	Pricing information Product availability Product specifications Datasheets Application notes Product delivery information Availability to conduct e-commerce

Online issues to address in distributor assessment include:

- Who to assess – the Internet-based third-party entities or the distributor that actually services the order.
- What to assess – the distributor functions or the website features.

Since the quality of the service obtained depends primarily on the distributor, the distributors should be assessed instead of the third-party company. Sometimes the website from which the order is transacted is owned but not operated by the distributor. In that case, the availability of the Web-based services can be counted in the distributor assessment. The participation of a distributor in the Internet service may affect the following distributor assessment criteria:

- delivery service quality
- customer service and support

8.5   Distributor assessment

A distributor can compromise the quality and reliability of electronic parts even when they are procured through the franchised channel. Some of the possible ways the distributor's actions can affect the part's quality and reliability include physical damage such as lead bending, ball shearing and scratching, mislabeling, packing in shipping containers of the wrong type and size, ESD damage, uncontrolled "time out of bag" for moisture-sensitive parts, and improper (or expired) dry packing material. Tracking supplier performance and other supply chain–related information helps OEMs to catch and handle errors quickly and effectively. Replacing the emotional relationship selection practice with qualitative data on performance helps to assess the distributors rigorously to reduce risks [3].

Distributor assessment is anchored in on-time delivery performance, the quality of packaging, handling and storage, the quality of products, and the level of responsiveness and flexibility in adjusting to demand and forecasting changes [3]. In keeping with the analogy of the part manufacturer assessment, distributor assessment is broken down into seven categories. These categories are described in the following sub-sections. Table 8.3 summarizes the categories and the acceptance criteria according to those categories. To gain acceptance, a distributor must attain excellence level 1 in each category.

In all the categories, compliance must be maintained through a verifiable audit by an external accreditation agency. This audit is performed through the implementation of ISO-9000 or an equivalent quality system. A quality system is the core policy for a company whose documented quality assurance manual defines and governs the quality policies and quality operating procedures in the company. It is maintained and developed by the corporate quality group. In all cases, the assessing team needs to ensure that the quality system manual specifically addresses the assessment category.

8.5.1   Delivery service quality

Delivery service quality is the confidence that orders will be filled accurately and on time. Distributors must have performance measurement and improvement methodologies that focus on increasing order and delivery accuracy, accelerate the information flow through the distribution channel, and improve the response time to customers' requests for quotes, orders, and inventory status.

8.5.2   Quality control

Policies and operating procedures that ensure the organization's quality awareness, quality policies, and procedures for implementation and continuous improvement constitute the quality control apparatus of a distributor. The distributor must have quality control policies, procedures, and measurements in place. All records that affect quality must be filed and retained for auditing, tracking, and reference purposes. Quality records include inspection data, audit data, corrective actions, sales orders, packing slips, supplier rating data, field quality reports, and purchase orders.

8.5.3   Process control

Process control consists of the activities established by the distributor to achieve and maintain part quality. Process control techniques must be in use to monitor operating processes over time, and operators must be qualified and trained.

Table 8.3: Summary – Distributor assessment criteria

Assessment category	Assessment criteria	
	Excellence level 0	**Excellence level 1**
Delivery service quality	Criteria for excellence level 1 are not satisfied	Distributor must have performance measurement and improvement methodologies that focus functions on increasing order and delivery accuracy, accelerate the information flow through the distribution channel, and improve the speed of response to customer requests for order and inventory status. Compliance is ensured through verifiable audits by an external accreditation agency.
Quality control	Criteria for excellence level 1 are not satisfied	Quality control policies, procedures, and measurements are in place. All records that affect quality are filed and retained for auditing, tracking, and reference purposes. Quality records include inspection data, audit data, corrective actions, sales orders, packing slips, supplier rating data, field quality reports, and purchase orders. Compliance is ensured through verifiable audits by an external accreditation agency.
Process control	Criteria for excellence level 1 are not satisfied	Process control techniques are in use to monitor operating processes over time, and operators are qualified and trained. Compliance is ensured through verifiable audits by an external accreditation agency.
Handling, storage, and shipping controls	Criteria for excellence level 1 are not satisfied	The distributor does not handle the parts in a manner that results in quality deterioration. Procedures are defined by the part distributors to help prevent part deterioration and damage during handling, storage, and shipping. Compliance is ensured through verifiable audits by an external accreditation agency.
Corrective and preventive actions	Criteria for excellence level 1 are not satisfied	Distributor actively seeks to locate and act upon nonconformities. Root cause analysis is used by the distributor to identify the cause of part nonconformities and to use preventive measures to eliminate the cause. Compliance is ensured through verifiable audits by an external accreditation agency.
Part traceability	Criteria for excellence level 1 are not satisfied	Distributor maintains part traceability within its processes and retains it for its suppliers and customers. Compliance is ensured through verifiable audits by an external accreditation agency.
Customer service and support	Criteria for excellence level 1 are not satisfied	Distributor maintains available support and service capability and plans potential solutions to meet customer and supplier requirements for basic and value-added services. Compliance is ensured through verifiable audits by an external accreditation agency.

8.5.4   Handling, storage, and shipping control

The distributor should not handle the parts in a manner that results in quality deterioration. Procedures[3] are defined and implemented by the parts distributors to help prevent damage during handling, storage, and shipping.

8.5.5   Corrective and preventive actions

Corrective actions are required whenever nonconformities are identified. Preventive actions are used for continuous improvement and to resolve potential nonconformities. The distributor should actively seek to locate and deal with nonconformities. Root cause analysis should be used by the distributor to identify the cause of part nonconformities and to take preventive measures to eliminate it.

8.5.6   Part traceability

Part traceability is the ability of the part distributor to trace a part to a manufacturer's incoming lot or from one region of the world to another. "Cross-referencing a component" is another related term that refers to the distributor's ability to substitute one supplier part for another. The distributor needs to maintain part traceability and cross-referencing within its processes and for its suppliers and customers.

8.5.7   Customer support and service

This is the ability to respond to and satisfy concerns and queries from customers. The distributor must maintain available support and service capability and plan potential solutions to meet customer and supplier requirements for basic and value-added services.

8.6   Summary

Consolidation of the distribution industry, coupled with the desire of OEMs to buy more parts from fewer sources, is giving rise to what market watchers term broad-line "mega distributors," which carry most, if not all, of the leading parts manufacturer lines in a given product category. Distributors are becoming bigger by forming alliances, mergers, or acquisitions with other smaller distributors. Mergers and acquisitions represent another way to broaden a distributor's scope and line card. Consolidation has caused shelf sharing for Asia's parts manufacturers to become extinct.[4]

With the constant shifting and consolidation of distributors, suppliers and customers are forced to consider the effect on the end customer. What will happen to the OEM suppliers when a semiconductor supplier drops or adds a distributor, disrupting a long-term relationship?[5] Part manufacturer practices like the minimum buy policy,[6] enforced through

---

[3] Some of these procedures include inventory storage (barcoding, ESD consideration, pick to belt system, computerized real-time inventory system–SIMS, military special lot code/date code, first-in/first-out system), advanced system checking, online inventory system, order tracking/management, product management system, packaging and labeling standards (established to prevent mechanical defects such as bent leads, chipped bodies, etc.), ESD, and solderability damage, as well as to clearly identify the product throughout processing.

[4] "Shelf-sharing policy" refers to the rule that a distributor signs an agreement with manufacturers about not carrying that manufacturer's competitors' line cards. Due to the growth of distributors as well as the development of consolidation and global market, it is becoming very difficult for both distributors and manufacturers to maintain this rule and satisfy customers' requirements at the same time. For example, Motorola has discarded its shelf-sharing policy, and more manufacturers are going to do so.

[5] In one example, when Xilinx and Marshall Industries severed their 10-year relationship, effective December 31, 1998, customers had approximately two months to find another supplier. For those customers with design-in contracts through Marshall, the impact was noticeable. Marshall did add Lucent to its line card, but customers cannot change system designs overnight. These changes may force distributors to provider alternative products to

their distributors, are squeezing OEMs that are already cash-poor due to weak end market demand [9]. Policies like this create attrition, and may severely affect the business of smaller OEMs that are forced to buy full orders rather than replenishing partially spent stocks.

The future directions of the distribution industry show justification for allocation of resources for distributor assessment in the part selection and management process. More parts will be bought from the distributors, and distributors will provide more support services traditionally provided by the manufacturers.

8.7   References

[1]   Brewer, G., "Independent Distributors Provide a Strategic Lift," *EBN (Electronic Buyers' News)*, August 2002, <www.ebnews.com/showArticle?articleID=2916347>, accessed April 21, 2003.

[2]   Robertson, J., "China's Redistribution Channel Is Being Reborn," *EBN (Electronic Buyers' News)*, June 2001, <www.ebnews.com/showArticle?articleID=2912400>, accessed April 18, 2003.

[3]   Shah, J. B., "Special Report: Distribution Quality – Assessment Strategies," *EBN (Electronic Buyers' News)*, p. 75, April 2000.

[4]   Spiegel, R., "Avnet Acquires China's Distributor," *Electronic Business News*, May 5, 2001, <www.ebnews.com/showArticle?articleID=2912239>, accessed April 18, 2003.

[5]   Sullivan, L., "Top EMS Players Rethink Sourcing Strategies," *EBN (Electronic Buyers' News)*, issue 1301, cover page, February 2002.

[6]   Sullivan, L., "Arrow, Avnet Eye Growth Prospects in China as Trade Vote Nears," *EBN (Electronic Buyers' News)*, issue 1228, cover page, September 2000.

[7]   Sullivan, L., "Avnet Introduces Web-Based Supply Chain Tool," *EBN (Electronic Buyers' News)*, June 2001, <www.ebnews.com/showArticle?articleID=2912510>, accessed April 18, 2003.

[8]   Sullivan, L., "Japan's Gate Open a Bit for Foreign Distributors," *EBN (Electronic Buyers' News)*, issue 1265, cover page, June 2001.

[9]   Sullivan, L., "Smaller OEMs Squeezed by Minimum Buy Policies," *EBN (Electronic Buyers' News)*, issue 1351, cover page, February 2003.

[10]   Sullivan, L., "Special Report: Independent Distributors," *EBN (Electronic Buyers' News)*, p. 80, January 2000.

[11]   Sullivan, L. and C. Serant, "Distributors and EMS Firms Compete for the Same Turf," *EBN (Electronic Buyers' News)*, June 2001, <www.ebnews.com/showArticle?ArticleID=2912403>, accessed April 18, 2003.

customers, but a sole-source relationship may require a buy/sell relationship between distributors and, in any case, causes the customer to make adjustments.

[6] A minimum buy policy compels customers to buy entire boxes or component reels.

Chapter 9

Tracking Part Changes Through the Part Supply Chain

Steven Murray, Michael Pecht, and David Erhart

The supply chain for electronic products has grown more complex over the past several decades, making the tracking of part changes through the supply chain more difficult. Electronic systems manufacturers must now take an increasingly active role in tracking part changes, and may need to establish contractual agreements to guarantee that they receive the information they need.

This chapter provides the information needed for system manufacturers to plan their change control policies. The relevant standards and current industry practices for change control and notification are discussed. Policies of major manufacturers, distributors, and contract manufacturers are analyzed. Examples of commonly made changes and the reasons why they are made are provided. Finally, policy recommendations for electronic systems manufacturers are made.

9.1   Introduction

The rapid growth of the electronics industry has spurred dramatic changes in the parts that make up electronic products and systems. Increases in speed, reductions in feature size and supply voltage, and changes in interconnection and packaging technologies are constantly generating new and improved electronic parts.

At the same time that changes have become more frequent in electronic products, most electronics supply chains have become more complex. The number of captive part suppliers has been greatly reduced, and an increasing amount of product development is being outsourced. Distributors and contract manufacturers are being more widely used. Today's supply chain looks less like a chain and more like a web.

The increasing complexity of supply chain structures has greatly complicated the change tracking process. As companies outsource more of their design and manufacturing functions and purchase components from open market rather than captive sources, the amount of control that they have over part and process changes greatly diminishes. Left without the power to dictate how changes are implemented, OEMs must focus on how to track change effectively.

Change tracking should be an integral part of an OEM's part selection and management program. The tracking of part and process changes is critical for ensuring that no changes are made that jeopardize product reliability or availability. Other components of this part selection and management program include supplier assessment [1]–[4], part assessment [5]–[6], and obsolescence management [7]–[8].

9.2   Manufacturers change control

Changes to parts are made throughout the life cycles of the parts. These changes are usually managed by a manufacturer's change control board. The policies of these boards generally vary from one manufacturer to another.

9.2.1   Why changes occur

The types of changes that are made to parts, as well as the motivations for making changes, depend on the life cycle stage of the parts. The life cycle for electronic parts is often described by a Gaussian curve when sales are plotted versus time [9]. The typical semiconductor part can be considered to go through sequential phases of introduction, growth, maturity, decline, and obsolescence. An illustration of a typical life cycle is presented in Figure 9.1.

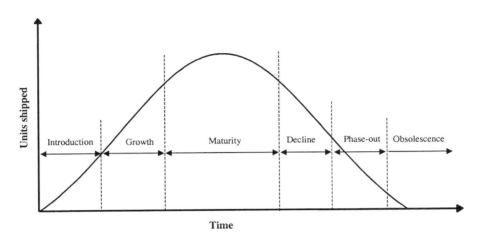

Figure 9.1: Typical life cycle of an electronic part (adapted from [7])

During the introduction stage, the changes implemented are mostly design improvements and manufacturing process adjustments. The part is continuously modified so that it can meet datasheet requirements, achieve economic yields, and meet reliability and quality requirements.

During the growth and maturity stages, a part is in high-volume production. Changes are implemented both to enhance the part and to minimize costs. Feature enhancements may be made to maintain competitiveness and generate new interest in the part. Material, fabrication, and assembly and testing locations may change to reflect changing business needs and capacity. Changes to improve yields and minimize costs may be necessary to maintain competitiveness in the marketplace.

During the decline stage of the part, sales levels start dropping, and manufacturers try to transition customers to newer parts and technologies. Part discontinuance usually occurs when the volume of sales for a part drops to the point where the part can no longer be profitably manufactured. However, it could also occur when a semiconductor company transitions its facilities to a new manufacturing technology.

After the part has been discontinued, it is in the obsolescence stage. Parts are no longer available for purchase, and OEMs must utilize previously purchased and stockpiled parts, obtain parts from an aftermarket source, find an equivalent substitute part, or redesign their products [8].

9.2.2   How changes occur

For most part manufacturers, the change process starts with the submission of a proposal to a change control board, sometimes called an "engineering control board" [10]. This board is usually composed of people from all major divisions within the company, including marketing, manufacturing, product engineering, and reliability engineering. Any division within the company can propose a change to the board.

Upon receipt of the change proposal, the board classifies the change according to its internal classification process. This classification involves deciding how significantly the form, fit, or function of the part would be affected by the change. Part characterization and integrity test results, contractual agreements with customers, and the number of parts affected by the change are also considered. If the board determines that the benefits of the change outweigh the risks, the change is approved.

9.2.3   Industry change control policies

Change classification systems and change control policies vary widely from one company to the next. Many companies have policies detailing the amount of testing that needs to be done to propose a change to a change control board. Many also have policies on how quickly the changes are phased into production.

The change control process for IBM Microelectronics is illustrated in Figure 9.2. All changes go through a single standardized review process, regardless of the type of change [11].

ON Semiconductor uses a two-tier system. Changes are classified as either major or minor. Major changes affect form, fit, function, quality, or reliability and require customer notification. Minor changes do not affect these factors and do not usually require notification.

Samsung has a three-tier classification system [12]. The most drastic changes, such as changes in production location, are classified as Class A and require full requalification data. More moderate Class B changes, such as changes in testing procedures, require only semiqualification data. Finally, less significant Class C changes, such as changes to packaging used for shipment or marking on components, require only a simple data review.

National Semiconductor has a four-tier classification system [13]. Level 1 changes do not represent a fundamental change to the process technique, material, or part functionality and require no formal qualification or customer notification. Level 2 changes are changes that have only a small chance of impacting device performance and require qualification testing but generally no customer notification. Level 3 changes represent a fundamental change and require formal qualification as well as customer notification. Level 4 changes consist of large-scale part or process transfers that require significant resources and coordination. These changes necessitate complete requalification, as well as customer notification. National Semiconductor also has three levels of change control boards, one of each to handle changes of levels 2–4, which incorporate increasingly senior employees.

9.3   Standards and authorities for notifying customers of part changes

Industry standards on change control and notification have been developed by the Joint Electron Device Engineering Council (JEDEC), the standards development arm of the Electronic Industries Association (EIA). These standards form the basis of most change notification procedures. The U.S. military also has specifications on change control for military electronics. However, it is increasingly adopting industry practices as it uses more commercial-off-the-shelf (COTS) parts [14].

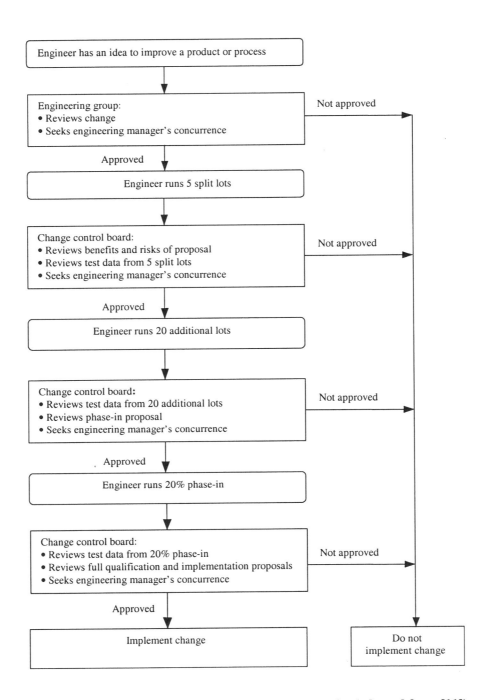

Figure 9.2: Change control process at IBM Microelectronics (adapted from [11])

9.3.1   EIA/JEDEC

EIA/JEDEC Standard 46-A [15] establishes guidelines for customer notification of part and process changes by semiconductor suppliers.[1] This standard has been republished in Europe by the International Electrotechnical Commission (IEC) as IEC/PAS 62166 [18]. Standard 46-A defines part changes, establishes change notification policy requirements, describes the content of written notifications, and suggests how these procedures can be customized for individual customers. Under this standard, documents called Part or Process Change Notifications (PCNs) form the basis of the change notification process.

Standard 46-A classifies changes as either major or minor. Major changes affect form, fit, function, quality, or reliability and necessitate the notification of all customers. Examples of changes that are considered major are provided in Table 9.1. Minor changes are defined as those that do not affect these factors. Notification policies for minor changes depend on individual company policies and requirements specified by customer contracts, but usually no notification is provided. Examples of minor changes are the replacement of an old transfer molding machine with an equivalently functioning new one, the re-sequencing of tests performed on parts following manufacture, and recalibration of equipment.

Table 9.1: Examples of major changes (adapted from [15])

Type of change	Examples
Fabrication	Manufacturing site, process flow, materials, wafer diameter, mask
Assembly	Assembly site, materials, marking, package style
Test	Test elimination
Electrical specification	AC or DC datasheet parameters
Mechanical specification	Case outline, package tolerances
Packing/shipping/labeling	Carrier tape dimensions, maximum storage temperature, dry pack requirements

Compliance to Standard 46-A requires a manufacturer to have a documented and controlled change notification procedure. This procedure should detail the methods of identifying customer notification requirements, notification timing, contents of change notification documents, methods of documenting delivery of notifications to customers, record retention periods, and methods of obtaining customer approval of proposed changes. The classifications of types of changes should also be defined in a controlled document. The specification recommends that 60 days' advance notice be provided to customers before they start receiving changed product, and that records on changes be maintained on file for a minimum of 5 years.

---

[1] Customer notification of product discontinuance is addressed by EIA/JEDEC Standard 48 [16], which has been republished in Europe as IEC/PAS 62167 [17]. Suppliers are requested to provide a minimum of 6 months' advance notice for discontinuance of multiple-source parts and 12 months' notice for single-source parts. This notice should be provided in writing to all affected direct customers who have purchased the parts being discontinued during the past two years, all direct sales channels, and all authorized distributors. At a minimum, the notification should include the last day for which purchase orders will be accepted and the affected part numbers. Upon request, the manufacturer should also provide information and/or technical data to help customers find replacement sources for discontinued parts.

There are two types of PCNs – advance and standard. Advance notifications are not required unless a special contractual agreement exists, but are sometimes used to communicate advance notice of changes being considered to major customers and to solicit feedback. Standard notifications issued publicly follow advance notifications and include additional details on implementation timetables.

Recommended content for notifications includes part identification (part numbers or product lines), a description of the change, PCN number, reason for the change, implementation date, effects on quality and reliability, methods for distinguishing old parts from new parts, qualification data, and the name of the person to contact with any questions. Customer part numbers may also be included to customize the notification for major customers.

Major customers may establish special guidelines with suppliers governing the change notification process. Such agreements built into contracts ensure that a company gets the maximum possible control over changes. Customer profiles (also called "user profiles") on record with part manufacturers keep track of any such special requirements and aid the part manufacturer in customizing the change notification process for each customer. These profiles detail for each customer the part numbers purchased, commodity types, approval requirements, contacts, advance notification requirements, and special contractual agreements.

9.3.2   U.S. military

U.S. military electronics manufacturers supplying parts to specifications MIL-PRF-38535, MIL-PRF-38534, and MIL-PRF-19500 require stricter change notification procedures than those required by EIA/JEDEC. They are linked with the Qualified Manufacturers List (QML)[2] and Government Industry Data Exchange Program (GIDEP).[3] QML dictates change notification and test requirements, while GIDEP serves as a path for change and discontinuance notifications. However, since the late 1990s, very few of the major semiconductor manufacturers have subscribed to the QML approach [20].

9.4   Change notification paths

There are four major paths for the propagation of a PCN through the supply chain to the OEM (see Figure 9.3). The choice of paths depends on the manufacturer and customers involved. Some manufacturers send out notifications along multiple paths for maximum assurance that the message will reach its destinations.

9.4.1   Direct to OEMs

For large customers who buy parts directly from the part manufacturer, direct notification to OEMs is the predominant notification method. This method may also be used for customers who have special notification requirements specified in contractual agreements.

---

[2] QML is documented as MIL-I-38535 [19]. The manufacturer must notify the qualifying agency prior to the implementation of any change to a product or quality assurance program that may affect performance, quality, reliability, or interchangeability. The manufacturer must submit engineering data, quality conformance inspection data, or a suggested test plan.

[3] Participants in GIDEP are provided electronic access to engineering data, failure experience data, metrology data, part information data, reliability and maintainability data, and PCNs. In addition, GIDEP provides Diminishing Manufacturing Sources and Material Shortages (DMSMS) notices of part discontinuances and information on alternate sources and aftermarket suppliers.

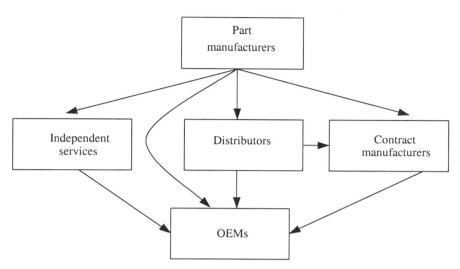

Figure 9.3: Paths for propagation of PCNs through the supply chain

9.4.2   Via distributors

Since most customers buy their parts through distributors and not directly from manufacturers, notification through distributors is the most commonly used change notification path. Manufacturers generally send all of their PCNs to their distributors. The distributors are then responsible for reviewing sales records and notifying affected customers of part changes.

9.4.3   Via contract manufacturers

OEMs that outsource manufacturing functions may receive some of their change notifications from contract manufacturers. In this case, the contract manufacturer receives a change notification from a part manufacturer or distributor, identifies the products that use the part, and passes that information on to the appropriate OEMs. OEMs may also receive notifications of assembly process changes on their products made directly by the contact manufacturers through the same line of communication.

9.4.4   Via change notification services

Most large manufacturers maintain PCN websites that can be directly searched by the public. For example, Texas Instruments has a website that lists part change notifications for each part type, categorized by date of issue and change type and with an abstract summarizing each change [21]. Many manufacturers also have e-mail notification lists to which the general public can subscribe.

Independent change notification services are also available, such as the Web-based PCN-Alert.com. These services collect change information from a large number of part manufacturers and redistribute it to OEMs. One can search PCNs, view the PCN history for various part numbers, and set up an e-mail alert based on specified filters. Information can be filtered against a bill of materials (BOM) or searched based on manufacturer or type of change (shipping/labeling, design change, discontinuance, etc.). GIDEP can be considered an independent notification service as well, but it is targeted to military system manufacturers.

9.5   Case study: Change notification policies in practice

In order to determine how closely change notification policies in industry follow EIA/JEDEC standards, representatives of 25 large semiconductor part manufacturers, distributors, and contract manufacturers were interviewed [1]. Change notification policies used in industry were found to vary in practice. Part manufacturers, distributors, and contract manufacturers each set their own policies, and may have different policies at different manufacturing locations or sales offices within the same company. These companies may also provide different levels of service to different customer populations.

9.5.1   Part manufacturers

Change notification policies of manufacturers vary, depending on the individual manufacturer, the company division or manufacturing location, the type of customer to whom the manufacturer sells most parts, and the geographical location of the customer.

9.5.1.1   Differences by manufacturer

Most large manufacturers provide change notifications in a manner compliant with the EIA/JEDEC specifications, despite the common practice of putting disclaimers on data-sheets saying that the manufacturer is free to make changes or discontinue parts without notice. Changes that could affect reliability or performance are usually sent out 60–90 days in advance, while minor changes such as marking changes are usually sent out 30–90 days in advance.   For additional assurance that customers do not miss a change, some manufacturers, including Fujitsu and Xilinx, also modify part number suffixes after all changes. If for some reason a company missed a change notification, its shipping dock would then likely catch the change when it started receiving parts under an unrecognized part number. Part discontinuance notices are generally sent out at least 6 months in advance for common parts and 1 year in advance for specialized parts that are likely to be single-sourced.

In general, smaller manufacturers are much more likely than their larger counterparts to not meet advance notification requirements, not have their policies documented, or not have their quality system audited by external accreditation agencies. In a study of 21 smaller component manufacturers, many of which were passive component manufacturers, only 9 were found to have documented and audited change notification policies [1].

9.5.1.2   Differences by company division or manufacturing location

A common company name does not necessarily guarantee that all parts made by that company will have the same change notification policy. Different company divisions or manufacturing locations can have different quality and policy manuals and therefore different policies. Vishay, a large passive device manufacturer, is one such company. While Vishay Sprague, in Sanford, Maine, has a documented policy in its quality manual to notify customers of part or process changes that affect customer requirements, Vishay Israel, in Holon, Israel, has no such policy [1]. Any time a production location change is made, OEMs should verify that the same notification policies will still apply to the parts that they buy.

9.5.1.3 Differences by customer type

Large customers generally have more influence with part manufacturers, and can demand stricter and more customized notification requirements from manufacturers than can a customer buying smaller numbers of parts. Differences also exist depending upon the industry in which customers operate. Customers manufacturing products that have large liabilities in the event of failure, such as automotive or avionic systems, tend to have stricter policies on the handling of change notifications from their suppliers.

9.5.1.4 Differences by geographical location

Change notification policies differ, depending upon the country in which the customer is located. In fact, manufacturers often tailor their policies to match the business culture of the country in which they are selling parts [22]. For example, Japanese customers are generally provided more advance notice of part or process changes than American customers. Where necessary, part manufacturers stock excess inventory in advance of changes to allow longer time periods to elapse before Japanese customers receive changed parts. Japanese companies also place much more emphasis on personal contact and relationships in business than do American companies. Common practice in Japan calls for PCNs to be translated into Japanese, typed on deluxe paper, and hand-delivered by agents of the part manufacturer to the customer.

Practices in Western Europe are similar to those in the United States. However, practices vary more and are generally less formal in Eastern Europe. Less emphasis is placed on formal written documentation and rigid change notification policies in Eastern Europe, and distributors use more discretion in deciding whether to pass on change notification information. In addition to cultural influences, these less formal practices are a product of the legal systems in these countries. Regulations requiring business documentation are generally weaker than in the West, and only enough documentation to obtain ISO certification is usually produced.

9.5.2 Distributors

Purchasing through distributors allows OEMs to order smaller numbers of parts, reduce lead times, and outsource some of their supply chain management functions. The National Electronic Distributors Association (NEDA) estimates that currently 35% of all North American electronic component sales are made through distributors and that this number is rapidly increasing [23]. Distributors are also expanding their businesses to offer a variety of value-added services to their customers, including device testing, component customization, light manufacturing, and business planning. Franchised distributors control most of the market, although smaller independent distributors are available. The franchised distributors Arrow and Avnet dominated the U.S. and European distribution industries in 2000, and are gaining market share elsewhere through partnerships and acquisitions.

There are no standards within the distribution industry that guide change and discontinuance notification policies. Although EIA/JEDEC standards 46-A and 48 are intended to cover franchised distributors as well as part manufacturers, quality of service varies widely in practice. Most distributors are good about passing product discontinuance notifications on to customers, but many are less diligent in communicating change notifications. Distributors generally have a central office that collects PCNs and forwards them to sales offices. However, it is often up to the individual sales offices to decide which customers receive the notifications. The requirements of EIA/JEDEC Standard 46-A may be exceeded or not met, depending on the distributor and the sales office within that distributor with which an OEM is working. Policies also vary based on product line and on any agreements that a distributor has with a particular part manufacturer. There are clear service levels within the distributor customer population, as customers who order large volumes of parts or pay higher prices in exchange for more services get better treatment in the part change and discontinuance notification process. The policies for distributors also vary from one country to another, reflecting the local culture and business customs, although these differences are gradually being reduced as distributors try to standardize their global operations to a greater degree [22]. The large number of acquisitions and mergers currently being made in the distribution industry increases the importance of verifying practices at the individual sales offices used by an OEM.

9.5.3   Contract manufacturers

Contract manufacturing had an estimated market penetration of 13% in 1999 and is forecast to grow at a 28% compound annual growth rate through 2004 [24]. The largest contract manufacturers in 2001 were Solectron, SCI, Celestica, and Flextronics. Some contract manufacturers are now providing supply chain management services to their customers and are taking a more active role in part selection and qualification than they have in the past.

Policies of contract manufacturers vary, depending on agreements with individual customers and the level of involvement of the contract manufacturer in part selection and management. Larger contract manufacturers active in part selection and management usually have centralized PCN tracking and notification systems, and have notification policies similar to those recommended for semiconductor manufacturers in EIA/JEDEC standards 46-A and 48. Reporting of part or process changes made by the contract manufacturer themselves usually follows similar guidelines, with at least 60 days' advance notice provided. The policies of smaller contract manufacturers vary more widely, but these companies are also generally more open to the preparation of special agreements with customers to address any concerns.

9.6   Case study: Examples of commonly made changes

Electronic parts generally have very short life cycles and undergo frequent change. Technology, packaging materials, and business environments are constantly evolving. New regulations, such as lead-free product initiatives, also bring about changes in electronic parts. PCN databases of major part manufacturers were reviewed to determine what types of changes are most commonly made to semiconductor parts. Examples of common part and process changes, and the reasons they are made, are provided in the following sections.

9.6.1   Wafer fabrication facilities changes

This change involves moving the manufacture of semiconductor wafers from one location to another. The change is often made in order to accommodate changing production volumes or to minimize costs [25]. In order to keep operating margins high, fabrication facilities need to be kept as close to full capacity as possible. These facilities may also be purchased or sold due to changing business needs, which can result in part changes or discontinuances. For example, when Texas Instruments closed its fabrication facility in Hatogaya, Japan, in late 2000, many parts sourced from the facility were discontinued [26]. Introduction of new product lines can also lead to fabrication changes, as production locations of existing product lines are rearranged.

Acquisitions also continue to be a popular strategy for corporate growth [27], leading to realignment of manufacturing capabilities and operations. These changes occur particularly frequently for semiconductor manufacturers who do not own their own facilities and utilize pure-play foundries. There were roughly 350 such companies in industry as of February 2001, mostly in the United States [28]. Occasionally a part may be shifted to a new fabrication process when wafer fabrication facilities are changed. For example, a part built in a bipolar fabrication process may be redesigned to use CMOS technology. Instead of issuing a PCN, the company generally discontinues old parts and introduces new parts under a new part number due to the severity of such a change.

9.6.2   Die revisions

Mask changes are defined as a layout modification on the die and are often a result of process changes. A mask change is often made to add new functionality to a device or improve performance, such as adding improved ESD protection circuitry [29]. It may also be made to allow for processes that increase throughput rates [30] and improve manufacturability [31]. Additionally, mask changes may be needed as a result of wafer fabrication facility changes, as a part must change to meet the capabilities and process flows of the wafer fabrication facility in which it is built [32].

Die shrinks are defined as the scaling down of a mask to produce a smaller die size. Die shrinks lead to higher manufacturer profits, as more dies can be fit on a single wafer, and often lead to improved electrical performance due to the shorter on-die interconnect lengths. Since design rule changes are generally required to accommodate the reduction in size, PCNs are usually triggered. Microprocessor manufacturers such as Intel are examples of companies that perform very frequent die shrinks. Having just finished converting all production to 0.25-μm technology, Intel moved rapidly to the next die shrink in 1999, converting virtually all of its production to 0.18-μm technology [33]. For convenience and cost savings, die shrinks may be performed simultaneously with mask changes [29].

9.6.3   Changes to assembly and test locations

Part manufacturers are increasingly outsourcing the assembly and testing portions of their manufacturing process. This outsourcing is projected to increase at a 25% compound annual growth rate [34]. Outsourcing allows part manufacturers to concentrate their investments in areas where they have competitive advantages, to gain access to specialized knowledge and innovations outside the boundaries of their companies, and to increase their ability to react quickly to innovations or changes in market demands [35]. As high-volume manufacturing of commodity products has traditionally been a low-margin business, outsourcing these functions can also lead to a more attractive balance sheet for a company [28]. For example, Motorola sold two package assembly and test units to the ASE Group in 1999 and then outsourced these functions [36]. Parts can also be moved from one manufacturing location to another in order to accommodate changing levels of demand [37].

Manufacturers are always seeking out new regions of the world to set up manufacturing operations in order to minimize costs. In earlier days, American companies built factories in areas of the United States with lower wages and manufacturing costs, such as Arizona and Texas. For many decades after that, East Asian nations such as Taiwan, Malaysia, and Korea were preferred. As costs in these regions are increasing, new locations are now being sought, including the Philippines [37], Thailand [37], and China [38]. Some manufacturers are now building manufacturing facilities in Eastern Europe and the former Soviet Union [39]. For example, ON Semiconductor has moved some manufacturing operations to the Czech Republic [40]. Asian manufacturers are also setting up manufacturing operations in the Balkans, mainly to lower the cost of products destined for European Union countries [41]. At the same time, however, manufacturing is starting to move back to developed countries such as the United States for some types of products. As the amount of automation used in the assembly of semiconductor products is increasing, the portion of manufacturing costs resulting from direct labor is decreasing. If the manufacture of a product is largely automated, the cost savings that could be saved in labor by manufacturing overseas are often more than offset by the additional transportation and logistics costs and by the decreases in overall company operational efficiency that would be incurred [28].

9.6.4   Changes to assembly materials

As a part moves from one assembly house to another, changes must sometimes be made to the part to adjust it to the process flow and capabilities of the new assembly house. For example, Texas Instruments changed the die attach and lead frame in their TO-220 package to match the process flow of one of their assembler and testers [42]. Changes may also be necessary in order to improve the reliability of parts if problems are encountered in the field. For example, Texas Instruments increased the solder ball size on their MicroStar Junior Ball Grid Array packages to improve solder joint and board-level reliability [43].

New resin materials, as well as new additives, are constantly being developed to lower costs, reduce moisture absorption, improve processability, and improve electrical and mechanical properties. Lower-stress and faster-flowing encapsulants are constantly being developed to improve device reliability and reduce costs, and are a necessity for some newer, smaller package styles such as ball grid arrays (BGAs) and chip scale packages (CSPs). Special encapsulants are even being developed that offer increased compatibility with the copper die metallization [44] which is starting to replace aluminum in some types of semiconductor parts.

Many changes in assembly materials will also occur as a result of the introduction of lead-free technologies. Changes to parts include lead plating changes and soldering compatibility changes. For example, Texas Instruments issued a PCN in 2000 offering NiPdAu lead finishes as a lead-free option on its parts, in addition to the NiPd lead-free option already offered [45]. Encapsulants must also be developed that can tolerate the higher processing temperatures required to melt lead-free solders for attachment to circuits boards. Texas Instruments has also qualified a new encapsulant material in preparation for the conversion to lead-free manufacturing [46].

9.6.5   Packing, marking, and shipping changes

Changes to the markings on a device and the packing materials used for shipping are frequently made. Although such changes do not impact part performance, they can wreak havoc on OEMs' receiving and manufacturing departments if proper advance accommodations are not made. Examples include changes between laser, ink, or paper device labeling [47], changes to make labeling more legible or to improve traceability [48], or changes to the number of devices packed in each reel or tray within a box [49].

Packing changes are also made for environmental reasons. Waste reduction and increased recycling are additional green efforts currently popular in industry. In order to promote reuse and recycling of the boxes used for shipping parts to customers, for example, Texas Instruments stopped placing its logo on boxes in 1999 [50]. Moisture sensitivity labeling has also been increasingly applied to packing materials as the use of lead-free technologies has grown [51].

9.6.6   Part discontinuance

The electronics industry is not only one of the fastest-growing industries, but also one of the most rapidly changing. Business and technology changes in semiconductor companies can result in part discontinuances, which will impact the OEMs using those parts. When ON Semiconductor was spun off by Motorola, for example, almost one third of its parts were discontinued.

Military-grade parts, which are certified to meet Mil-Spec tests, have also been discontinued in large numbers over the past decade. The military no longer buys enough parts for semiconductor manufacturers to manufacture Mil-Spec parts economically and maintain the additional infrastructure required to ensure compliance with the specifications [14], [52]. Parts made with through-hole configurations have also been discontinued frequently and replaced by surface mount parts.

9.7   Conclusions and recommendations

Change is an inevitable aspect of part manufacturing. The development of new technologies and manufacturing processes, constantly changing business forces, and the emergence of new environmental regulations all necessitate change for a manufacturer to remain competitive. The manner in which a manufacturer manages change can have a large impact on economic success and customer satisfaction. If changes are not implemented in a controlled manner, changes that adversely affect part reliability are more likely to be inadvertently made, damaging the reputation of a manufacturer and increasing the risk of liability. If changes are made frequently or if insufficient notice or reason is provided for changes, manufacturers can also receive negative reactions from customers. Effective change notification requires semiconductor manufacturers to communicate with their customers frequently and openly, so that a bond of understanding can develop. In addition to careful crafting of communications and management of business relationships, judgment calls are often made in change control. The complete effects of changes are often unknown, and the distinction between major and minor changes is often fuzzy. Change control is therefore not only a science but also an art.

For OEMs, change tracking is becoming increasingly complicated. As captive parts suppliers are divested, the amount of control OEMs have over the change control process has diminished. An increasing number of companies are also purchasing through distributors and contract manufacturers, increasing the number of paths for the flow of change notification information through the supply chain. OEMs must therefore take an active role in the change tracking process and establish contractual agreements with the manufacturers, distributors, and contract manufacturers from which they purchase parts to ensure that they receive the change notifications they need. Larger OEMs that have the benefit of being able to work directly with part manufacturers should clarify what types of changes result in notifications and make special arrangements to address any omissions from this list that may affect their products. A request to be included on advance notification lists allows the most advance warning of impending changes to be received as possible, often early enough so that feedback to the part manufacturer that may influence the implementation of the change can be provided.

OEMs should take particular care when relying on distributors for change notification, since these companies often have varying levels of service from one office to another and from one customer to another. While the services provided by a distributor can streamline the change control activities of OEMs, customers must clearly establish with their sales representatives at distributors the level of service they expect. Companies working with contract manufacturers would be wise to do the same. As with any other type of supplier, the compliance of distributors and contract manufacturers with change control standards and change notification agreements should be audited on a regular basis. When in doubt, OEMs can track changes themselves using e-mail notification lists and Web-based PCN databases to ensure that they are receiving complete and timely notification of all changes being posted.

9.8   References

[1]   Syrus, T., U. Ramgopal and M. Pecht, "Manufacturer Assessment Procedure and Criteria for Parts Selection and Management," *IEEE Transactions on Electronics Packaging Manufacturing*, Vol. 24, No. 4, October 2001.

[2]   Foucher, B., et al., "Why a New Parts Selection and Management Program?" *IEEE Transactions on Components, Packaging, and Manufacturing Technology*, Part A, Vol. 21, No. 2, pp. 375–382, June 1998.

[3]   Jackson, M., A. Mathur, M. Pecht and R. Kendall, "Part Manufacturer Assessment Process," *Quality and Reliability Engineering International*, Vol. 15, pp. 457–468, 1999.

[4]   Schwach, C. A., A. Mathur, R. Solomon, P. Sandborn and M. Pecht, "Equipment Supplier Intervention Techniques," *Future EMS International*, No. 3, pp. 97–100, 2000.

[5]   Das, D., N. Pendse, M. Pecht, L. Condra and C. Wilkinson, "Deciphering the Deluge of Data – Understanding Electronic Part Data Sheets for Part Selection and Management," *IEEE Circuits and Devices Magazine*, Vol. 16, No. 5, pp. 26–34, September 2000.

[6]   Syrus, T., M. Pecht and D. Humphrey, "Part Assessment Guidelines and Criteria for Parts Selection and Management," *IEEE Transactions on Electronics Packaging Manufacturing*, Vol. 24, No. 4, October 2001.

[7]   Solomon, R., P. Sandborn and M. Pecht, "Electronic Part Life Cycle Concepts and Obsolescence Forecasting," *IEEE Transactions on Components and Packaging Technologies*, Vol. 23, No. 3, pp. 707–717, December 2000.

[8]   Condra, L., A. Anissipour, D. Mayfield and M. Pecht, "Electronic Components Obsolescence," *IEEE Transactions on Components, Packaging, and Manufacturing Technology*, Part A, Vol. 20, No. 3, pp. 368–371, 1997.

[9]   Pecht, M. and D. Das, "The Electronic Part Life Cycle," *IEEE Transactions on Components and Packaging Technologies,* Vol. 23, No. 1, pp. 190–193, March 2000.

[10]   Phillips, C., "Controlling and Managing Engineering Changes," *Manufacturing Systems*, Vol. 5, No. 12, pp. 40–43, December 1987.

[11]   Pennington, S., "Before We Improve It …We've Got to Prove It: The Role of the Technical Review Board," *IBM Micro electronics*, pp. 32–34, First Quarter 1999.

[12]   Samsung Semicondutor. (2000), "Customer Support," <http://www.intl.samsungsemi.com/top/support/qualityassurance/customersupport.htm>.

[13]   National Semiconductor. (2000), "Change Management System," <http://www.national.com /quality/0,1790,90,00.html>

[14] Pecht, M., "Issues Affecting Early Affordable Access to Leading Electronics Technologies by the U.S. Military and Government," *Circuit World*, Vol. 22, No. 2, pp. 7–15, 1996.

[15] EIA/JEDEC, Publication EIA/JESD46-A, "Guidelines for User Notification of Product/Process Changes by Semiconductor Suppliers," Alexandria, VA, August 1997.

[16] EIA/JEDEC, Publication EIA/JESD48, "Product Discontinuance," Alexandria, VA, September 1997.

[17] IEC, Publication IEC/PAS 62167, "Product Discontinuance," Geneva, Switzerland, August 2000.

[18] IEC, Publication IEC/PAS 62166, "Guidelines for User Notification of Product/Process Changes by Semiconductor Suppliers," Geneva, Switzerland, August 2000.

[19] United States Department of Defense, Publication MIL-PRF-38535E, "Integrated Circuits (Microcircuits) Manufacturing, General Specification for," Washington, DC, December 1, 1997.

[20] Pecht, M., J. Fink, E. Hakim and J. Wyler, "An Assessment of the Qualified Manufacturer List (QML)," *IEEE Aerospace and Electronic Systems Magazine*, Vol. 12, No. 7, pp. 39–43, July 1997.

[21] Texas Instruments, (2000), "Product/Process Change Notifications (PCN)," <https://mist.ext.ti.com/pcn>.

[22] Sullivan, L., (May 17, 2001), "Going Global: Around the World, Everywhere the Same." *Electronic Buyers News* <http://www.ebnews.com/story/OEG20010511S0027>.

[23] National Electronic Distributors Association (1999), "We're Established," <http://nedafoundation.org/industry/index.html>.

[24] Carbone. J. (October. 19, 2000), "Growth Means More Consolidation, More Services," *Purchasing,* <http://www.manufacturing.net/magazine/purchasing/archives/2000/pur1019.00/102elec.htm>.

[25] Texas Instruments, "PCN #19991105003: Qualification Notification for the SN65ALS180 and SN75ALS180, Die Revision K," Dallas, TX, November 17, 2000.

[26] Texas Instruments, "PCN #19990908001: Texas Instruments HAT0 Wafer Fabrication Closing: Lifetime Buy Notice," Dallas, TX, June 16, 2000.

[27] Ahuja, G. and R. Katila, "Technological Acquisitions and the Innovation Performance of the Acquiring Firms: A Longitudinal Study," *Strategic Management Journal*, Vol. 22, pp. 197–220, 2001.

[28] Kline, J., "Outsourcing in the Electronics Industry," *HDI Magazine*, Vol. 4, No. 2, pp. 40–42, February 2001.

[29] Texas Instruments, "PCN #199990624001: Qualification Notification for Select-SN74CBT Technology Devices," Dallas, TX, August 30, 1999.

[30] ON Semiconductor, "Change Notification #10001: New Mask Set for LM317T/D27 from Tesla," Phoenix, AZ, August 28, 1999.

[31] Texas Instruments, "PCN #19991216004: Qualification Notification for the SN65ALS176, SN75ALS176, SN75ALS176A, SN75ALS176B, SN65ALS1176, SN75ALS176, SN75ALS176A, SN75ALS176B, SN65ALS1176," Dallas, TX, January 7, 2000.

[32] Xilinx, "PCN2000-08: Additional Equivalent Wafer Foundries for Fabrication of Virtex Comm (C-Grade and T-Grade) Product Family," San Jose, CA, December 29, 2000.

[33] Electronic News, February 8, 1999, "Process Pedal to the Metal," <http://www.electronicnews.com/enews/Issue/1999/02081999/antenna.asp>.

[34] McCray, K., "Are Individual Domestic Test and Assembly Houses in for a Change?," *HDI Magazine*, Vol. 4, No. 2, pp. 20–21, February 2001.

[35] Nielsen, A. P., "Outsourcing and the Development of Competencies," *PICMET'99*, pp. 72–77, July 1999.

[36] CNET July 6, 1999, "ASE Acquires Two Motorola Chip Plants," <http://news.cnet.com/news/0-1006-200-344451.html>, Pensacola, FL.

[37] Texas Instruments, "PCN #20000606002: Qualification of 5DBV Package at Hana," Dallas, TX, February 13, 2001.

[38] ON Semiconductor, "Change Notification #10340: Seefull China Fabrication/Assy Site Qual for General Purp Rectifiers," Phoenix, AZ, October 24, 2000.

[39] Carbone, J., March 22, 2001, "Buyers are Putting the Pieces Together," *Purchasing*, <http://www.manufacturing.net/magazine/purchasing/archives/2001/pur0322.01/032 mfg.htm>.

[40] ON Semiconductor, "Change Notification #10135: Assembly/Test Site Change from Motorola KLM to Seznam, Czech Republic," Phoenix, AZ, February 10, 2000.

[41] Farish, R., "East Meets East," *Business Eastern Europe*, Vol. 24, No. 44, pp. 17–19, October 1995.

[42] Texas Instruments, "PCN #19990623003: Notification of Assembly Material Changes in the TO-220 Package Built at PSI Technologies, Inc. in Manila, Philippines," Dallas, TX, November 9, 1999.

[43] Texas Instruments, "PCN #20010817000: HIJI/TITL MicroStar Junior BGA Solder Ball Size Change," Dallas, TX, November 30, 2001.

[44] Design News, November. 15, 1999, "Epoxy Adhesives," <http://www.manufacturing.net/magazine/dn/archives/1999/dn1115.99/products.htm>

[45] Texas Instruments, "PCN #20000718001: Notification of Additional Lead Finish," Dallas, TX, March 29, 2001.

[46] Texas Instruments, "PCN #20000418001: Qualification of the Sumitomo EME5050 Mold Compound," Dallas, TX, December 20, 2000.

[47] Intel, "PCN #1068-00: Product Marketing Change of Embedded Pentium II Processor
– Low Power Module at 333 MHz and 256K L2 Cache on Die," Santa Clara, CA,
December 12, 2000.

[48] Intel, "PCN #932: Intel Flash Memory Package Topside Lasermark Change for all
Easy BGA Packages," Santa Clara, CA, February 18, 2000.

[49] Texas Instruments, "PCN #20000202006: Informational Notification of Quantity
Change for Select Devices Previously Offered by Harris," Dallas, TX, February 17,
2000.

[50] Texas Instruments, "PCN #19990120001: Informational Notification for the Removal
of Texas Instruments Logo and Texas Instruments Name on Shipping Materials,"
Dallas, TX, February 23, 1999.

[51] Texas Instruments, "PCN #19990401001: All Sites/All SBE's: Universal Change to
the Box/Bag/Reel Label," Dallas, TX, May 19, 1999.

[52] Philips Semiconductor, "Product Discontinuation Notice Number DN-
35MILITARY," Los Gatos, CA, December 1, 1996.

Chapter 10

Parts Selection and Management to Avoid Counterfeit Electronic Parts

Sanjay Tiku, Diganta Das, and Michael Pecht

The agreement on "Trade-related Aspects on Intellectual Property Rights," the TRIPs Agreement [23], defines counterfeiting as follows:

> Counterfeit trademark goods shall mean any goods, including packaging, bearing without authorization a trademark which is identical to the trademark validly registered in respect of such goods, or which cannot be distinguished in its essential aspects from such a trademark, and which thereby infringes the rights of the owner of the trademark in question under the law of the country of importation.

Counterfeiting is an infringement of the legal rights of an owner of intellectual property. It is the deliberate cheating of consumers by anyone, including manufacturers, distributors, and retailers, who reproduces known trademarks, packaging, and product configurations to market goods that look identical to those produced by brand owners.[1] Since manufacturers of counterfeit parts do not pay for research and development, staff, or marketing costs for their products, they can make higher profits. Moreover, due to the difficulties in detecting counterfeiting and apportioning blame to the primary perpetrators, detection and prosecution has been weak.

10.1 Business and national security implications of counterfeiting

Counterfeiting is a global problem estimated to account for 5% to 7% of all worldwide trade (amounting to US$350–500 billion), and this number is expected to grow [5]. Figure 10.1 shows the value and percentage of the different types of counterfeit merchandise seized by U.S. Customs in 2000. The International Anti-counterfeiting Coalition (IACC) has reported that in 2001 U.S. Customs seized about $57.4 million in counterfeit or pirated products, including $4 million in computer hardware. Out of the total, $26.5 million came from China. IACC has also reported that Fortune 500 companies each spend between $2 million and $4 million annually to combat global counterfeiting, and some companies are reported to be spending up to $10 million [18].

The economic repercussions of counterfeit parts reach far beyond the cost of merely replacing the devices and include, costs of safety, loss of performance, repair or replacement, and reputation. For example, a component that may be worth only $2 can cost as much as $20 to replace if it is detected to be counterfeit after it is mounted on a board

---

[1] The Federal Aviation Administration (FAA) defines an unapproved part as a part, component, or material that has not been manufactured in accordance with the approval procedures in FAR § 21.305 or repaired in accordance with FAR Part 43; that may not conform to an approved type design; or may not conform to established industry or U.S. specifications (standard parts). FAA-unapproved parts include, but are not limited to counterfeit or fraudulently marked parts, components, or materials. Avionics manufacturers are obliged to report counterfeit parts as "suspected unapproved parts" to the FAA. The FAA forbids the installation of unapproved parts on a type-certificated product unless a determination of airworthiness can otherwise be made [10].

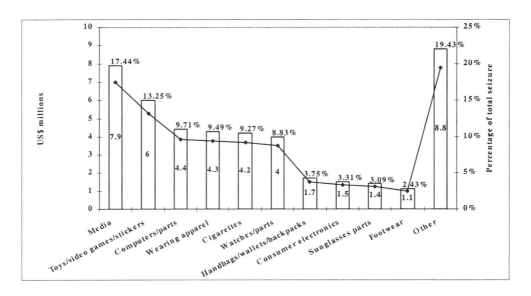

Figure 10.1: Counterfeit goods seized by U.S. customs in 2000 [5]

[21]. More importantly, counterfeit parts have been blamed for causing disasters, e.g., the September 8, 1989, crash of a chartered Convair 580 while approaching the coast of Denmark.[2] Another issue requiring national attention is that counterfeiting could be used for sabotage, and could pose a serious challenge to national security and public well-being.

10.2   Examples of counterfeit electronic parts

For any equipment manufacturer, it is difficult to identify counterfeit parts among the thousands of parts used unless there is a sustained and deliberate attempt to do so. The detection of counterfeit parts usually occurs when field failures in products are traced to a part and the subsequent investigation reveals that the defective part is counterfeit. Yet, failures caused by counterfeit parts may not be detected at all or may be attributed to other causes. Therefore, counterfeit parts that have been identified and publicly reported are in no way representative of the entire electronic parts counterfeiting problem but only provide examples. Listed below are examples of counterfeit parts that have been reported in the years between 1998 and 2003.

1.   In January 2001, a shipment of 4250 semiconductors, packaged as SMCG10A, was purchased from the parts broker International Semiconductor, Inc. (ISI), by Hamilton Sundstrand and sent to New Jersey Micro-Electronic Testing Inc., a testing laboratory. Examination of the marking and the logo on the parts established General Semiconductor (GSI) as the manufacturer. However, there were some problems in testing, and discussions with GSI about the marking and

---

[2] When the FAA reviewed its accident/incident database, it found that unapproved parts played a role in 174 aircraft accidents or crashes from May 1973 to April 1996, resulting in 17 deaths and 39 injuries, although none involved major commercial carriers [15].

the date code revealed that that particular lot of parts had been scrapped by GSI [24].

2. In January 2001, Linear Technology Corp., in Milpitas, California, began receiving a significant number of part returns (voltage regulators, phase comparators, and operational amplifiers) with IC failures. Linear found the failed ICs to be counterfeits. Linear's examination of the returned parts revealed mismatched part numbers and production dates etched on them. Some parts contained Linear's die, but the parts appeared to have been removed from boards and then cleaned and reworked, whereas other parts had the correct part number but had competitors' die. Between January and April 2001, Linear received nine such batches consisting of more than 180,000 parts [21].

3. Micros S.C., in Guangdong, China, sold 50,000 pieces of International Rectifier part number IRF540S to a customer. The customer, after experiencing a high failure rate of its product and tracing the problems to the IRF540S, sent the samples of the parts directly to International Rectifier (IR) for inspection. IR determined that the parts were counterfeit. IR could not cross-reference the lot code appearing on the parts (Lot Code F3666) with any D2Pak packaged devices it assembled. IR also discovered three different Gen5 dies used in this part. The IRF540S is supposed to be a Gen3 device, not a Gen5 device [9].

4. On May 13, 2002, Hynix Semiconductor Inc. disclosed that it had found counterfeits of its 128-megabyte, 133-MHz SDRAM module being distributed in China. It determined that the counterfeit parts are packaged in Taiwan with a Hynix label. Construction analysis of parts by Hynix engineers revealed that the die itself bears the marking of various DRAM manufactures that were probably rejected parts that had found their way into the black market counterfeit channels. In this disclosure, Hynix urged DRAM buyers to acquire their parts only through bona fide channels to avoid obtaining counterfeit parts [17].

5. On April 16, 2002, GIDEP (Government-Industry Data Exchange Program)[3] issued an alert regarding unauthorized distribution/sale of unqualified IC-memory, SRAM/FIFO, part number CY7C199 of Cypress Semiconductor Corp. The U.S. Army's Brilliant Anti-Armor Submunition (BAT) program had experienced numerous failures of Cypress SRAM CY7C199-20VI, and failure analysis revealed a number of anomalies including shorts, missing bond wires, and broken/separated bond wires, as well as memory, timing, and parametric failures. It was determined that Raytheon (Andover, Maryland), which builds circuit card assemblies for the BAT program, had purchased parts from Second Source, a parts broker. At the same time, Cypress had become aware of an ongoing security problem in the summer of 2000 (possibly going as far back as January 1999), with unqualified products (screening failures and material rejects) being smuggled out of the Philippines assembly facility (and possibly other sites) and sold on the gray market to an unknown number of brokerage firms and parts dealers. This security breach

---

[3] GIDEP is a cooperative activity between government and industry participants (in the United States and Canada) seeking to reduce or eliminate expenditures of resources by making maximum use of existing information. The program provides electronic access to engineering data, failure experience data, metrology data, product information data, and reliability and maintainability (R&M) data for systems, facilities, and equipment in the database.

was thought to have resulted in the return of products from various customers exhibiting problems. Cypress has since increased security at its assembly sites and taken measures to strictly control the disposal and destruction of scrap products. It has cautioned customers to buy products only from Cypress or Cypress-authorized distributors and to accept products in sealed, original Cypress external packaging only [12].

6. In May 2002, Hewlett-Packard Co. (HP), working in cooperation with Chinese authorities, seized $1.2 million in counterfeit computer parts during a raid on an illicit operation in Guangdong province. The raid turned up secondhand and gray market Compaq Computer Corp. parts allegedly used to build fake Intel-based servers. HP found that the counterfeiters had the capability to produce a complete server and computer including the packaging material, labels, and warranty cards. The raids also revealed that the monitor screens and the casings were purchased from the secondary market, and the hard drives were either substandard or counterfeit [18].

7. Independent distributor Classic Components Corp. rejected 12 shipments of counterfeit components coming from mainland China worth about $75,000 from September 2000 to February 2001 [20].

8. In 1998, counterfeiters were repackaging 266-MHz Pentium IIs as 300-MHz chips since 300-MHz Pentium II chips cost $375 per processor, while 266-MHz Pentium II chips cost $246 per processor. If a 266 is operated at 300 MHz, it runs, but reliability becomes an issue since it becomes hotter at 300 MHz and can give incorrect answers to instructions [2].

9. TVK Aviation Supply, Inc. (TVK), in Miami, Florida, allegedly sold 158 counterfeit and/or suspect unapproved Allison 501 gas turbine engine thermocouple probes, used to detect an overheated engine, to North American Turbines, in Miami, Florida, an FAA-certified repair station [7].

10. Three thousand pieces of DS12887A (Dallas Semiconductor) were purchased from Shenyang Elite Electronics Co., Ltd, in Liaoning, China, by a member of Electronic Resellers Association International. When the parts started failing, they were sent to Dallas Semiconductor for inspection. Dallas Semiconductor indicated that the parts were used and had an odor that is caused by a defective battery or excessive temperatures in operation. It was determined that these failed parts had been re-circulated after initial failure [9].

Table 10.1 shows a sample list of counterfeit parts of different manufacturers that the Electronic Resellers Association International (ERAI) has identified [8]. The list shows that parts of any manufacturer are prone to counterfeiting and that parts varying from passives to Digital Signal Processors have been counterfeited.

Table 10.1: Electronic parts of different manufacturers found to have been counterfeited by ERAI [8]

Manufacturer	Part number	Part type	Lot code	Date listed
AVX	08051C103JA1045M	Multilayer ceramic capacitor	8042XX605, 8040X034-M	6/13/2001
Vishay Sprague	593D475X9050D2T	Solid tantalum chip capacitor	—	4/19/2001
Analog Devices	AD849JN	Operational amplifier	—	4/25/2001
Dallas Semiconductor	DS 12887A	Real-time clock	—	3/30/2001
Motorola	DSP96002RC-33	Digital signal processor	—	5/3/2001
Texas Instruments	OPA111AM	Operational amplifier	—	4/02/2001
Linear Technology	LT1016CS8	High-speed comparator	—	2/23/2001
Philips	PCF8584T	Integrated circuit	—	5/13/2001
NEC	PS2805-4	Photocoupler	N982 / N992 / N997	3/30/2001
Kemet	T491X2256K035AS	Tantalum capacitor	0940245815 (not valid)	4/19/2001

10.3   Legislative initiatives and organized activities against counterfeiting

At the international level, government-planned action for intellectual property protection started with groups of countries joining together to form Intellectual Property (IP) treaties like the Patent Cooperation Treaty (PCT), all of which were drafted to form the WIPO (World Intellectual Property Organization) Copyright Treaty. WIPO is one of the 16 specialized agencies of the UN. With the forming of the World Trade Organization (WTO), protection of IP rights was included in the General Agreement on Tariffs and Trade (GATT) in the TRIPS agreement.

Listed below are the U.S. laws, in chronological sequence, that are directed towards prevention of infringement of intellectual property rights with penalties for such activities [14].

- Trademark Infringement – Lanham Act, 15 U.S.C. §1114(1)(a)
- Trademark Infringement – Lanham Act, 18 U.S.C. §2318-2320
- Anti-Counterfeiting Consumer Protection Act of 1996, 18 U.S.C. §2311
- Trademark Dilution Act, 15 U.S.C. §1125(c)

- Economic Espionage Act of 1996, 18 U.S.C. §1831(a).

In addition to these laws, Bill H.R.3597, the "Keep America Secure Act," was introduced in the House of Representatives on December 20, 2001 [1].[4] The bill proposes to prohibit the U.S. Secretary of Defense from purchasing any equipment that contains electronic components that are not manufactured in the United States. However, the bill is not practical, since the majority of affordable leading-edge electronics technologies come from outside the United States.

Besides legislative actions, a number of public and private agencies are also involved in the prevention, detection, and reporting of counterfeit electronic parts. The IACC is a nonprofit trade association devoted to combating product counterfeiting and piracy. The Department of Energy (DoE)[5] Lessons Learned Program can be used to develop and disseminate DoE Suspect / Counterfeit Items lessons learned. Lockheed Martin Idaho Technologies Company (LMITCO)[6] includes a comprehensive table of suspect indications and another tabulation of suspect components as aids for detecting or avoiding suspect or counterfeit parts. The U.S. Department of Defense maintains a Diminishing Manufacturing Sources and Material Shortages (DMSMS) database which is described in the GIDEP [4]. Other government organizations involved in such activities include the Defense Industrial Supply Center (DISC) and the Naval Supply Systems Command (NAVSUP). ERAI [8], an affiliation of electronics distributors, is playing an active role in the reporting of counterfeit parts. In April 2003, independent distributors launched a nonprofit trade association called the Independent Distributors of Electronics Association (IDEA) to "combat the stigma of illegitimacy and inhibit unscrupulous brokers from selling inferior parts" [19]. The intention is to promote the quality of the parts supplied by the member companies by educating them about achieving high business standards through seminars, panel discussions, conferences, and newsletters.

10.4 Preventing counterfeiting of parts: Recommendations for electronic part manufacturers

Part manufacturers have a vested interested in solving the counterfeiting problem to protect their reputation and to avoid potential liability claims, since it is their parts that are being counterfeited. Some of the recommendations that part manufacturers can use to prevent their parts from becoming counterfeits are provided below.

- Part manufacturers outsource for strategic business reasons to take advantage of the core competencies of the outsourcing facilities. However, in the process, they may lose direct oversight of the manufacturing process, which provides opportunities to counterfeiters. The counterfeiters may use these facilities to produce counterfeit parts by running a "third shift" of production. A total exclusion of outsourcing may not be a viable solution; however, a better oversight and control mechanism is needed.

- Technologies such as chemically altered dyes and inks can be employed to label products invisibly. Such overt or covert product protection makes counterfeiting harder and more expensive. Effective overt authenticating technologies enable the public to recognize, avoid, and report instances of counterfeiting, and covert

---

[4] The bill was referred to the House Committee on Armed Services on December 20, 2001, and an "Executive Comment" has been requested from DOD on January 21, 2002.
[5] Described in DoE Standard DOE-STD-7501-95, "Development of DoE Lessons Learned Program," May 1995 [23], and in DoE Handbook DoE-HDBK-7502-95, "Implementing U.S. Department of Energy Lessons Learned Program," August 1995 [24].
[6] LMITCO Internal Report INEL-95/227, "Guidelines for Identifying Suspect/Counterfeit Material," September 1995 [11].

technologies can alert company representatives and enforcement authorities to counterfeiting activity. Anti-counterfeiting technologies also provide evidential support in a court of law, where issues of product genuineness and liability may have to be determined [6].

- In two of the case studies, scrapped parts at the original manufacturing facility itself found their way back into the supply chain. Electronics parts that are scrapped must be destroyed so that they cannot be cleaned, reworked, and returned to the supply chain. There must be a proper disposal procedure for such scrappings in order to preclude their resurfacing. All electronic parts manufacturers should have adequate security at their manufacturing sites, and they must develop procedures for the disposal and destruction of scrap products.

10.5   Preventing supply of counterfeit parts: Recommendations for OEMs

Management of the parts supply chain governs decisions like which manufacturers and distributors should be used for what parts, when new orders should be placed, how part performance should be tracked, and what level of inventory should be maintained. Managing the supply chain involves deploying resources across it to get high-quality parts as inexpensively as possible and simultaneously satisfying the schedule's requirements. Supply chains of parts used by OEMs in their products are networks of companies and can involve part manufacturers, authorized distributors (franchisees),[7] independent distributors,[8] and part brokers.[9] The last two part providers can be sources of concern.

In this world of fast technological innovations, parts quickly become obsolete, and there may be a life cycle mismatch between the product and the parts that go into it. OEMs may make themselves vulnerable to the adoption of counterfeit parts through lack of planning for obsolescence. This is evident from the Urgent Data Requests (UDRs) issued by GIDEP on a regular basis for components that are difficult to find.[10] The note of urgency in the request and the subsequent distress purchase provides an opportunity for counterfeiters to slip in their parts without the customer's being able to establish the authenticity of the parts. Several obsolescence management strategies that rely on obtaining parts from sources other than the part manufacturer or distributor (e.g., part substitution, part emulation, reverse engineering, buying from aftermarket sources, and reclamation) also provide a window of opportunity for counterfeiters [16].

One tempting but impractical method used to prevent the introduction of suspected unauthorized parts into the supply chain had been to create a licensing procedure for all members of the parts supply chain. In 1994, the FAA was asked to consider this option for

---

[7] The term "franchise" refers to a continuing commercial relationship between the franchisee and the franchiser [11]. Franchisee distributors are those who have signed selling and marketing contracts with part manufacturers for the distribution of goods or services identified by the franchiser's trademark or trade name. Manufacturers provide franchised distributors with technical support and sales referrals and give the distributors certain safeguards, including the right to return a certain percentage of slow-moving inventory for credit, the right to send back products that have become technologically obsolete, and price protection in the event that the supplier reduces prices. In return, distributors act as an extension of their suppliers' sales force, marketing to a broad customer base that suppliers cannot serve economically. Part manufacturers, however, reserve the right to make direct sales to listed customers.

[8] Independent distributors are aftermarket sources of parts that offer end users parts and service. They make a one-time purchase of parts without a continued commercial relationship with the manufacturer/supplier [11].

[9] Part brokers are scouting agencies who typically handle hard-to-find replacement parts and components. They may hold inventory of possible sources of parts and often search for parts when the need arises.

[10] Example: Gidep UDR number HK3-U-02-02, dated May 28, 2002, regarding the Source of Supply Request For Microcircuit, Non Rad-Hard, states: "Seakr Engineering is in urgent need of twenty (20) pieces of National Semiconductor 54actq16244fmqb or 5962-561901qxa with matching National Stock Number (NSN) of 5962-01-434-9081 or 5962r9561901vxa. The date code must be 0132 or earlier for the non rad-hard version."

avionics.[11] It rejected the idea as not being an effective deterrent and being impractical to implement because of the large number of suppliers (estimated at several thousand entities) and the FAA's limited resources to conduct oversight [22]. In addition, in spite of National Electronic Distributors Association's (NEDA) efforts to promote authorized distribution [3], independent distribution and part brokering are unavoidable since such facilities provide the OEMs with difficult-to-find parts and help them to keep their assembly lines running. To prevent counterfeit parts from entering their final products, OEMs can make use of the following recommendations.

- Following a structured, risk-informed part selection and management methodology, which incorporates an "eyes-on, hands-off" approach, is the first defense against counterfeit parts to be used during product development [13]. This methodology has been developed to evaluate the risks inherent in the use of an electronic part and then facilitate informed decision making regarding its selection and future management activities based on availability and cost constraints so that discrepancies with respect to unusual part availability and cost can be detected. The manufacturer, part, and distributor assessment steps of this methodology include several points where counterfeit risks can be detected. The manufacturer identification process identifies the actual source of the part. The handling, storage, and shipping controls, part traceability, and change notification process assessment criteria provide confidence that the manufacturer keeps track of its products and processes. The distributor assessment criteria of quality control, process control, part traceability, and customer support and service reduce the risk of getting counterfeit parts by eliminating dubious entities that may not be legitimate businesses from becoming suppliers of parts.

- If a counterfeit part is located in the product stream in the postdesign phase, a part selected through a proper part selection process will provide benefits. The user can trace the part back to determine what the source should have been and what the source is. All the agencies involved in the manufacture, dispersion, and use of a counterfeit part can also be alerted. This process and the associated risk management steps can help clarify the possible supply chain failures or weaknesses in the chain that contributed to its penetration by counterfeits. This will not only solve one counterfeiting issue, but will also provide directions for further action.

- Packaging and labeling of electronics parts can help track a part from manufacture through sales. OEMs must develop a knowledge base of the part manufacturers' part traceability tools and technologies and have the ability to trace the parts from these identification marks. Since Electronic Manufacturing Services (EMS) and test labs often purchase parts, it is not enough for the OEMs alone to have these counterfeit tracking methods, tools, and practices. An industrywide process involving the EMS and the labs needs to be developed to ensure that they also have such measures in place.

- The aftermarket sources for electronic parts (that are no longer readily available from their manufacturers) are not reliable sources of genuine parts. Electronic parts not purchased from the manufacturers or their authorized distributors have the highest probability of being counterfeit, particularly if they are purchased from Asia. The sources from where the parts are procured or the final selling entities must be evaluated, as rigorously as distributors during product development, for their ability to supply the genuine parts of a manufacturer. Asking them to indicate the source of

---

[11] During Senate hearings on the DoT's fiscal year 1994 appropriation, the FAA was asked to report on the possibility of regulating and licensing brokers, distributors, and other parties engaged in the sale of parts for aeronautical products.

their purchases and verifying the purchase of such parts with the part manufacturer for proper dates and lot codes is advisable.

• The OEMs should particularly avoid purchasing from part brokers since they have no direct relation or any commitment to the manufacturer or the buyer of the parts. Part brokers have negligible control over their supply and can be duped into purchasing and selling counterfeits. Furthermore, brokers can close shop at any time after supplying the parts, leaving the customer without the possibility of any follow-up action.

10.6   Summary

Counterfeiting is an infringement of the legal rights of an owner of intellectual property. High profits, low risk of detection, and weak prosecution contribute to the supply of counterfeit parts. Counterfeiting of electronic parts causes potential safety hazards and loss of profits to companies, as well as maligning the reputation of manufacturers and distributors. Counterfeit parts can also be a cause of security concerns, since these parts can be weapons of sabotage. All types of parts and part manufacturers are susceptible to counterfeiting, as illustrated by the case examples and the list of identified counterfeit parts provided in this chapter.

A number of laws have been enacted in the United States to penalize counterfeit activities and other IP violations. Several private and public organized groups have also taken notice of and created technological and information-sharing tools to help the industry detect and avoid the use of counterfeit parts.

The emerging market economies of the world are more likely to be havens for counterfeiting activities. This problem may be reduced when these economies adopt fair trade practices and conform to international business norms and enforcement standards. Part manufacturers can also help reduce counterfeiting of their parts by using overt and covert anticounterfeiting technologies and by having effective parts disposal procedures.

OEMs run the to risk of counterfeit parts entering their products by their failure to use proper obsolescence prediction and management methods. However, in today's world of fast technological changes, reliance on nonauthorized sources is an unpleasant inevitability. Licensing of all suppliers of electronics parts or promotion of authorized distribution only is an onerous and impractical task. This chapter provides a set of recommendations for OEMs to use during the product development stage and the post-design phase of their products to avoid their falling prey to counterfeiters.

10.7   References

[1]   "Bill Summary and Status for the 107th Congress," Thomas Legislative Information on the Internet, <http://thomas.loc.gov/cgi-bin/bdquery/z?d107:h.r.03597>, accessed August 12, 2002.

[2]   "How to Spot Pentium II Fakes", *CNET Networks,* April 1998, <http://news.com. com/2100-1001-210597.html?legacy=cnet>, accessed April 5, 2002.

[3]   "NEDA White Paper on Distribution's Role in the Supply Chain," National Electronic Distributors Association (NEDA), <http://www.nedassoc.org/Supplychain.htm.>, accessed July 5, 2002.

[4]   "Other Significant Activities to Address S/CI Issues," <http://twilight.saic.com/ qawg/scitrend/Sci297/sci_sec4.htm>, accessed April 5, 2002.

[5]   "The International Anti-Counterfeiting Directory 2002, Counterfeiting Intelligence Bureau," <http://www.iccwbo.org/ccs/cib_bureau/CIBDir.pdf>, accessed April 5, 2002.

[6]   Bastia, S., "Next Generation Technologies to Combat Counterfeiting of Electronic Components," *IEEE Transactions on Components and Packaging Technologies*, Vol. 25, Issue 1, pp. 175–176, March 2002.

[7]   Department of Defense, Office of the Inspector General, Defense Criminal Investigative Service, Central Field Office, "Crime Awareness Newsletter," Vol. 01–05, p. 3, July 2001.

[8]   Electronic Resellers Association International (ERAI), "Counterfeit Part Listing," <http://www.erai.com/scripts/counterfeit.asp>, accessed July 2, 2002.

[9]   Electronic Resellers Association International, "Counterfeit Part Alert!" (April 05, 2001), <http://www.yuga.com.sg/eg/3eg.htm>, accessed April 5, 2002.

[10]   Federal Aviation Administration, "Detecting and Reporting Suspected Unapproved Parts," <http://www.asy.faa.gov/safety_products/unapprovedparts.htm>, accessed April 5, 2002.

[11]   Federal Trade Commission, "Informal Staff Advisory Opinion 97-6," <http://www.ftc.gov/bcp/franchise/advops/advis97-6.htm>, accessed July 11, 2002.

[12]   GIDEP (Government-Industry Data Exchange Program) Alert, Document No. 6L-A-02-02A, dated May 20, 2002.

[13]   Jackson, M., P. Sandborn, M. Pecht, C. H. Davis and P. Audette, "A Risk-Informed Methodology for Parts Selection and Management," *Quality and Reliability Engineering International*, Vol. 15, pp. 261–271, September 1999.

[14]   Marrs, S. D. (Beirne, Maynard & Parsons, L.L.P.), "Winning the Counterfeiting Wars," <http://www.bmpllp.com/Articles/IP-CntrftgWar.htm>, accessed April 5, 2002.

[15]   Maynard, W. L. and P. D. Moak, "Repaired and Reconditioned Products in the Aviation Industry," ©1997 Beirne, Maynard & Parsons, <http://www.bmpllp.com/Articles/IP-aviation.htm>, accessed April 5, 2002.

[16]   Pecht, M., P. Sandborn, R. Solomon, D. Das and C. Wilkinson, *Life Cycle Forecasting, Mitigation Assessment and Obsolescence Strategies*, CALCE EPSC Press, University of Maryland, College Park, MD, 2002.

[17]   Robertson, J., "Hynix Says Counterfeit DRAMs found in China," *EE Times*, May 2002, <http://www.eetimes.com/story/OEG20020513S0047>, accessed May 14, 2002.

[18]   Sullivan, L., "HP Cracks Down on Counterfeit PC Parts in China," *Electronic Business News*, June 2002, <http://www.ebnonline.com/story/OEG20020626S0013>, accessed July 2, 2002.

[19]   Sullivan, L., "Independent Distributors Look to Rally Around Trade Association," *Electronic Business News*, April 2003, <http://www.ebnews.com/business/opinion/showArticle.jhtml?articleID=8600151>, accessed April 21, 2003.

[20] Sullivan, L. and J. Graham, "China Seen as Key to Counterfeiting Problem," *The Electronics Design, Technology and News Network,* February 2001, <http://www.edtn.com/story/biz/OEG20010216S0069-R>, accessed April 5, 2002.

[21] Sullivan, L. and J. Graham, "Fake Parts Plague Industry," *Electronic Business News,* February 2001, <http://www.ebnonline.com/printableArticle?doc_id=OEG20010 212S0054>, accessed April 5, 2002.

[22] The FAA Suspected Unapproved Parts' Task Force, "Suspected 'Unapproved Parts' Program Plan," U.S. Department of Transportation (Federal Aviation Administration), October 6, 1995, <http://www.faa.gov/avr/supfnl11.pdf>, accessed May 22, 2002.

[23] Vithlani, H., "The Economic Impact of Counterfeiting," Organization for Economic Co-operation and Development, 1998, <http://www1.oecd.org/dsti/sti/industry /indcomp/prod/Fakes.pdf>, accessed May 21, 2002.

[24] Written communication from Victor G. Mosca, Manager, Supplier Quality Assurance, Hamilton Sundstrand, to Joe Garafola, General Semiconductor, dated November 13, 2001.

Chapter 11

Equipment Supplier Intervention Techniques

Anant Mathur, Rajeev Solomon,
Peter Sandborn, and Michael Pecht

Changes in the electronics marketplace have led to discontinuances of many parts. In some cases, specialized parts are available, but their manufacturers may not meet the equipment supplier's quality and integrity requirements. Considering the high costs associated with alternate source development and the cost of missed deliveries to the customer, the equipment supplier may still have to buy parts from such a manufacturer. To ensure the quality and integrity of such parts, some form of intervention may be required. The role of equipment supplier intervention in the overall scheme of parts management, methods, cost, and schedule implications, together with a case study, are discussed in this chapter.

11.1 The parts selection and management process

The purpose of part manufacturer, part family quality and integrity, and distributor assessment is the mitigation of risk in using a part in a product. The risk mitigation process includes the evaluation of the manufacturer's policies, specific part family data, and the distributor's processes. The parts selection and management process establishes procedures and guidelines for performing these evaluations. The final decision to use a part is a company-specific trade-off governed by company resources, policies, culture, and goals and customer demands [1].

If the part manufacturer, the part family, and the distributor conform to the minimum acceptability levels on each of the assessment criteria, the part family is considered acceptable from the quality and integrity (part manufacturer's qualification) standpoint and can be further evaluated for parts selection and management flow [2]. However, there can be instances when a manufacturer, the part family, or the distributor does not conform to one or more of the assessment criteria. Depending on the criteria that are not met, some intervention by the equipment supplier may help raise the part family to an acceptable level. Such action taken by the equipment supplier in order to make a part family acceptable is called "equipment supplier intervention." The responsibility of ensuring the effectiveness of intervention lies with the parts management team.

11.2 Why use equipment supplier intervention?

The resources to establish adequate quality systems and appropriate testing facilities may not be available to every part manufacturer. For example, some manufacturers produce parts whose volumes are simply not significant, and they may lack statistical quality control methods such as process control. This means that when evaluated per the assessment matrices, these manufacturers and their products may fail to conform to the equipment supplier's requirements. However, considering the high costs associated with alternate source development and the cost of missed deliveries to the final customer, the equipment supplier may still have to buy parts from such a manufacturer.

To ensure the quality and integrity of parts from manufacturers and distributors who do not have all the supporting information or do not conform to certain assessment criteria, equipment supplier intervention may be required. Moreover, significant investment in terms of both time and money may have already been made by the equipment supplier in assessing the manufacturer. In some cases, alternative solutions may not be timely enough for the equipment supplier, forcing it to intervene in the quality control methods of the manufacturer. Equipment supplier intervention is motivated by the needs of some equipment suppliers due to changes in the market and supply chain relationships [3].

11.3 Methods of equipment supplier intervention

Methods of equipment supplier intervention vary, depending on the specific criteria on which the part manufacturer or the part is found lacking. Equipment supplier intervention methods can be relatively straightforward, such as establishing a special contract requirement for adherence to change notification procedures, or as complex and costly as conducting part qualification assessment. Equipment supplier intervention may be applied to the manufacturer, part family, and distributor, and are discussed in the sections that follow.

11.3.1 Equipment supplier intervention for the part manufacturer

If the part manufacturer does not provide information required for facilitating a manufacturer's assessment, the equipment supplier must assume that no data are available from the part manufacturer and/or there is a lack of customer support and responsiveness. If an alternative part source is not available, past experience with the part manufacturer, as well as the results of other part evaluations, must be considered during the decision process. However, intervention without data or manufacturer responsiveness is almost certainly doomed to failure.

If the part manufacturer is responsive but does not have a quality function, the part manufacturer's quality system is less than ideal. The part manufacturer should be informed about the assessment and the criteria found lacking. Possible business implications of not improving the quality function should be communicated to the part manufacturer. Assistance to the part manufacturer in establishing the quality functions can be provided via the following mechanisms:

- The equipment supplier may advise the part manufacturer of the part quality acceptability standards by timely information exchange and by periodic monitoring and feedback on the part's average outgoing quality levels. The equipment supplier may indicate specific reasons for finding the part manufacturer unacceptable and provide the part manufacturer with suggestions on specific quality function improvements.
- The equipment supplier may write quality manuals to be used by the part manufacturer in accordance with the equipment supplier's quality acceptance levels. These documents establish the policies and guidelines defining the quality standards that should be maintained for a part to be considered acceptable by the supplier. For example, assistance can be provided in developing an ISO 9000 plan by providing quality system expertise and auditor service.
- The equipment supplier may provide a technical specialist to assist the manufacturer in correcting its faulty processes or loan test equipment to test the quality and

integrity of the parts and identify where (in which part of the manufacturing flow) corrective action must be taken. The equipment supplier may provide training and expertise in quality control methods to assist the manufacturer in applying statistical process control techniques, as well as root cause failure analysis to raise the manufacturer's part quality and reliability to acceptable levels.

- The equipment supplier may indicate possible business implications for the part manufacturer if the quality function(s) is not improved and provide incentives to the manufacturer for making improvements.
- The equipment supplier may perform frequent periodic monitoring of the part manufacturer's performance of the particular quality function if the manufacturer does not provide any assurance about quality function improvements.
- The equipment supplier may provide financial intervention by sharing development costs with the part manufacturer to raise the manufacturer's productivity and quality levels or by granting loans to the manufacturer to improve its processes and increase the quality of its parts.

If the part manufacturer is found to be below minimum acceptability levels required by the equipment supplier, on-site audits of the part manufacturer's processes may be performed. Audits provide firsthand knowledge of the part manufacturer's processes and may inspire confidence in the part manufacturer and the product. The audit teams, however, must be qualified to obtain useful information from the audit, and must focus on exploring the part manufacturer's specific quality function and identifying the part manufacturer's improvement and implementation plans.

A resource investment is required when employing audit intervention, both to educate the auditors and to finance the audit. Third-party surveillance houses may be used; however, resource commitments are also necessary in this case, both to fund the activity of the house and to evaluate the capabilities of the third party conducting the audit. Part manufacturer receptiveness to an audit must also be considered. A part manufacturer unwilling to allow audits demonstrates an unwillingness to enter into a certain type of relationship that may be necessary to ensure that the equipment supplier's needs are met.

If the part manufacturer's change notification process is unacceptable to the equipment supplier, the equipment supplier may attempt to include the requirement for change notifications in the contract. The willingness of a part manufacturer to sign such a contract is a measure of its willingness to cooperate in a relationship that may be necessary to ensure that the equipment supplier's needs are met. The decision may depend on the volume of parts purchased.

If the part manufacturer does not have an acceptable documented and implemented storage, shipping, and handling system, the equipment supplier may make it a contractual requirement for the part manufacturer to ensure that specific procedures (to be provided by the equipment supplier) are implemented during storage, shipping, and handling of the parts specified in the contract. One hundred percent incoming inspection may be performed by the equipment supplier, but such action represents a significant resource commitment.

11.3.2 Equipment supplier intervention for the part family

The following equipment supplier intervention methods may be used for the part family. As with supplier intervention for the part manufacturer, the selection of an intervention method must be balanced by the cost and schedule implications and the assumption of associated responsibility.

If the average outgoing quality (AOQ) data for a part or part family are unacceptable, the equipment supplier may perform 100% inspection for incoming parts. This involves allocation of resources in terms of inspection facilities, personnel, and equipment. Additionally, depending on the number of parts, the time taken for 100% inspection could have a significant impact on the project's schedule.

If AOQ levels are acceptable but the trends show lack of consistency over previous periods and demonstrate significant swings in part family outgoing quality, the equipment supplier can conduct frequent periodic monitoring. This will help the equipment supplier keep track of the part's quality on a regular basis and take relevant action if the product quality falls below acceptable limits. If the AOQ levels are below acceptable limits and demonstrate lack of consistency, frequent periodic monitoring can be conducted, coupled with 100% inspection of incoming parts.

If continuous monitoring tests are not acceptable or if the test results are not acceptable, the equipment supplier may consider screening all incoming parts as a possible intervention step. Such intervention, however, is cost and resource intensive. Additionally, screening could introduce detrimental effects on part quality and reliability.

If the part manufacturer does not perform part qualification testing, if the qualification test data are not available, or if the selected qualification tests are insufficient to demonstrate part integrity, the equipment supplier may consider one or more of the following intervention options:

- The equipment supplier may ask the part manufacturer to perform qualification tests on the part(s). Qualification testing may be feasible only when the equipment supplier's purchase volumes justify the part manufacturer's added costs. If the part manufacturer agrees to perform qualification testing on the parts, the equipment supplier may also consider asking the part manufacturer to qualify the part not only to the part design specifications, but also as per the equipment supplier's application requirements.
- If the above option is not feasible, the part family may still be accepted if it meets *all* the following conditions:
    - The part family and the specific part have been used in the market for at least five years.
    - During this period of time no changes that may affect the form, fit, or function have been made to the part and part family design or the manufacturing processes.
    - The parts in the part family have demonstrably met the part manufacturer's specifications in the field.

If the above conditions are met, qualification requirements may be waived, provided that other part family assessment criteria are met.

The equipment supplier may perform qualification tests for the parts; these tests can be carried out by the equipment supplier or by a contract facility. However, such a step entails considerable investment. Also, by qualifying the part, the equipment supplier assumes the responsibility of ensuring that the part meets its design specifications.

11.3.3   Equipment supplier intervention for the distributor

The following equipment supplier intervention methods apply to the distributor. While selecting any of the intervention methods, it is important for the equipment supplier to minimize not only the cost and schedule implications, but also the assumption of responsibility, which could be potentially detrimental to the equipment supplier's interests.

If the distributor does not have a management system of acceptable quality, the equipment supplier may consider one or more of the following possible interventions:

- If the distributor's quality system is unacceptable due to lack of a specific quality function, the equipment supplier can inform the distributor about the assessment and explain the possible business implications of not improving the specified quality function.
- The equipment supplier may also conduct 100% inspection of parts procured from the distributor. This is a costly action and should be resorted to only when no alternative distributors can be found. Due to the cost implications, 100% inspection should be considered only for provisional acceptance of the distributor, not for long-term practice.

If the distributor does not have written and implemented procedures to prevent damage or deterioration during storage, shipping, handling, marking, testing, and other distributor processes, the equipment supplier can contractually require the distributor to implement specific procedures for the parts to be procured by the equipment supplier. These can include specific packaging/handling instructions or electrostatic discharge (ESD) damage prevention procedures.

If the distributor does not have acceptable problem notification and resolution procedures, the equipment supplier may contractually require the distributor to provide notification of all problems and institute time-bound resolution methods.

If the distributor does not have an acceptable product/process change notification process, the equipment supplier can contractually require the distributor to forward all change notifications received from the part manufacturer to the equipment supplier within a fixed time period. In some cases, the part manufacturer may not provide change notifications to the equipment suppliers who are serviced entirely through distributors. In such cases, it is important for the equipment supplier to ensure that the distributor will forward the change notifications as a matter of policy. A distributor that does not provide such an assurance cannot be considered an acceptable source.

11.4   Cost implications of equipment supplier intervention

Since equipment supplier intervention involves some action on the part of the equipment supplier, most equipment supplier interventions have some associated cost and/or schedule implications. Costs may be direct, such as the cost of 100% inspection if the part family AOQ level is below acceptable limits, or indirect, such as the cost of a detailed review of the manufacturer's documents if the manufacturer is not ISO 9000 (or equivalent) certified. These actions also have schedule implications; for example, 100% inspection is a time-intensive operation. The equipment supplier should evaluate the merits of intervention in light of its compatibility with the project's schedule. Clearly, intervention practices involve substantial cost and time commitments, and should be undertaken only if extended business relations are anticipated.

In some cases, such as when the equipment supplier conducts qualification testing of the parts, the responsibility for part integrity may shift from the manufacturer to the equipment supplier. A direct economic result of establishing equipment supplier intervention methods may be missed deliveries of the equipment supplier's product to the customer. This leads to late market entry of the equipment supplier's product. Costs are also associated with developing alternate sources for the same part. Implementing no equipment supplier intervention and developing alternate sources may also result in loss of "preferred customer" status on the part of the equipment supplier in dealing with the manufacturer. Such trade-offs must be evaluated prior to establishing the correct equipment supplier intervention.

11.5 Provisional acceptance

"Provisional acceptance" refers to acceptance of the parts after either a 100% or a sample inspection of the part family in accordance with the equipment supplier's acceptability criteria. Periodic monitoring of the part family's average outgoing quality levels and performance monitoring of the parts already out in the field determine whether approval can be granted to continue the use of the part in the equipment supplier's product without further intervention.

11.6 Case study

This case study is taken from Rockwell Collins. The part in question was a complex RF amplifier, which was required in a product estimated to return $35 million per year in sales. The part had a failure rate that varied from 5% to 35%, resulting in equipment supplier costs that were double those originally anticipated. The part was subsequently obtained from a second source. However, the second source faced financial difficulties in producing the part even after a four-year effort and a $1 million investment by the equipment supplier. The following equipment supplier intervention techniques were employed:

- An auditor was dispatched by the equipment supplier to the second source to review the company finances and make recommendations about possible financial help. After the review, cash was infused to stabilize the financial situation of the second source.
- Technical support was provided to improve quality levels of the part family.
- Test facilities were loaned to the second source.
- Part requirements were reevaluated by the equipment supplier in the product application.
- The end customer also participated in the intervention by redefining the product's requirements.

11.7 Summary

In order to preserve and promote effective supply chains in today's environment of rapid technology transformation and diminishing sources, the parts management team will be required to apply equipment supplier intervention. Developing the skill of knowing "how much" and "when" is an essential element of an effective risk mitigation process.

11.8 References

[1]   Foucher, B., R. Kennedy, N. Kelkar, Y. Ranade, A. Govind, R. Solomon and A. Mathur, "Why a New Parts Selection and Management Program?" *IEEE Transactions on Components, Packaging, and Manufacturing Technology*, Part A, Vol. 21, No. 2, pp. 375–378, June 1998.

[2]   Jackson M., A. Mathur, M. Pecht and R. Kendall, "Part Manufacturer Assessment Process," *Quality and Reliability International*, Int. 15, pp. 457–468, May 1999.

[3]   Berry, D., D. W. Towill and N. Wadsley, "Supply Chain Management in the Electronics Products Industry," *International Journal of Physical Distribution and Logistics Management*, Vol. 24, No. 10, pp. 20–32, 1994.

Chapter 12

Determination of the Life Cycle Environment

Niranjan Vijayaragavan, Daniel N. Donahoe, and Michael Pecht

The part life cycle environment is closely related to the part requirements. The part life cycle environmental conditions affect parts selection and management, as well as design and development decisions, qualification and specification processes, quality assurance, safety, warranty costs, field service and support, and regulatory conformance. IEEE Standard 1332-1998, "IEEE Standard Reliability Program for the Development and Production of Electronic Systems and Equipment" [9], states, "The supplier shall determine the customer's requirements and product needs," and "The supplier, working with the customer, shall include the activities necessary to ensure that the customer's requirements and product needs are fully understood and defined, so that a comprehensive design specification can be generated." IEC 60134 (July 1994), "Rating Systems for Electronic Tubes and Valves and Analogous Semiconductor Devices," states that "[A part] should be designed so as not to exceed the absolute-maximum value for intended service under the worst probable operating conditions with respect to supply voltage variation, equipment component variation, equipment control adjustment, load variations, signal variation, environmental conditions, and variation in characteristics of the device." One of the factors for fulfilling this requirement is the determination of the life cycle environment of the part.

The part life cycle environment includes the assembly, storage, handling, and scenario for the use of the part, as well as the expected severity and duration of these environments, and thus contains the necessary load input information for failure assessment and the development of design guidelines, assembly guidelines, screens, and tests. Specific load conditions may include steady-state temperature, temperature ranges, temperature cycles, temperature gradients, humidity levels, pressure levels, pressure gradients, vibrational or shock loads and transfer functions, chemically aggressive or inert environments, acoustic levels, sand, dust, electromagnetic radiation levels, and stresses caused by power, current, and voltage. These conditions, either individually or in various combinations, may influence the performance and reliability of the part.

A formal method is necessary to capture all environmental information and to develop an environmental profile. This chapter describes the methods and considerations for developing a life cycle environment profile for electronic equipment.

12.1 Defining the life cycle environmental profile

A life cycle environment profile (LCEP) is a forecast of events and associated environmental conditions that equipment will experience from manufacture to end of life [1], [2].[1] MIL-STD-810E defines the life cycle profiles as follows:

---

[1] In some cases, the environmental factors experienced by constituents of the system begin before manufacturing (e.g., storage of components from a lifetime buy of a large quantity purchased far in advance of their use in manufacturing) [3].

A time history of events and conditions associated with an item of equipment from its release from manufacturing to its removal from service. The life cycle should include the various phases that an item will encounter in its life, such as: handling, shipping, and storage prior to use; mission profiles while in use; phases between missions, such as stand-by or storage, transfer to and from repair sites and alternate locations; and geographical locations of expected deployment. [13]

The useful life of a product depends on the magnitude of the stresses, rate of change of stresses, and spatial variation of the stresses that are generated by the loads acting during its life cycle. An LCEP helps to identify all possible load combinations so that the stresses acting on the product can be identified and their effects can be accounted for in the product's design, test, and qualification process to ensure the reliability of the electronic equipment for its entire life [4]–[8].

12.2   Steps in developing an LCEP

The steps in developing an LCEP are as follows:

1. **Describe expected events for an item or equipment from manufacture through end of life.**

   This process involves identifying the different environmental phases, that the equipment will pass through. Typical environmental phases include testing and qualification, storage at the test facility, transportation to the place of installation, storage at the place of installation, transportation to the specific site of installation, installation, operation, and field service during scheduled maintenance. It also involves identifying system requirements such as who will use the system, what platform[2] will carry it, and the operational requirements, deployment, and transportation concepts.

2. **Identify significant natural[3] and induced[4] environmental factors or their combinations for each expected phase.**

   This process involves identifying the environmental factors which act in each of the identified environmental phases. These environmental factors can be due to the ambient conditions or the specific usage conditions/functionality of the equipment. For example, the operation phase for electronics used in an aircraft engine includes high steady-state temperature, temperature cycling, low pressures, and random vibrations. A failure modes and effects analysis is usually performed to identify key loads acting on the system that can influence its performance and hence its design.

3. **Describe environmental load conditions (in both narrative and statistical form) to which equipment will be subjected during the life cycle.**

   This process involves the quantification of load conditions identified as a result of the previous two steps. Data should be determined from real-time measurements but may be estimated by simulation and laboratory tests. For example, the vibrations experienced by the electronic equipment during shipping could be obtained by a mock shipping experiment wherein sensors are kept with the equipment to record vibration

---

[2] The platform is the next higher-level assembly for the equipment under consideration. For example, for an electronic component the platform is the PCB.

[3] The natural environment is the product's natural ambient conditions, e.g., temperature, pressure, and humidity.

[4] The induced environment is the product's environmental conditions related to the specific functionality of the product; for example, electronics on the drilling tool experience mechanical vibration during the drilling process.

data. The loads should be quantified in a narrative manner to identify specific conditions used during measurement of the load and in a statistical manner to identify the typical range and variability of the load.

12.3 Considerations and recommendations

The following are recommendations for using readily available sources of environmental data and issues which need to be addressed while developing an LCEP.

12.3.1 Extreme specifications-based design (global and local environments)

Extreme environmental conditions in the location of deployment of the equipment are often used for design.[5] Extreme conditions are unlikely to be encountered by the equipment in its lifetime. Moreover, the duration of maximum conditions is typically short. Further, the environment in the vicinity of the part can be modified by its functionality (local environment). Hence, the use of extreme-based specifications for the design of electronic equipment can lead to overdesign or underdesign (due to a change in the local environment).

The part's local environment, i.e., the environment in the immediate vicinity of the part, often varies from the system's global environment, i.e.,the environment in the larger vicinity of the part.[6] The variation between the global environment and the local environment may be a function of the part's isolation from the global environment, the existence of fans within the system, the heat generated by nearby parts, and insulating air between the part and the system environment.

For example, the lowest recorded ambient temperature in Greenland is −70°C. To meet needed performance and reliability objectives, the local environment of parts in systems located in Greenland must be thermally insulated or regulated (i.e., through the use of heaters). The design procedure should incorporate extreme specifications as a baseline, along with their probability of occurrence, and should modify it according to the expected local environments.

12.3.2 Standards-based profiles

Standards-based environmental data like MIL-STD-210 [12] and MIL-STD-810 [13] can be used as a baseline for deciding environmental criteria for design of electronic equipment. MIL-STD-210 is a database of regional and worldwide climatic data. The data are divided into three groups − worldwide surface environment, regional surface environment, and worldwide air environment − and include details about basic regional types: hot regions, cold regions, severe cold regions, and coastal/ocean regions. The environmental factors discussed for each of the groups include temperature, humidity, pressure rainfall rate, wind speed, blowing snow, snow load, ice accretion, hail size, ozone concentration, sand and dust, and freeze-thaw cycles in terms of extreme values, nominal (average) values, and frequency of occurrences. In spite of the details provided in MIL-STD-210, climatic data derived from this standard should not be used directly for design criteria. Rather, they should be used to derive design criteria for each item based on the

---

[5] The highest temperature recorded on Earth is 57.77°C in Al Aziziyah, Libya, in September 1922. Death Valley, California, recorded 56.77°C in July 1913. The place that has the world's highest average temperature is Dakol, Ethiopia, in the Danakil Depression, with a mean temperature of 34.44°C. Places in Pakistan (e.g., Pad Idan) have recorded temperatures up to 50.55°C [10]. The lowest recorded temperature on Earth to date is −89.44°C in Vostok, Antarctica. Temperatures in Greenland have gone as low as −70°C [11].

[6] For example, the local environment of certain parts in a desktop computer, given the heat generated from the power dissipation of the parts on the board, will be significantly hotter than the regulated office environment.

response of the item to both the natural environment and the forcing functions induced by the platform on or within which the item is located (local environments).

MIL-STD-810 provides guidelines for conducting environmental engineering tasks to tailor environmental tests to end-item equipment applications. It contains test methods for determining the effects of natural and induced environments on equipment performance and is mainly focused on system-level design. The conditions and procedures described in MIL-STD-810 can be used for deriving the LCEP for electronic products. Other standards, like those of the EIA and SAE also provide environmental data, which can be used to derive the LCEP.

Standards-based designs can be used as a starting point for developing an LCEP for electronic equipment. Direct use of the data given in these standards without proper scrutiny is, however, not recommended.

12.3.3   Combined environmental conditions

Combined environments (incorporating two or more environmental factors) may affect equipment reliability differently than a single environmental factor. If the combined effect of the environmental factors proves to be more harmful than that of a single environmental condition, then the equipment must be designed for failures arising from the combined effects. Some examples of the possible effects of pairs of environmental factors appear in Table 12.1. Pairs of environmental factors can be classified as follows based on their effects.

12.3.3.1   Intensified deterioration

The combined effect of environmental factors on electronic equipment is greater than that caused by each environmental factor alone. For example, the electronics used along with aircraft engines experience high temperature, shock, and vibrations. Each of these factors could intensify the effect of the others. High temperatures may make the material weaker, and associated shock and vibrations can then cause cracking-type failures in the assemblies.

An increase in one environmental factor can also lead to an increase in another, thereby intensifying the net effect. This is different from the above case, where both environmental factors affect the failure mechanism directly. For example, high temperatures accelerate the growth of some fungi and microorganisms. This can be a problem with permanently installed downhole fixtures, which are at high temperatures. With a small amount of humidity present, microorganisms can grow on electronic assemblies and the organic processes can cause chemical changes and contamination, resulting in loss of performance.

12.3.3.2   Coexistence with no synergistic effects on deterioration of the equipment

Two environmental factors act independently on electronic equipment and do not influence each other's effect. For example, acoustic vibrations to which electronic equipment might be subjected do not have any significant additive effect on the potential hazards caused by fungal activity in the vicinity of electronic parts.

12.3.3.3   Weakened net effect

The two environmental factors diminish the effect of each other. For example, high temperature can increase outgassing of constituents of the structural material of electronic parts, while high pressure generally decreases it. Permanently installed downhole gauges typically experience high-temperature and high-pressure conditions.

The increase in one environmental factor can also lead to the reduction of another; consequently, the net effect is reduced. For example, low temperature generally retards growth of fungi; therefore, the effects of the presence of fungi are reduced with low temperature.

12.3.3.4 Incompatible environmental factors

The coexistence of the environmental factors is not possible. Certain combinations of environmental factors, e.g., snow and high temperature, cannot exist.

12.3.4 Change in magnitude and rate of change of magnitude

Failure mechanisms in electronic equipment can be caused by steady-state loads or changes in the magnitude of the load (absolute change or rate of change). Therefore, the nature of the application of the loads (steady state or dynamic) should be determined. For example, failure by intermetallic growth at the wire bonds in integrated circuit (IC) packages is dominated by steady-state temperature conditions, while failures by die fracture in IC packages depend more on the rate of temperature change [20], [21].

Table 12.1: Examples of generic effects of combined environments on electronic products (adopted from [2]).

Combined environments	Classification of effects	Possible effects
High temperature and salt spray	Intensified deterioration	High temperature tends to increase the rate of corrosion caused by salt spray and thereby increase the net effect [14].
High temperature and high relative humidity	Intensified deterioration	High temperature increases the rate of moisture penetration and the rate of corrosion. Thus the combination can aggravate failures caused by humidity (e.g., corrosion, delamination) [14], [15].
High temperature and high pressure	Intensified deterioration	Each of these environmental factors leads to deterioration in the strength of the material and can cause structural failure in electronic assemblies.
High temperature and fungi	Intensified deterioration	High temperatures provide a congenial environment for growth of fungi and microorganisms. Thus high temperatures aggravate failures caused by fungal growth of electronic assemblies (typically from 25°C to 71°C) [17].
High temperature and acceleration	Intensified deterioration/ weakened net effect	Both acceleration and high temperature affect material properties. The combination, however, can reduce failure caused by fatigue/fracture because the material stress relaxes at high temperatures and the material becomes more pliable. Failures caused by solder joint fatigue and cracking are diminished by the combination [16]. In the case of brittle materials, however, this combination can lead to early failures because the material becomes weak at high temperatures and can easily fracture.
High temperature, sand, and dust	Intensified deterioration/ weakened net effect	The erosion caused by sand may be accelerated by high temperature, which can cause wear of structural parts due to abrasion. High temperature also reduces the penetration of sand and dust, thereby decreasing failures that occur from dust penetration [2].

Combined environments	Classification of effects	Possible effects
High temperature, shock, and vibration	Intensified deterioration/ weakened net effect	Vibration, shock, and high temperature affect material properties and cause deterioration of mechanical properties. The combination, however, reduces failure caused by fatigue/fracture, because the material stress relaxes at high temperatures and the material becomes more pliable. Failures caused by solder joint fatigue and cracking are diminished by the combination [16]. In case of brittle materials, however, this combination can lead to early failures because the material becomes weak at high temperatures and can easily fracture.
Low temperature and humidity	Intensified deterioration	Relative humidity increases as temperature decreases (especially in moist conditions), and lower temperature may induce moisture condensation. If the temperature is low enough, frost or ice may result. Hence low temperatures can aggravate failures caused by humidity, frost, or ice (e.g., corrosion).
Low temperature and high pressure	Intensified deterioration	The combination can cause structural failure, such as leakage through seals and airtight enclosures.
Low temperature and salt spray	Weakened net effect	Low temperature reduces the corrosion caused by salt spray; the combination causes weakening [14].
Low temperature, sand, and dust	Intensified deterioration	Low temperature increases dust penetration and can aggravate failures caused by wear of assemblies and alteration of electrical properties.
Low temperature and fungi	Weakened effect	Low temperature reduces fungus growth. At subzero temperatures, fungi remain in suspended action, thereby weakening the net effect [17].
Low temperature, shock, and vibration	Intensified deterioration	Low temperature tends to intensify the effects of shock and vibration, because certain materials (like aluminum) tend to become brittle at lower temperatures. However, this is a consideration only at very low temperatures.
Low temperature and acceleration	Intensified deterioration	Acceleration produces shock, vibration, or both. Hence, low temperature and acceleration intensify the effects of acceleration because of brittleness at low temperatures.
Humidity and high pressure	Intensified deterioration	The effect of this combination varies with the temperature. High temperature can aggravate the deleterious effects caused by humidity and high pressure, indirectly increasing the net effect on an electronic assembly.
Humidity and salt spray	Intensified deterioration	High humidity may dilute the salt concentration and could affect the corrosive action of the salt by increasing its mobility and spread, thereby increasing the conductivity. Corrosion failures, like shorts and opens in the metallization and interconnects, are typically aggravated [14].

Combined environments	Classification of effects	Possible effects
Humidity and fungi	Intensified deterioration	Humidity helps the growth of fungus and microorganisms but adds nothing to their effects [17].
Humidity, sand and dust	Intensified deterioration	Sand and dust have a natural affinity for water, and this combination increases deterioration by corrosion.
Humidity and vibration	Intensified deterioration	This combination tends to increase the rate of breakdown of electrical material and connections.
Humidity, shock, and acceleration	Intensified deterioration	The periods of shock and acceleration, if prolonged, aggravate the effects of humidity, because humidity tends to cause deterioration of material properties. The combination can lead to early structural failure.
High pressure and vibration	Intensified deterioration	This combination intensifies structural failures in electronic and electrical equipment.
High pressure, shock, and acceleration	Intensified deterioration	This combination intensifies structural failures in electronic and electrical equipment.
Salt spray and dust	Intensified deterioration	Sand and dust have a natural affinity for water, and this combination increases deterioration by corrosion.
Salt spray, shock, or acceleration	Coexistence without any synergistic effects on deterioration of the equipment	These combinations produce no added effect.
Salt spray and vibration	Intensified deterioration	This combination tends to increase the rate of breakdown of electrical material and connections.
Salt spray and explosive atmosphere	Incompatible	This is considered an incompatible combination.
Sand, dust, and vibration	Intensified deterioration	Vibration increases the wearing effects of sand and dust.
Shock and vibration	Coexistence without any synergistic effects on deterioration of the equipment	Since shock is a form of vibration, this combination does not produce any added effects.
Vibration and acceleration	Intensified deterioration	This combination produces increased effects when encountered with high temperatures and low pressures (typically in applications like oil suction).
High temperature and low pressure	Intensified deterioration	As pressure decreases, outgassing of constituents of materials increases. As temperature increases, outgassing increases. Hence, each tends to intensify the effects of the other.
High temperature and explosive atmosphere	Coexistence without any synergistic effects on deterioration of the equipment	Temperature has minimal effect on the ignition of an explosive atmosphere but does affect the air-vapor ratio, which is an important consideration.

Combined environments	Classification of effects	Possible effects
High pressure and explosive atmosphere	Intensified deterioration	High pressure aggravates the effects of explosion and thereby enhances the hazards of an explosive atmosphere.
Low temperature and low pressure	Intensified deterioration	This combination can accelerate leakage through seals and airtight regions. It can cause material deterioration and loss of functionality in hermetic parts.
Low temperature and explosive atmosphere	Coexistence without any synergistic effects on deterioration of the equipment	Temperature has minimal effect on the ignition of an explosive atmosphere but does affect the air-vapor ratio, which is an important consideration.
Humidity and explosive atmosphere	Weakened net effect	Humidity has no effect on the ignition of an explosive atmosphere, but high humidity will reduce the pressure of an explosion.
Low pressure and salt spray	Intensified deterioration	This combination can lead to increased penetration of moisture into the equipment and thus enhance the rate of material deterioration and corrosion-related failure mechanisms.
Low pressure and fungi	Coexistence without any synergistic effects on deterioration of the equipment	This combination does not add to overall effects.
Low pressure and explosive atmosphere	Intensified deterioration	At low pressures, an electrical discharge is easier to develop, but the explosive atmosphere is harder to ignite.

12.4 Life cycle phases and environmental factors

This section discusses typical life cycle phases and the environmental factors that are usually critical in each phase. However, since the LCEP is strongly application and product dependent, a thorough analysis of environmental factors in each of the phases is necessary during the design of any electronic equipment. Typical environmental factors in most phases include temperature, vibration, shock, pressure, and humidity. However, factors like radiation, fungi/microorganisms, fog, freezing rain, snow, hail, sand and dust, salt spray, and wind should not be overlooked.

12.4.1 Assembly

Assembly of electronic parts and/or modules usually has temperature and vibration as the primary environmental loads. Welding and soldering operations lead to significant thermal stresses in the parts and/or assemblies. The mechanical handling, placement, and assembly procedures also induce significant vibration and shock loads. Other environmental factors which might be critical are radiation, chemical and ionic contamination (plasma machining or welding), humidity, and pressure, depending on the assembly process.

12.4.2 Qualification and testing

The environmental loads the electronic equipment is subjected to during qualification and testing are strongly application specific. Standard tests on electronic parts and modules include, high-temperature bake, high-temperature operating life, vibration, shock, temperature humidity bias (THB), highly accelerated life testing (HALT), and highly accelerated stress testing (HAST). Depending on the application, special tests for radiation, sand and dust, snow, and other such environmental factors may be performed on the electronic equipment. Depending on these tests, the LCEP of the product for this phase will change.

12.4.3 Storage

Storage typically has temperature (diurnal cycles) and humidity as the prime environmental factors. However, depending on the quality of the storage spaces, environmental factors such as fungi, sand and dust, and radiation might also come into the picture. These environments should also be considered when the levels are high enough to cause a potential reliability problem. Chemical gases are also an issue when equipment is stored in chemically aggressive climates.

12.4.4 Transportation

Transportation phases are usually characterized by high vibration, shock, and temperature loads. Transportation by road will cause shock and vibration due to rocky and uneven paths, internal vibrations in the automobile, and accidents of the vehicle. The equipment will also be subjected to diurnal temperature cycles as well as to heat generated by the operation of the automobile. Transportation by air similarly will subject the equipment to severe shock while taking off and landing (unless packaged appropriately), as well as to severe temperature cycling due to the differences in ground and airborne temperatures. Apart from these specific problems, electronic equipment can also experience sand and dust, rain, snow, gases, humidity and radiations.

12.4.5 Installation

The installation process is typically characterized by vibration and shock loads. Due to the brief duration of this phase, other factors may not be significant but can be an issue in specific cases. For example, in deployment of permanent monitoring equipment for oil wells, the equipment suddenly encounters very high temperatures when it comes in contact with the hot oil inside the tubes during deployment, and temperature shocks become a serious issue.

12.4.6 Operation

The environmental loads during operation are specific to the application. For example, the electronic equipment in the under hood of a car encounters a great deal of temperature cycling and vibration, whereas the electronics inside a computer rarely do so. On the other hand, humidity might be a consideration in both of these applications.

12.4.7 Maintenance

Maintenance, in some cases, can subject the equipment to severe loads due to mishandling of the equipment. Shock and vibration are typical loads associated with maintenance procedures. Electrostatic discharge can also be an issue when proper care is not taken during maintenance.

12.5 Environmental factors and their effects

Table 12.2 provides some of the environmental factors and their generic effects on electronic parts and subsystems. Some of these parameters are discussed in more detail in the following sections.

12.5.1 Temperature

Temperature has a major influence on the electrical, mechanical, chemical, and physical deterioration of materials for two main reasons: the physical properties of nearly all known materials are altered by changes in steady-state temperature, temperature gradients, and temperature extremes; and the rate of a chemical reaction between two or more reactants is dependent on the temperature of the reactants. Some of the adverse effects of temperature include the expansion or contraction of materials due to temperature changes, causing problems with fit between system interfaces and interconnections, outgassing of corrosive volatile products due to application of heat, local stress concentrations due to nonuniform temperature, and the collapse of metal structures when subjected to cyclic heating and cooling due to induced stresses and fatigue caused by repeated flexing. Many IC failure mechanisms are caused and/or accelerated by temperature [20].

12.5.2 Humidity

Water vapor in the air or any other gas is called "humidity"; water in solids or absorbed in liquids is usually designated "moisture". There are various ways of expressing humidity:

- *Relative humidity (RH or ϕ)*: the ratio of actual vapor pressure (p_V) to saturation vapor pressure (p_S) at the prevailing temperature. It is usually expressed as a percentage.
- *Absolute humidity (W)*: the mass of water vapor (m_V) per unit mass of dry air (m_A) in the sample volume at the prevailing temperature
- *Vapor pressure (p_V)*: the part of the total pressure contributed by the water vapor
- *Dew point temperature (T_{dp})*: the temperature to which a gas must be cooled at constant pressure to achieve saturation

Humidity or moisture can play a major role in accelerating failures in parts and systems. Failure mechanisms such as metallization corrosion, conductive filament formation, swelling of polymer-based structural elements or potting, and connector plating corrosion are all adversely impacted by the presence of moisture. In addition, moisture can cause mated systems to lock together, especially when water condenses on them and then freezes. Similarly, many materials that are normally pliable at low temperatures can become hard and brittle due to absorption of water, which subsequently freezes at low temperatures. The volume increase due to freezing of water can also separate parts, materials, or connections. Condensed moisture can also act as a medium for the interaction between several otherwise relatively inert materials. The chlorine released by polyvinyl chloride (PVC), for example, will form hydrochloric acid when combined with moisture. Moisture can cause shorts or leak paths between board traces.

Although the presence of moisture may cause deterioration, the absence of moisture can also cause reliability problems. Many nonmetallic materials become brittle and crack when they are very dry. The properties of these materials depend upon an optimum level of moisture. Similarly, fabrics wear out at an increasing rate as moisture levels are lowered, and fibers become dry and brittle. Environmental dust, which is usually held in suspension by moisture, can cause increased wear and friction on moving parts. Freed dust can clog filters due to the absence of moisture.

Table 12.2: Environmental factors and their effects on electronic equipment for downhole applications

Environmental factors	Principal effects		Possible failures
Temperature (natural/induced)	High	Thermal aging Oxidation Structural change Chemical change Softening and melting Viscosity reduction/ evaporation Physical expansion	Insulation failure because of melting
			Alteration of electrical properties owing to changes in resistance
			On-chip failures (metallization migration, Kirkendall voiding in wire bonds, slow trapping, time-dependent dielectric breakdown) [4], [19], [20], [21]
			Melting of solder joints [19], [22]
			Unequal expansion leading to fracture [19]
			Ionic contamination [19]
	Low	Physical contraction Brittleness	Alteration of electrical properties owing to changes in resistance
			Unequal expansion between components and board leading to fracture because of coefficient of thermal expansion (CTE) mismatch
			Increased brittleness of metals
Relative humidity/moisture (natural/induced)	High	Moisture absorption [23] Chemical reaction Corrosion Electrolysis	Metallization corrosion (on-chip) [4], [14], [24]
			Delamination [4], [24], [25], [26]
			Loss of electrical properties owing to corrosion and chemical reactions
			Cracking of electronic parts owing to moisture absorption [1], [4], [24]
			Reduction in electrical resistance because of conduction through moisture
	Low	Desiccation Embrittlement Granulation	Loss of mechanical strength
			Structural collapse of components
			Alteration of electrical properties
Pressure (natural/induced)	High	Compression	Structural collapse of assemblies including electronic components
			Penetration of seals
			Interference with function
	Low	Expansion Outgassing	Explosive expansion of assemblies
			Alteration of electrical properties
			Loss of mechanical strength
			Insulation breakdown and arc over

Environmental factors	Principal effects	Possible failures
Wind (natural)	Force application Deposition of materials Heat loss (low velocity) Heat gain (high velocity)	Structural collapse Interference with function Loss of mechanical strength Mechanical interference and clogging [1] Accelerated abrasion [1] Accelerated low-temperature effects (low velocity) Accelerated high-temperature effects (high velocity)
Salt spray (natural)	Chemical reactions Corrosion Electrolysis	Increased wear of electronic parts and assemblies Alteration of electrical properties Interference with function Surface deterioration Increased conductivity Metallization corrosion [1], [4], [14], [24]
Sand and dust (natural)	Abrasion Clogging	Increased wear of electronic parts owing to material degradation Interference with function Alteration of electrical properties
Rain (natural)	Physical stress Water absorption and immersion Erosion Corrosion	Structural collapse Increase in weight Electrical failure Structural weakening Removal of protective coatings Delamination and cracking [4], [15], [24] Surface deterioration Enhanced chemical reactions like corrosion
Ionized gases (natural)	Chemical reactions Corrosion Change in conductivity	Change in electrical properties Deterioration in material properties
Air pollution (natural)	Chemical reactions Clogging	Interference in functionality because of clogging Deterioration in material properties owing to chemical reactions Corrosion

Environmental factors		Principal effects		Possible failures
Freezing rain/frost/snow (natural)		Low temperature Moisture ingress [25], [26] Corrosion/chemical reactions Clogging		Mechanical stress caused by CTE mismatch between structural components Increase in weight Change in electrical properties owing to change in resistance/conductivity Delamination [4], [24]–[26] Material deterioration On-chip failures (metallization, corrosion, delamination) [4], [14], [24]
Fungi (natural)		Clogging		Change in electrical characteristics owing to shorts and alteration in electrical resistance Oxidation of structural elements of the circuit
Static electricity electrostatic discharge (natural/induced)		Change in electrical response Electrical overstress		Interference in function owing to changes in electrical properties (resistance, voltage) Shorts and opens in circuit caused by electrical overstress and electrostatic discharge [4]
Chemicals (induced)		Chemical reactions Reduced dielectric strength		Alteration of physical and electrical properties Insulation breakdown and arc over [4] Metallization corrosion [14], [20], [24]
Explosion (induced)		Severe mechanical stress		Rupture and cracking Structural collapse of assemblies and parts
Shock (induced)	Thermal	Mechanical stress		On-chip failures (die fracture, cracking, electromigration, wire flexure fatigue, shear fatigue) [4], [20], [21], [24] Solder joint fatigue [16], [27], [28] Unequal expansion between components and board leading to fracture because of CTE mismatch
	Mechanical	Mechanical stress Fatigue		Loss of mechanical strength Interference with function Increased wear of electronic assemblies Solder joint fatigue [16], [27], [28] Structural collapse of assemblies
Vibration (induced)		Vibration/ acceleration	Mechanical stress Fatigue	Loss of mechanical strength Interference with function Increased wear of electronic assemblies Solder joint fatigue [16], [27], [28] Structural collapse of assemblies

Environmental factors	Principal effects		Possible failures
	Rotation	Mechanical stress Torsional acceleration	Twisting of electronic assemblies Intermittent interconnections Loss of mechanical strength
	Bending	Mechanical stress Fatigue	Bending failure of electronic components and assemblies Cracking

Some design techniques that can either be used separately or combined to counteract the effects of moisture are:

- eliminating moisture traps by providing drainage or air circulation
- using desiccant systems to remove moisture when air circulation or drainage is not possible
- applying protective coatings
- providing rounded edges to allow uniform coating of protective material
- using materials resistant to fungi, corrosion, and other moisture-related effects
- hermetically sealing components, gaskets, and other sealing systems
- impregnating or encapsulating materials in moisture-resistant waxes, plastics, or varnishes
- separating dissimilar metals or materials that might combine or react in the presence of moisture or of components, which might damage protective coatings

Apart from devising measures to keep moisture out of parts, the design team must also consider possible adverse effects caused by specific methods of protection. Hermetic sealing, gaskets, and protective coatings may, for example, increase moisture by sealing moisture inside or contributing to condensation. The gasket materials must be evaluated carefully for outgassing of volatile vapors or for incompatibility with adjoining surfaces or protective coatings.

12.5.3 Vibration and shock

Vibration is defined as periodic motion about an equilibrium position. Vibrations result from dynamic forces that set up a series of motions within a system. The forced motions may be linear, angular (torsion), or a combination of both. A vibratory system includes, in general, a means for storing potential energy (spring or elasticity), a means for storing kinetic energy (mass or inertia), and a means by which energy is gradually lost (damping or resistance).

Some of the common faults that may be caused by vibration include bent shafts, damaged or misaligned drives and bearings, fretting corrosion, onset of cavitation, and worn gears. In addition, vibration and shock can harmfully flex leads and interconnects, cause parts to strike each other or the housing, dislodge parts from their positions in the system, cause acoustical and electrical noise, and lead to structural instabilities. Protective measures against vibration and shock are generally determined by an analysis of the deflections and mechanical stresses produced by these environmental factors. This involves the determination of natural frequencies and evaluation of the mechanical stresses within components and materials produced by the shock and vibration environment. If the mechanical stresses so produced are below the acceptable safe working stress levels of the materials involved, no direct protection methods are required. If the stresses exceed the safe levels, corrective measures such as stiffening, reduction of inertia and bending moment effects, and incorporation of further support members, as well as possible uses of isolators, may be required. If such approaches do not reduce the stresses below the acceptable safe levels, further reduction is usually possible by the use of shock-absorbing mounts.

Another failure mode that may result from vibrations is fatigue, which is the tendency of a material to yield and fracture under cyclic stress loads considerably below its tensile strength. This type of failure includes high cycle fatigue, acoustic fatigue, and fatigue under combined stresses such as temperature extremes, temperature fluctuations, and corrosion.

In addition to using proper materials and configurations, it is necessary to control the amount of shock and vibration experienced by the system. Damping systems are used to reduce peak oscillations and special stabilizers can be employed when unstable configurations are involved. Typical examples of dampers are viscous hysteresis, friction, and air damping. Vibration isolators are commonly identified by their construction and material used for resilient elements like rubber, coiled spring, and woven metal mesh. Shock isolators differ from vibration isolators in that shock requires a stiffer spring and a higher natural frequency for the resilient element. Isolation mounting systems are of the type installed underneath, the over-and-under type, and inclined isolators. In some cases, however, even though an item is properly insulated and isolated against shock and vibration damage, repetitive forces may loosen the fastening systems. If the fastening systems loosen enough to permit additional movement, the system will be subjected to increased forces and may fail. Many specialized self-locking fasteners are commercially available to counter this occurrence.

12.5.4 Solar radiation

Solar radiation contributes several types of significant loads to the life cycle environment. The solar flux provides radiant heating, ionizing radiation including ultraviolet exposure, and visible wavelengths that can interfere with optics.

The maximum solar load outside the atmosphere occurs on January 2nd of each year when the Earth is closest to the sun. The solar flux is taken at an average of 1367 W/m^2, with a January peak of 1413 W/m^2 and the July 4th minimum at 1332 W/m^2. The sun can be modeled as a black body radiator at 6000 K. Therefore, the sun emits ultraviolet (UV) radiation. Objects in orbit receive this flux projected onto their area unless shadowed by the Earth. The Earth's atmosphere attenuates and scatters much of the incident solar energy. The solar radiation on objects on the surface of the Earth is the sum of the projected area normal to the Earth-sun line flux, a function of the time of day and location, energy incident by a scattered path, and energy reflected off other surface objects.

The primary effect of sunlight is heating. The surface temperature in orbit is directly dependent upon the ratio of solar absorbtivity to infrared emissivity. This ratio is important on the surface of the Earth, although convection plays a dominant role in determining the surface temperature.

The sun's light also provides damaging UV radiation on earthborn electronics assemblies. Organics used in plastics and paints, wiring, cables, and connectors are especially vulnerable to damage by UV radiation. UV radiation inhibitors are required for exposure. Optical components like security cameras are vulnerable to damage by heat, direct solar exposure, thermal loading, and functional interference by glint or overexposure.

12.5.5 Electromagnetic radiation

Electronic products stored near nuclear reactors, isotropic nuclear sources, accelerators, or nuclear detonations must be designed to tolerate the two basic effects of nuclear irradiation: mechanical and electrical failures. The mechanical failure mechanism causes time-dependent wearout failures through an embrittlement phenomenon that increases the hardness and decreases the ductility of metals. However, the electrical failure mechanism of random overstress when a single radiation particle interacts with the LSI/VLSI circuitry is a more critical problem.

In general, metals are quite resistant to radiation damage in the space environment. Semiconductor devices may be affected by gamma rays, which increase leakage currents. The lattice structure of semiconductors can be damaged by high-energy electrons, protons, and fast neutrons, which cause permanent effects through atomic displacement and damage to the lattice structure. Organic materials are particularly susceptible to physical changes in cross-linking and scission of molecular bonds. Radiation-induced formation of gas, decreased elasticity, and changes in hardness and elongation are some of the predominant changes in plastics which have been subjected to radiation of the type encountered in the space environment. Electronic circuits are affected by a lowering of input and output impedances [1].

Protection against the effects of electromagnetic radiation has become an engineering field by itself: electromagnetic compatibility design. The most direct approach to protection is to entirely avoid the region in which high radiation levels are found. When exposure cannot be avoided, shielding and filtering are the protective measures used. In other cases, material design changes or operating procedural changes must be instituted in order to provide protection or to minimize the effects on normal operation of the system. Table 12.3 shows the radiation susceptibility of various semiconductors.

Table 12.3: Allowable radiation dosage levels for various semiconductors

Radiation environment	Discrete bipolar transistors	Silicon-controlled rectifiers	TTL devices	Low-power Schottky TTL	CMOS devices	NMOS devices	Light-emitting diodes
Neutrons (c/nm^2)	10^{10} to 10^{12}	10^{10} to 10^{12}	10^{14}	10^{14}	10^{15}	10^{15}	10^{13}
Total radiation dose (rads)	$> 10^4$	10^4	10^6	10^6	10^3 to 10^4	10^3	$> 10^5$
Transient dose rate (upset)	—	10^3	10^7	5×10^7	10^7	10^5	—
Transient dose rate (survival)	10^{10}	10^{10}	$> 10^{10}$	$> 10^{10}$	10^9	10^{10}	$> 10^{10}$

12.5.6 Pressure

Pressure is defined as the normal force per unit area exerted by a fluid (either a liquid or a gas) on any surface. The surface is typically a solid boundary in contact with the fluid. Finding the component of the force normal to the surface is sufficient for determining the pressure. Pressure can be expressed in four ways:

- *absolute pressure* — the same as the definition given above. It represents the pressure difference between the point of measurement and a perfect vacuum where the pressure is zero.
- *gage pressure* — the pressure difference between the point of measurement and the ambient pressure.
- *differential pressure* — the pressure difference between two points, one of which is chosen to be the reference.
- *stagnation pressure* — the pressure due to fluid flow.

In high-vacuum conditions (such as space), materials having a high vapor pressure will sublimate or evaporate rapidly, particularly at elevated temperatures. In some plastics, the loss of the plasticizing agent by evaporation will cause cracking, shrinking, or increased brittleness. Inorganic coatings with low vapor pressures can be used to protect metals such as magnesium, which would normally evaporate rapidly (1 g/cm^2/year at 250°C).

In a high vacuum, adjoining solid surfaces can become cold-welded after losing adsorbed gases from their surfaces. Some form of lubrication is therefore necessary. Conventional oils and greases evaporate quickly. Graphite becomes unsatisfactory and actually behaves as an abrasive because of the loss of absorbed water. However, thin films of soft metals, such as lead, silver, or gold, are effective lubricants in a high vacuum. Thin films of molybdenum disulfide are often sprayed over chrome or nickel plating, forming easily sheared layers. The film also releases sulfur at the interfaces during sliding of adjacent parts, performing the same function that water vapor does for graphite.

12.5.7 Chemicals

The Earth's environment contains numerous chemically active elements, such as oxygen, carbon dioxide, nitrogen, snow, ice, sand, dust, salt-water spray, and organic matter, which have the ability to corrode and deteriorate materials. Hence, system specifications often specify limits on temperature, humidity, altitude, salt spray, fungi, sunshine, rain, sand, and dust that the system can tolerate.

A material or structure can undergo a chemical change in a number of ways. Among these are interactions with other materials, such as metal migration and diffusion, and modifications in the material itself, such as recrystallization, stress relaxation, phase change, or changes induced by irradiation. In addition to the deterioration problems associated with the external environments to which systems are subjected, adhesives, batteries, and certain types of capacitors are susceptible to chemical aging and biological growths due to biochemical reactions.

Materials widely separated in the electrochemical series are subject to galvanic action, which occurs when two chemically dissimilar metals are in contact in an electrolytic liquid medium. The more active metal dissolves, and an electric current flows from one metal to the other. Coatings of zinc are often applied to iron so that the zinc, which is more active, will dissolve and protect the iron. This process is commonly known as "galvanization." Galvanic action is also known to occur within the same piece of metal if one portion of the metal is under stress and has a higher free-energy level than the other. The part under stress will dissolve if a suitable liquid medium is present. Stress-corrosion cracking occurs in certain magnesium alloys, stainless steels, brass, and aluminum alloys. It has also been found that a given metal will corrode much more rapidly under conditions of repeated stress than when no stress is applied. Proper design of a system therefore requires trade-offs in:

- selecting corrosion-resistant materials
- specifying protective coatings if required
- avoiding the use of dissimilar metallic contacts
- controlling metallurgical factors to prevent undue internal life cycle conditions
- preventing water entrapment
- using high-temperature resistance coatings when necessary
- regulating the environment through dehydration, rust inhibition, and electrolytic and galvanic protective techniques

12.5.8   Sand and dust

In addition to the effect of reduced visibility, sand and dust can degrade any system by abrasion, leading to increased friction causing increased wear and heat, as well as clogging of filters, small apertures, and delicate systems. Thus, a system having moving parts requires particular care when designing for sand and dust protection. Sand and dust will abrade optical component surfaces, either by impact when being carried by air or by physical abrasion when the surfaces are improperly wiped during cleaning. Dust accumulations have an affinity for moisture and may lead to corrosion or even biological growth.

In relatively dry environments, such as deserts, fine particles of dust and sand can readily be agitated into suspension in the air, where they may persist for many hours, sometimes reaching heights of several thousand feet. Thus, even though there is virtually no wind present, the speed of vehicles that may be housing an electronic system and moving through these dust clouds can also cause surface abrasion by impact.

Although dust commonly is considered to be fine, dry particles of earth, it also may include minute particles of metals, combustion systems, and solid chemical contaminants. These other forms may cause direct corrosion or fungal effects on systems, because this dust may be alkaline, acidic, or microbiological. Other hazards due to dust can be soiling and discoloration, environmental pollution, harm to human health, and economic losses.

Dust reduction methods are mainly of two types: active and passive. Active methods include installation of fans to increase the flow of air; use of filters and shelters, and so on. Passive methods include measures such as planting trees, and improving pollution standards. Dust protection must be planned in conjunction with protective measures against other environmental factors. For example, specifying a protective coating against moisture, if sand and dust are present, is useless unless the coating is carefully chosen to resist abrasion and erosion. When systems require air circulation for cooling or for removing moisture, the issue is not whether to allow dust to enter, but rather to control the size of the dust particles. The problem becomes one of filtering the air to remove dust particles above a specific nominal size. For a given working filter area, these filters decrease the flow of air or other cooling fluids through the filter, while the ability of the filter to stop smaller and smaller dust particles is increased. Therefore, there must be a trade-off between the filter surface and the decrease of flow of the fluid through the filter or the allowable particle size.

12.5.9   Human factors

Although humans are not directly a part of the environment of any system, they can contribute to failures and affect the system in several ways. Humans are active participants in the operation of most systems, and this interaction must be weighed against safety, reliability, maintainability, and other system parameters to assess system reliability, maintainability, time performance, safety analyses, and specific human engineering design criteria.

12.6 Quantification of environmental loads

The environmental loads acting on the system during each stage of its life cycle should be quantified. The quantification is necessary to provide a basis for designing for reliability. The methods for quantification are discussed here.

12.6.1 Real-time/in-situ monitoring

Real-time monitoring of the environmental loads using sensors can be used to quantify the loads due to environmental factors. Sensors are placed at the location of installation or on similar equipment (in a similar location). The sensors should be so placed that they do not significantly change the local environment at the location where the sensing is performed.

The following consideration should be addressed for real-time monitoring of environmental factors.

12.6.1.1 Extent and frequency of monitoring

Once the parameters to be monitored have been selected, the monitoring frequency and extent should be decided. Each of the parameters may require different frequencies of monitoring. For an environment where there is substantial heat generation, monitoring the temperature might be an almost continuous requirement. A schedule should be devised that specifies the monitoring frequency of each parameter.

To decide the extent of monitoring, it is essential to first know what effect the particular parameter will have at every level of the system. Highly localized heating, for example, might require continuous monitoring, but its effect on other components in the vicinity might be nominal enough to not require installation of a temperature sensor for the other parts.

12.6.1.2 Operating environment extremes

The conditions of operation of the system determine the monitoring requirements for the system. A system operating on a submarine is subjected to conditions that are very different from the conditions encountered by a system in a microwave tower in terms of the bounds of factors like temperature, vibration, and humidity. Identification of these extremes is necessary to avoid over/underdesign when selecting the monitoring system. It is not necessary to install a monitoring system that can monitor across the entire range of operation capabilities of the part if the local environment will not attain that state. It is adequate to have a monitoring system that can monitor across the limits of the greatest extremes of the actual environment, with some tolerances for unprecedented changes. The system for monitoring should be chosen so that it can operate within the maximum specified environmental extremes.

12.6.1.3 Deciding which sensors to use

The choice of the appropriate sensor to use to determine the environment depends on a variety of factors, including:

- sensing technologies available
- cost limitations on the selection of a sensor for a given application
- space constraints for the sensing device

- environmental and operating conditions in which the sensing device is to be used
- physical limitations during the monitoring process
- ease of monitoring and accessibility of the sensing device

12.6.1.4   Sensing methods for selected environmental factors

Commonly used sensing methods for environmental factors discussed in Section 12.5 are as follows.

Temperature

The most widely used thermal sensors are resistive temperature detectors (RTDs), thermistors, and thermocouples. Thermistors and RTDs are based on changes of mobility and carrier density, resulting in a change of resistance with temperature, which is represented by temperature coefficients that may be constants or nonlinear functions of temperature.

A thermocouple is defined as any pair of electrically conducting and thermoelectrically dissimilar elements coupled at an interface. Its operation is based on the Seebeck effect (one of the three thermoelectric effects – Seebeck, Peltier and Thompson [29]), which is the appearance of a thermo-EMF (electromotive force) in an electric circuit composed of two heterogeneous conductors whose junctions are at different temperatures. Two different materials (usually metals) are joined at one point to form a thermocouple. The reference junction is held at a known temperature, such as the ice-water equilibrium point (i.e., known thermocouple voltage), and the difference between this fixed voltage and the thermocouple voltage at the measurement junction is measured by a voltmeter. This EMF is calibrated to directly obtain the temperature at the measuring junction.

Humidity

The most commonly used mode of humidity sensing is electric sensing. Electric relative humidity (RH) sensors are of two types: capacitive and resistive. Capacitive sensors have nonconducting elements that absorb or give up moisture until a state of equilibrium is reached, whereas resistive elements use conductive organic polymers. A change in the RH results in a change in the amount of water vapor absorbed by the sensing layer, whose resistance and/or capacitance varies with the humidity.

The resistive RH sensor is a thin wafer of a water-absorbent polymer printed with two interlocking combs of conducting metal or polymer. The electrical resistance, measured through or across the surface of the polymer, varies with the water content. This sensor also needs an AC excitation voltage, not for the measurement but to prevent it from being destroyed by causing a one-way electrolytic ion movement in the polymer.

Vibration and shock

Vibration is usually measured using accelerometers. Piezoelectric, piezoresistive, and capacitive types are the most common accelerometer types. Piezoelectric accelerometers utilize a mass in direct contact with the piezoelectric element (usually lead zirconate titanate ceramic [PZT] and quartz). When a varying motion is applied to the accelerometer, the element experiences a force excitation, causing a proportional electric charge to be developed across it. Since the applied force is proportional to the inertial load, measurement of the induced electric charge yields information regarding the applied force.

Piezoresistive accelerometers are essentially strain gages with large gage factors. In piezoresistive accelerometers, a piezoresistive material (e.g., a silicon crystal) is deposited over the sensor. When the sensing structure is stressed due to an inertial load, the resistance of the sensing area containing the piezoresistive material changes; in turn the change, is proportional to the acceleration to be measured. Most piezoresistive accelerometers use two or four active gages arranged in a Wheatstone bridge for carrying out the measurement.

Capacitive sensors measure the change in capacitance between two electrodes when one of them is displaced or deformed due to an inertial force introduced by the applied acceleration. The sensor is usually approximated as a multiple parallel plate capacitor.

Solar radiation

Solar radiation measuring instruments can be broadband, which measure the combined solar intensity (irradiance) at all wavelengths, or spectrally selective, which measure the intensity in different wavelength bands. Broadband instruments are more common. The two most widely used broadband instruments in the field are:

- *Pyrheliometers*: These are instruments that measure direct radiation. Instruments measuring direct radiation usually include radiation coming out to an angle of about 3 degrees away from the sun's disk. The sensor is a temperature-compensated thermopile (a number of thermocouples in series) placed at the bottom of a blackened collimator tube that limits the angular acceptance of solar radiation to about 5–6 degrees (total). The instrument is oriented such that the direct radiation from the sun is parallel to the axis of the collimator tube.

- *Pyranometers*: These are instruments that measure global and diffuse radiation. They have a shading disk to prevent direct solar radiation from reaching the sensor. The measurement for diffuse radiation involves correcting for the portion of the radiation shielded from the sensor by the shading disk. The sensor is a thermopile with alternate blackened junctions heated by the sun. The unheated junctions are near ambient temperature, which may be ensured by putting the unheated junctions in thermal contact with a white surface. Heating by the sun is accomplished by placing the junctions in contact with a black surface (with high thermal conductivity) or by placing a black coating on the junctions. The instrument is installed in a level position, the sensor facing up towards the sky. Less expensive pyranometers may use a photovoltaic sensor to measure solar radiation.

Electromagnetic radiation

There are three types of electromagnetic radiation sensors: gas-filled detectors, scintillation devices, and semiconductor detectors. These can also be divided into two groups, depending on whether they measure the energy of the emitted rays (γ-rays, X-rays, etc.) or particles (neutrons, α- particles, β-particles, positrons, etc.). Semiconductor devices are used for their energy-resolving capability (photoelectric absorption), whereas scintillators and gas-filled devices are used to obtain spectral information.

Gas-filled detectors can be separated into three distinct types: ionization chambers, Geiger-Müller counters, and proportional counters. All of these utilize the ions and electrons created by the interaction of an incident photon with the detector gas, directly or indirectly, to produce an output signal. They differ in the characteristics of the electric field and the nature of the output signal.

Proportional counters use gas multiplication to amplify the charge of the initial ion-pair population. In these counters the charge associated with the amplified pulse is proportional to the number of ion pair produced initially. Geiger-Müller counters are the most widely used means of sensing X- and γ-radiation. They are similar to proportional counters in that they also use gas multiplication to generate an output pulse, but the output is independent of the number of initial ion-pairs.

Scintillation devices have the property of converting the energy absorbed from the incident X- or γ-radiation beam into visible or near-visible light. The detector consists of a scintillator element in which γ-radiation is converted into optical photons and a photocell (or a device similar to it) to convert these photons into an electrical signal for processing by other electronics. The detecting elements may be organic or inorganic.

The last category of sensors are semiconductor detectors, which may be either silicon or germanium. These are principally used due to their combination of efficient absorption and high-energy resolution. The energy of the electromagnetic ray is primarily transmitted to a small number of electrons in the semiconductor. The energetic electrons, in turn, interact with the other electrons in the valence shell to create mobile pairs of electrons and holes. In a detector with a sufficiently large volume, the average number of electron-hole pairs varies linearly with the energy of the ray. The motion of the ionized charge in the detector gives an indication of the strength of the radiation.

Pressure

Since pressure is defined as force per unit area, the most direct way of measuring pressure is to isolate an area on an elastic material for the force to act on and then measure the deformation of the sensing material. The most common pressure sensors are piezoresistive, piezoelectric, and capacitive sensors.

Piezoresistive pressure sensors (also called "strain-gage" sensors) are the most common type of pressure sensor in use. In these sensors, piezoresistive materials (such as silicon crystals) are incorporated into diaphragms and placed on the edges of the diaphragm, where the strains are the largest. A pressure differential across the diaphragm causes deflections that induce strains in the diaphragm, modulating the resistance values. The change in the resistance is an indication of the magnitude of the applied pressure. In piezoelectric pressure sensors, the strains associated with the deformation of a sensing element are converted into electrical charge by a piezoelectric crystal.

Capacitive pressure sensors consist of a pair of electrodes: a metal or silicon diaphragm (the pressure-sensing element) serving as one electrode and a stationary metal electrode formed by depositing a metal layer on a ceramic or glass substrate. The applied pressure defects the diaphragm, which in turn changes the gap spacing and the capacitance.

Chemicals

Common microfabricated chemical sensors include electrochemical sensors, tin-oxide (SnO_2)-based sensors, and calorimetric devices, of which electrochemical devices are the most common. Electrochemical sensors are essentially electrochemical cells consisting of two or more electrodes in contact with a solid or a liquid electrolyte. Electrochemical sensors can be classified according to their mode of operation, e.g., conductivity sensors, potentiometric sensors, and voltammetric sensors. A conductivity sensor measures the change in conductivity of the electrolyte in an electrochemical cell due to the presence of a chemical species. Potentiometric sensors use the effect of the concentration of the chemical species on the equilibrium of the redox reactions occurring at the electrode-electrolyte interface in an electrochemical cell as their sensing principle. These sensors sense the potential developed at the electrodes in the presence of the species. Voltammetric sensors utilize the concentration effect on the current-potential characteristics of the electrochemical system for their sensing operation.

Sand and dust

Dust concentration is usually expressed in mg/m^3. Commercial dust sensors monitor the flow of particulate emissions through a filter in a dust collector or bag house. Common sensing mechanisms include optical sensing and electrical sensing. Optical sensing utilizes the principle of near-forward light scattering to measure the concentration of airborne dust particles. This principle utilizes an infrared light source positioned at a 90 degree angle from a photodetector. As airborne dust particles enter the infrared beam, they scatter the light. The amount of light received by the photodetector is directly proportional to the dust concentration.

Electrical sensing works on the principle of tribo-electric charge transfer, which is the interaction between charged emissions particulates and a highly sensitive triboelectric probe mounted within the stack of dust. This interaction results in a charge transfer, which generates a minute current flow within the probe.

12.6.2 Simulation and modeling

Simulation and modeling can be used for virtual quantification of loads. It is useful to get down to subsystem level conditions from an overall system-level environmental load. For example, to estimate the temperature in the under hood of the car near the engine, the entire system can be simulated with nominal ambient temperature conditions as the global loads and the heat generation from different parts in the under hood to get an estimate of the temperature profile near the engine area. The influence of equipment used to modify the local environments (like fans and cooling systems) can also be incorporated. In order to use this method, the global or system-level environmental loads should be known to serve as an input to these calculations.

12.6.3 Human factors – what users really do

Although real-time monitoring, simulation, and modeling are methods used to get a more accurate idea of the environmental loads, they are not employed by users in most situations. Given the involving nature of the previous two methods, equipment designers usually resort to easier (and often less accurate) estimates of the environmental loads. Following are some of the methods usually followed.

12.6.3.1 Market studies as sources of data

Market surveys and other reports done independently by some agencies[7] or done by industries as a part of their design process are used as the basis for environmental load characterization. These kinds of data are derived most often from a remotely similar kind of environment and give a very coarse estimate of the actual environmental loads which the targeted equipment will experience. Care should be taken to avoid the use of such data as an estimate for environmental loads. Even if these methods become inevitable (due to constraints on time and cost), the date obtained should be used after considerable similarity analysis.

12.6.3.2 Field trial records, service records, and failure records

Field trial records are also sometimes used to get estimates on the environmental profiles. These are environmental loads encountered by previous and/or prototype equipment during field trials. The data available are for a short duration of time, and are extrapolated to get an idea of the environmental conditions that the actual equipment will face. Service records and failure records usually document the causes for unscheduled maintenance and/or failure in the equipment, which might have been due to certain environmental conditions. These conditions are sometimes used to estimate the kinds of environments the equipment might be subjected to. These data give an idea of the critical conditions but should not be used as a basis for developing the entire life cycle profile.

---

[7] These agencies include focus groups in organizations and standards committees like those that develop military standards.

They should be used only to accommodate for the extreme conditions which the equipment might encounter.

12.6.3.3   Data on histories of like parts, assemblies, or systems

Similarity analysis is a common technique used for developing LCEPs when data on similar products with a sufficient field history are available. However, care should be taken to ensure that the mature field equipment from which the data are taken is operating under very similar conditions. Changes and discrepancies in the conditions should be critically analyzed to ensure good accuracy on the life cycle of the new product.

12.7   Summary

This chapter presents an overview of the methods available for the determination and monitoring of the environmental parameter of a part. Characterizing the actual life cycle environment of a part together with physics-of-failure models will enable more accurate reliability assessments, more robust and cost-effective hardware design, and more realistic accelerated tests. This will help eliminate costly overdesign or problematic underdesign, unnecessary or inadequate qualification testing, and avoidance of new technologies or the selection of inappropriate technologies.

12.8   References

[1]   Pecht, M. and J. Pecht, *Long Term Non-Operating Reliability of Electronic Products*, CRC Press, New York, NY, 1995.

[2]   NASA, "NASA Preferred Reliability Practices – Environmental Factors," Practice Number PD-EC-1101, Lewis Research Center, <http://www.hq.nasa.gov/office/codeq/relpract/np1101.pdf>, accessed June 2, 2002.

[3]   Pecht, M., P. Sandborn, R. Solomon, D. Das and C. Wilkinson, *Life Cycle Forecasting, Mitigation Assessment and Obsolescence Strategies*, CALCE EPSC Press, University of Maryland, College Park, MD, 2002.

[4]   Pecht, M., R. Radojcic and G. Rao, *Guidebook for Managing Silicon Chip Reliability*, CRC Press, Boca Raton, FL, 1999.

[5]   Followell, D. A., "Enhancing Supportability Through Life Cycle Definitions," *IEEE Proceedings of the Annual Reliability and Maintainability Symposium*, pp. 402–409, 1995.

[6]   Mulkern, J. H. and G. T. Lommasson, "Out of This World Products – Designing for Space," *Proceedings of the Fourteenth Annual Applied Power Electronics Conference and Exposition*, Vol. 1, pp. 129–134, 1999.

[7]   Weidman, E. and S. Lundberg, "Life Cycle Assessment of Ericsson Third Generation Systems," *Proceedings of the IEEE International Symposium on Electronics and Environment*, pp. 136–142, 2000.

[8]   Cluff, K., D. Robbins, T. Edwards and D. Barker, "Characterizing the Commercial Avionics Thermal Environment for Field Reliability Assessment," *Journal of the Institute of Environmental Sciences*, Vol. 40, pp. 22–28, 1997.

[9]   IEEE Standard 1332-1998, IEEE Standard Reliability Program for the Development and Production of Electronic Systems and Equipment, ISBN 0-7381-1411-1.

[10] "Science Question of the Week," NASA Goddard Space Flight Center, <http://www.gsfc.nasa.gov/scienceques2001/20020118.htm>, accessed May 17, 2002.

[11] "Science Question of the Week," NASA Goddard Space Flight Center, <http://www.gsfc.nasa.gov/scienceques2001/20020524.htm>, accessed May 29, 2002.

[12] United States Department of Defense, "Military Standard: Climatic Information to Determine Design and Test Requirements for Military Systems and Equipment - MIL 210C," Washington, DC, 1987.

[13] United States Department of Defense, "Military Standard: Environmental Test Methods and Engineering Guidelines - MIL 810E," Washington, DC, 1989.

[14] Pecht, M. and W.C. Ko, "A Corrosion Rate Equation for Microelectronic Die Metallization," *International Journal for Hybrid Microelectronics*, Vol. 13, No. 2, pp. 41–51, June 1990.

[15] McCluskey, P., R. Munamarty and M. Pecht, "Popcorning in PBGA Packages During IR Reflow Soldering," *Microelectronics International*, No. 42, pp. 20–23, January 1997.

[16] Dasgupta, A., P. Sharma and K. Upadhyayula, "Micro-Mechanics of Fatigue Damage in Pb-Sn Solder Due to Vibration and Thermal Cycling," *International Journal of Damage Mechanics*, Vol. 10, pp. 101–132, April 2001.

[17] Janinska, B., "Environmental Requirements for Fungi," *Journal of Thermal Envelope and Building Science*, Vol. 23, No.4, pp. 339–348, 2000.

[18] Webster, J. G., *The Measurement, Instrumentation and Sensors Handbook*, CRC Press, Boca Raton, FL, 1999.

[19] McCluskey, P., R. Grzybowski and T. Podlesak, *High Temperature Electronics*, CRC Press, New York, NY, 1997.

[20] Pecht, M., P. Lall and E. Hakim, *Influence of Temperature on Microelectronics and System Reliability*, CRC Press, New York, NY, 1997.

[21] Pecht, M., P. Lall and E. Hakim, "The Influence of Temperature on Integrated Circuit Failure Mechanisms," *Quality and Reliability Engineering International*, Vol. 8, pp. 167–175, 1992.

[22] Nowottnick, M., W. Scheel, K. Wittke, U. Pape and J. Schulz, "Development of Temperature Resistant Joints – Testing and Influences," *Proceedings of the European Microelectronics Packaging and Interconnection Symposium*, International Microelectronics and Packaging Society, Prague, pp. 78–83, 2000.

[23] Gupta, V., R. Hernandez and P. Charconnet, "Effect of Humidity and Temperature on the Tensile Strength of Polyimide/Silicon Nitride Interface and Its Implications for Electronic Device Reliability," *Materials Science and Engineering A (Structural Materials: Properties, Microstructure and Processing)*, Vol. A317, No. 1–2, pp. 249–256, October 31, 2001.

[24] Pecht, M., L. T. Nguyen and E. Hakim, *Plastic Encapsulated Microelectronics*, John Wiley & Sons, New York, NY, 1994.

[25] Ardebili, H., C. Hillman, M. Natishan, P. McCluskey, M. Pecht and D. Peterson, "A Comparison of the Theory of Moisture Diffusion in Plastic Encapsulated Microelectronics with Moisture Sensor Chip and Weight-Gain Measurements," *IEEE Transactions on Components and Packaging Technologies*, Vol. 25, No. 1, pp. 132–139, March 2002.

[26] Pecht, M., "Moisture Sensitivity Characterization of Build-Up Ball Grid Array Substrates," *IEEE Transactions on Advanced Packaging*, Vol. 22, No. 3, pp. 515–523, August 1999.

[27] Barker, D., A. Dasgupta and M. Pecht, "PWB Solder Joint Life Calculations Under Thermal and Vibrational Loading," *Journal of the IES*, Vol. 35, No. 1, pp. 17–25, February 1992.

[28] Dasgupta, A., "Failure Mechanism Models for Cyclic Fatigue," *IEEE Transactions on Reliability*, Vol. 42, No. 4, pp. 548–555, December 1993.

[29] Barrett, R. C., W. Nix and A. Tetelman, *The Principles of Engineering Materials*, Prentice-Hall, Englewood Cliffs, NJ, 1973.

Chapter 13

Performance

Diganta Das, Michael Pecht, and Neeraj Pendse

The goal of performance assessment is to select parts capable of performing in the application environment. To do this, designers must select parts whose function and performance specifications meet the product circuit requirements. The performance specifications are provided in the datasheet and in associated documents (see Chapter 4).

There is often a minimum and a maximum limit beyond which the part will not function properly. Product manufacturers need to adapt their designs so that the parts do not experience conditions beyond their absolute maximum ratings, even under the worst possible operating conditions (e.g., supply voltage variations, load variations, and signal variations) [1]. It is the responsibility of the parts selection and management team to determine that the electrical, mechanical, and functional performance of the part is suitable for the operating conditions of the particular product. If mismatches exist between the manufacturer-specified environmental ranges and the local environment, the parts selection and management team must take additional steps to ensure proper operation of the part in the product.

13.1   Methods of mitigating the discrepancy between system needs and part ratings

When faced with the problem of finding a part that is necessary for the functional design but whose ratings do not meet the performance specifications for the system, the designers can evaluate several alternatives. These alternatives include:

- reevaluation of the required operational conditions
- using alternative parts that fit the bill (looking harder)
- modification of the equipment operating procedures
- environmental management
- uprating of parts

13.1.1   Reevaluation of the required operational conditions

When faced with a possible mismatch between part ratings and part requirements, one of the first questions to ask is, "do we really know the operating environment or are we just following some standards or precedents without justification?" The local conditions for each part should be reevaluated, including the assumptions regarding the external temperatures for the equipment in which the parts operate. Analysis of the environment in which the part's experiences may not be known may be necessary to estimate it from the equipment environmental specifications. Environmental sensor data collected for the same or similar equipment or site of use can also be used to determine the actual required operating conditions. For example, thermal analysis or sensor data can indicate a narrower temperature specification, and the part can be working in the specified temperature range. This type of reevaluation will not necessarily narrow the required temperature range; it may even result in an expansion of the temperature range of operation. Still, it is a valuable exercise because such information will result in a design that is more robust.

The stated environmental range for the equipment may not truly represent the operational conditions and may simply be a "boilerplate" requirement. For example, a specification for a new cockpit control panel may come with a boilerplate set of requirements that calls for operation over a –55 to +70°C range. Since other cockpit equipment may have been designed for a more benign temperature range (recognizing that humans must occupy the nearby space), there may be a large number of parts used in the other cockpit equipment that would not satisfy this requirement. If specifying a narrower temperature range (e.g., –20 to +55°C), is an option, and if that range is acceptable to regulatory agencies, known customers, or anticipated market applications, then the problem situation could be eliminated [1].

For customer requirements, an agreement with the customer is needed for exploring the trade-offs that may be necessary to meet the stated requirements. Any solution to this problem will be better if it has the customer's participation, whether it involves relaxing the equipment specification or special qualification of components. If the requirements come from internal product line's definition and not from a specific customer, then the product line's needs should be evaluated with a set of requirements spanning the anticipated customers.

13.1.2   Using alternative parts

Using alternative parts is the process of selecting a part which is a direct functional replacement of the original part. Part substitution provides equivalent functionality and an opportunity for technological upgrading of the system, such as a newer package style. The main disadvantage of using a substitute part is that quality and integrity assessments will have to be performed on the substitute part as per the part selection and management procedure [2], [4]. Substitution may also involve board layout redesign due to different pin-out pattern.

Manufacturers on the qualified manufacturer list (QML) are possible sources for parts, but most companies such as Motorola, Intel, AMD, and Philips are not on the QML. Aftermarket suppliers such as Lansdale Semiconductor, Rochester Electronics, and White Electronic Design Corporation can be sources for discontinued or upgraded parts. Emulated or re-engineered parts from the Generalized Emulation of Microcircuits (GEM) program and other sources (including some QML companies) may be available in required environmental conditions that are not catered to by the major commercial semiconductor manufacturers.

13.1.3   Modification of the equipment operating procedures

Operating procedures can sometimes be amended so that the part environment extremes are constrained without having any effect on the equipment operating capability. For example, many airborne electronic systems stay in narrow (~0-70°C) temperature ranges during steady-state operation. If sufficient warm-up/cool-down time is allocated, electronic parts rated for the industrial temperature range may be used in many avionic applications.

In aircraft operations, warm-up can be provided by an increased start-to-operation time. This can be accomplished in two ways: either by directly changing the specifications or indirectly by allocating more of the specified time to device warm-up and less to software initialization routines. Allowing reduced performance on ground (where hot), with full performance when airborne (where cool), is an example of cool-down.[1]

By the same token, the error detection and correction (EDAC) routines that help prevent failures caused by single-event upsets (a radiation effect) may be able to work more efficiently when circuits run at a lower speed. Critical systems can be run at lower speeds while operating in areas of high probability of radiation occurrence (e.g., higher altitudes).

13.1.4 Environmental management

When implemented correctly, environmental management can maintain proper device operation even when the outside ambient conditions may not meet the specifications on a part. When the necessary power quality for a part is more rigorous than the quality the default system conditions can provide, local power filters may be included in a circuit to provide a filter for incoming power.

Entire pieces of equipment can also be placed in a controlled local environment; this is called "cocooning." Thermal cocooning has been successfully used in ground-based navigation systems, the Mars Rover, and the Boeing Sikorsky MEP cooling subsystem [5]. Cocooning against electromagnetic effects and radiation is also possible. Cocooning efforts usually come with size and weight penalties.

13.1.5 Uprating of parts

Uprating is a process used to assess the ability of a part to meet the functional and performance requirements of an application in which the part is used outside the manufacturer's recommended operating conditions (see Chapter 4). In an ideal world, there would not be a need for uprating. The part manufacturers would supply all needed parts with appropriate ratings for all products and systems. One would not have to be concerned about using any part beyond its ratings, and the cost of the part, assembly, test, sustenance, and maintenance would not be factors of concern. However, that world does not exist, and it is unlikely that it ever will.

The CALCE Electronic Products and Systems Center at the University of Maryland has developed methods for uprating[2] [6] and has helped the International Electrotechnical Commission (IEC) develop processes for a guide on methods of using electronic parts outside the manufacturer's specified temperature range [7]. To ensure that all relevant aspects of product liability responsibility are considered, it is necessary to follow documented, controlled, and repeatable processes integrated with the parts selection and management programs [8], [9] while using parts outside manufacturer-specified ratings.

In fact, electronic parts are being used outside their specifications in many applications and industry sectors [28]. Parts are used outside their ratings in oil and gas exploration and recovery, home appliances, and commercial and defense electronics. Uprating has been successfully implemented in several government and air transport products since at least 1994 [40]. Uprating of electronic parts is here to stay, and the industry should devote time and resources to follow processes that best address the requirements of the equipment.

---

[1] D. Farwell of Raytheon Systems Company made this observation in a presentation to the U.S. Department of Defense.

[2] The term "uprating" was coined by Michael Pecht [6] to distinguish it from "upscreening." Uprating is defined as a process to assess the ability of a part to meet the functionality and performance requirements of the application in which the part is used outside the manufacturer's specification range. Upscreening is the practice of attempting to create a part equivalent to a higher quality level by additional screening of a part (e.g., screening a commercial part to a space qualification part).

The Federal Aviation Authority (FAA) of the United States accepts the use of parts outside the manufacturer's specifications (in case of temperature). It states that "If the declared installation temperature environment for the EEC (electrical and electronic components) is greater than that of the electronic components specified in the engine type design, the applicant should substantiate that the proposed extended range of the specified components is suitable for the application" [10].

The electronic part manufacturers are skeptical about and critical of the use of parts outside their ratings. National Semiconductor, for example, states that they are aware that some customers may be upscreening or retesting semiconductor components. They warn that using components in applications or environments for which they were not intended can lead to component or system failure. National Semiconductor strongly recommends that their products be used only within the electrical and environmental limits published in their respective datasheets [11]. Intel also states that they will not accept any legal responsibility for failures that occur due to the deliberate misuse of their products, including any damages resulting from the practice of uprating. Similarly, there are warnings from other manufacturers and trade associations such as Analog Devices [1], Xilinx [12], Texas Instruments [13], and the Semiconductor Industry Association (SIA) [14].

There is a growing mismatch problem between ratings of parts and the application requirements of products. In particular, this problem is becoming acute in the operating temperature ratings of the parts. The problem is caused not by technological shortcomings but by the nature of the electronic part marketplace. It cannot be solved economically in all circumstances by mandating the availability of parts with required ratings, obtaining parts from specialized sources, or environmental management techniques. The situation demands thinking outside of the box and developing an integrated approach to assessing parts, considering the integration of a part with the operational and environmental requirements of a product in a holistic manner. This process may result in use of a part outside of its manufacturer ratings.

13.2   Methods of uprating

The performance assessment step of the parts selection management process assesses if a part will work in its intended application. If there is a mismatch between the part specifications and the local environment for the part in the application, then there are several options that engineers could explore to mitigate the mismatch. Since the application-dependent assessment processes are performed in parallel, any one of the processes can potentially stop further consideration of the candidate part for use in the application. It is also possible that the application requirements will change during the design and development process and that the part needs to be reassessed.

Uprating analysis is integral to the performance assessment step of part selection and management. It is carried out after the part, the part manufacturer, and the distributor or any other supply chain link is assessed [15]–[18]. The analysis is carried out by the user of the part (the OEM) in accordance with his parts management and quality procedures. The analysis must be based on objective evidence, which will include the manufacturer's datasheet, application notes, and any other published data that the user may find to be

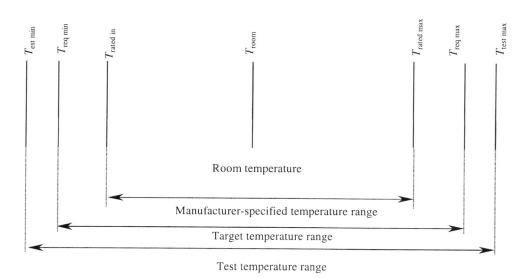

Figure 13.1: Schematic diagram showing the minimum required temperature test
points for parameter conformance and parameter recharacterization

relevant. It is usual for part manufacturers to be reticent in providing data for the use of
their parts outside published datasheet limits, and often they actively discourage the
practice.

Three methods for part uprating were developed by the CALCE Electronic Products
and Systems Consortium at the University of Maryland (See Figure 13.2). The processes are
described in detail in the following three sections. In all the methods, there is some level of
testing involved over a temperature range shown in Figure 13.1. These methods are accepted
as industry best practices by the International Electrotechnical Commission and by the
Electronics Industry Association for a guidebook [7]. Publications for the U.S. Department
of Defense acknowledge these methods as effective and rigorous [19]. Many small and large
companies already use the methods; for example, Applied Data Systems states that it is
using the CALCE methods for its embedded systems [20].

Some datasheets (or associated documents) can include electrical parameter data
beyond the recommended operating temperature limits. These kinds of data may be useful
in preassessing if a part can be uprated.

13.3 Parameter conformance

Parameter conformance is a process for thermal uprating in which the part is tested to
assess if its functionality and electrical parameters meet the part manufacturer's
specifications over the target temperature range. The electrical tests are of the go/no-go
type. Electrical parameters are not characterized over the target temperature range, and
electrical parameter specifications in the datasheet are not considered for modification.

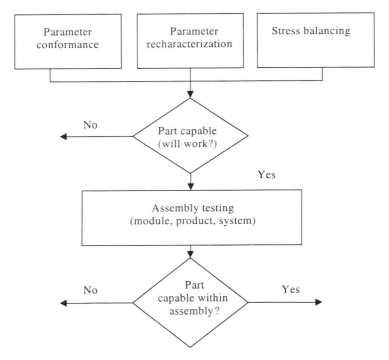

Figure 13.2: Approaches to thermal uprating of electronic parts

13.3.1   Electrical testing

Electrical testing of the part using the part manufacturer's datasheet specifications at the target temperature range is the key to the parameter conformance process. The tests consist of both functional and parametric go/no-go tests.

Appropriate functional coverage for the application must be ensured. Parameters that are critical to the end product's performance and operation must also be tested. The values of these parameters are selected from the part manufacturer's datasheet.

The ideal case is to test all electrical parameters and functional components of the system in which the part is used. However, this is not always practical and may even be impossible. For example, performing a complete functional test on a microprocessor could take years. Nonavailability of detailed internal information about the parts also limits the scope of the test.

The datasheet values for the maximum and minimum specifications, test limits, and test conditions should be used for all parameters except temperature. The test temperature range should be determined based upon the target application and the desired temperature margins.

13.3.2   Test temperature range

The temperature range in which electrical testing is to be performed depends on the temperature range the part is expected to experience during operation. This information can be based on the equipment's environmental requirements and the results of thermal analysis within the application.

The analysis and test conditions consider airflow rate and direction, proximity of other parts with significant power dissipation, and heat sinks. This will ensure creation of a test environment that is similar to the actual operating condition of the part in the system. The method of temperature measurement (ambient, case, or junction) used in the part ratings should be considered while selecting the temperature range for testing. The interaction between different parts operating at temperatures beyond the manufacturer's specifications must also be assessed.

13.3.3   Test margins

Margins should be added on the test parameters to account for any possible variability. The margins can be added to electrical parameters or to the target temperature extremes. Accordingly, there are two types of tests in parameter conformance, as shown in Figure 13.3.

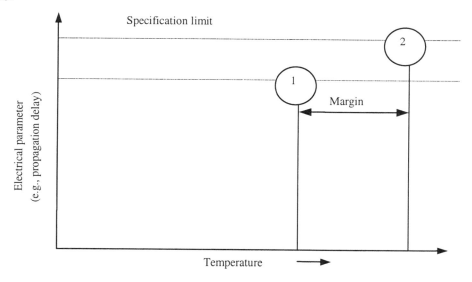

Figure 13.3: The two types of go/no-go tests in parameter conformance

In type 1, the test is conducted at the target temperature, with margins on the electrical parameters. In type 2, the test is conducted at the electrical parameter specification limit, with margins on the target temperature. One may also choose to combine both types of margins. Inaccuracies in the thermal environment can be accounted for by adjusting the margin on the test temperature extremes.

The pass-fail criterion for each parameter is based on the datasheet-specified limits. These parameter limits can be made tighter than the datasheet specifications, especially if a sample of parts is subject to parameter conformance testing. Putting margins on the datasheet electrical parameters will give equipment manufacturers confidence in the applicability of the test results to all parts.

To provide confidence in test results, one can experimentally determine the temperature at which the part fails. To find the temperature margin by test, the test temperature is incrementally increased (or decreased) beyond the target application temperature range ($T_{req-max}$, $T_{req-min}$) until at least one of the part parameters no longer meets the datasheet performance specifications. This is called the "fallout temperature" for that part. Once the fallout temperature for all the parts under test has been found, a distribution can be plotted, with the temperature on the x-axis and the percentage of fallout on the y-axis. The temperature margin at $T_{\text{req-max}}$ and $T_{\text{req-min}}$ can then be determined with confidence levels.

For successful uprating with parameter conformance, the maximum target temperature must be less than the temperature at which the first part parametrically or functionally failed to meet the datasheet specifications; the minimum target temperature must be greater than the lowest temperature at which the first part parametrically and functionally failed to meet the datasheet specifications.

The difference is essentially the temperature margin available. Based on the distribution of the electrical parameters failed beyond the target temperature range, a confidence interval on those parameters can be estimated at the extremes of the target temperature range. The number of samples depends on the desired temperature margin and the confidence level in that margin. Large temperature margins may use small sample sizes (and vice versa).

Finding the number of samples involves finding an approximate mean and standard deviation of the fallout temperature. The final sample size may change as the test progresses, due to continual recalculation of the mean and the standard deviation as more samples are tested, until the desired margin has been achieved. As a result of this exercise, it is possible that a sufficient margin cannot be achieved. In that case, a different part or a different method of uprating should be chosen.

13.4   Parameter recharacterization

Part manufacturers characterize their parts to determine and ensure the electrical parameter limits. They typically test a sample of parts at the recommended operating temperature extremes of the datasheet and at room temperature (typically 25°C) to determine the typical and limiting values of electrical parameters. Figure 13.4 shows a characterization curve for a digital logic part [21]. In the curve, a sample of 10 parts was used to characterize the output rise time of the logic part after a design change. The figure shows the mean values of the parameters and standard deviations at each temperature at which characterization is done. This result was used to determine electrical parameter values for the temperature ranges of –40 to 85°C and –40 to 125°C.

From the results of the electrical tests, the part may be determined to satisfy the parameter limits of the datasheet. In some cases, it may be necessary to develop new parameter limits. In fact, this practice is used by manufacturers for creating specifications for parts [22].

Electrical part datasheets show that the electrical parameter limits may be the same for parts characterized for different temperature ranges or may vary for different temperature ranges. For example, the Motorola MC74HC242 quad three-state bus transceiver is rated for –55 to 125°C, but its datasheet shows three different electrical parameter limits for three different temperature ranges. In this case, the manufacturer chose to share with the users the different parameter capabilities that can be achieved at different temperature ranges that it found through characterization. In another example, a Xicor NOVRAM X25401 part has the same electrical parameter limits for the three temperature ratings in which the part is available. In this case, the manufacturer provides the conservative parameter limits that can be guaranteed at all different temperature ranges.

During the characterization process of the part, the part manufacturer determines the parameter limits over the temperature range for which the part is sold. When performing a recharacterization, new parameter limits may be required.

Some electrical parameters may be known to be independent of temperature, and those parameters may be excluded from recharacterization. It may also be possible that electrical characterization data are available for one or more of the electrical parameters for the complete target temperature range and recharacterization is not necessary. Also, limits on many parameters, such as the logic voltage levels, are often defined based on the supply voltage choice. Recharacterization of all functions at all voltage conditions may not be required.

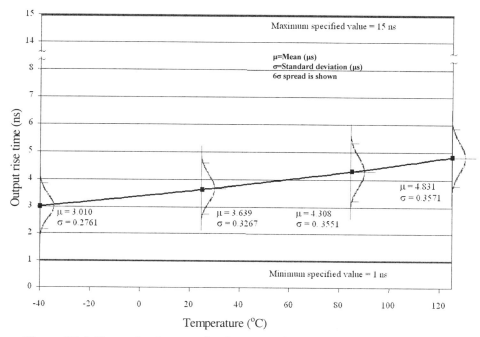

Figure 13.4: Example characterization curve from Fairchild Semiconductor

Functional testing of a part is required for parameter recharacterization because in certain temperature conditions a part may fail to function even though it meets all the parameter limits. In this case, the fault coverage of the test vectors becomes an issue. Gate-level design information is required for developing software that achieves specific fault coverage. In many cases, a full set of test vectors for a part will not be available for the uprating tests, and the percentage of fault coverage will be difficult to determine without detailed knowledge of semiconductor device architecture. In this situation, functional tests should include test vectors that exercise the part in a manner similarly to the application for which it is being considered.

When a part is considered for parameter recharacterization in an application, a set of requirements and limitations on the electrical parameters of the part needs to be developed. Acceptable margins for the electrical parameters under operational conditions in a product need to be established. When applicable, these margins need to be determined for both the upper limit ($M_{UL\text{-req}}$) and lower limit ($M_{LL\text{-req}}$) of the electrical parameters. The robustness of the overall design to individual part parameter fluctuation is the prime consideration in determining the level of margin required. Determination of this margin is an application-specific decision. The limits[3] thus determined will be compared with the actual margins determined through electrical testing to determine if parameter limit modification is warranted for the new temperature range of the part (see Figure 13.5).

13.4.1   Choice of test temperature range

The first determining factor for the test temperature range is the target temperature range. This information is available from the assessment of the application environment. The equipment manufacturer must consider the type of temperature (ambient, case, or junction) used in the part ratings when determining the temperature range for testing. The margin of uncertainty of the thermal assessment methodology also needs to be accounted for in determining the temperature range for testing.

Parametric and functional testing should be performed up to temperatures 15°C to 25°C above and below the target temperature range. Between the part manufacturer's temperature limit and the test temperature extreme ranges, data should be collected at close intervals so that within the extended temperature ranges unsatisfactory results can be identified. These tests will give equipment manufacturers confidence in the applicability of the parts over the target temperature range. If electrical margin loss beyond acceptable limits is observed at temperatures between the target temperature range and the test temperature range, a statistical analysis of the temperature-dependent electrical margin loss can be used to determine the margin of safety[4] of the uprating process. The tail of the electrical margin distribution can be used to estimate the safety margin in the target temperature range of interest.

13.4.2   Assessment of electrical test results

The process of parameter limit modification involves statistical analysis of the experimental results and the requirements of the application.[5] The first step is to estimate how much margin exists between the part parameters at the test temperature and the data-sheet specifications for the parameters. Figure 13.6 schematically shows a representation of parameter limit modification to account for changes in electrical parameters with temperature.

13.4.3   Determination of margins

After the electrical test data are collected over the target temperature range, the margins on the parameters from the part manufacturer-specified parameter limits at the temperature points of interest are assessed. This assessment process takes into account the confidence interval on the parameters, the measurement uncertainties, and the spread of the electrical parameters.

When the margin on a particular parameter is considered inadequate, new limits on the electrical parameters must be determined. The parameter limit modification process is the reverse process of determining the level of margin on electrical parameters. First, the required margin at a temperature extreme is selected. This may be either the same margin

---

[3] While considering a part uprated by parameter recharacterization for use in a product, the maximum (UL_{Max}) and minimum (LL_{Min}) limits on the parameters acceptable for operation of the product need to be determined in case parameter limits are to be modified. These limits provide the upper and lower bounds on parameter limit modifications for the particular application.

[4] "Safety" refers to the comfort level of the equipment manufacturer, not to equipment safety.

[5] A part can be rejected for uprating based on the test results before parameter limit modification is considered. Discontinuities in any of the parameters within the target temperature range or failures in functional tests are causes for rejection of a part for parameter recharacterization.

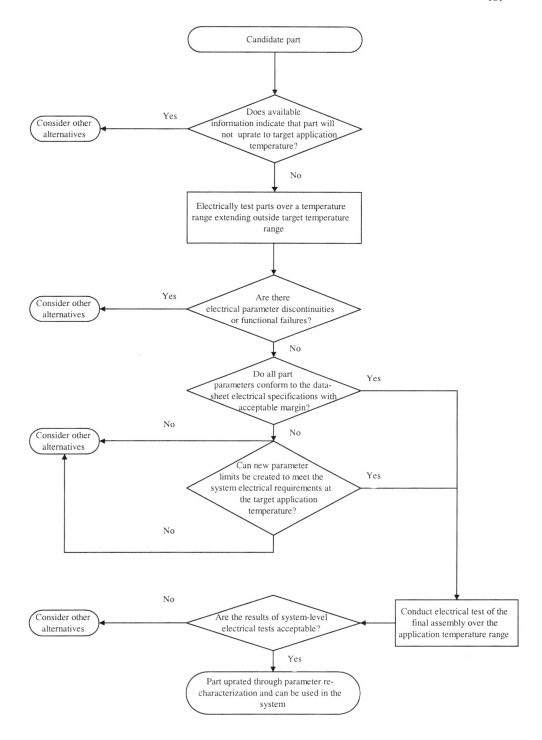

Figure 13.5: Flow diagram for the parameter recharacterization process

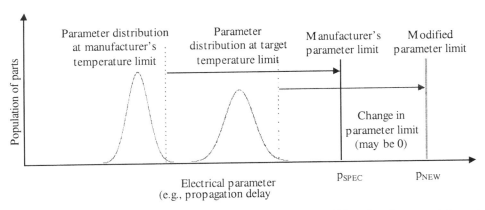

Figure 13.6: Schematic representation of parameter limit modification for use at target temperature range

used to determine the adequacy of electrical parameters or the margin(s) at the extremes of the part manufacturer-specified temperature range.

13.5   Stress balancing

Stress balancing is a process for thermal uprating of a part when there is an interest in using a part beyond the manufacturer's specified maximum recommended ambient or case operating temperature. In this method the part's electrical parameter is kept below its maximum allowable limit to reduce heat generation, thereby allowing operation at a higher ambient or case temperature than that specified by the semiconductor part manufacturer.

The process of stress balancing exploits possible power-temperature-functional performance trade-offs for a part. For active electronic parts:

$$T_J = T_A + P \cdot \theta_{JA} \tag{13.1}$$

where, T_J is the junction temperature, T_A is the ambient temperature, P is the power dissipation, and θ_{JA} is the junction to ambient thermal resistance. Equation (13.1) shows that the ambient operating temperature can be increased beyond the manufacturer's recommended values if the power dissipation is reduced while keeping the junction temperature constant. Because the power dissipation of the part will be a function of some electrical parameters (e.g., operating voltage and frequency), a trade-off can thus be made between increased ambient temperature and a change in some electrical parameters. If the electrical parameters can be selected to meet the application requirement, then the trade-off can be accepted. Figure 13.7 shows a flowchart of the stress balancing process.

In stress balancing, electrical testing of the part is performed at the application conditions to ensure operation of the part with modified electrical parameters. In other methods of uprating (parameter conformance and parameter recharacterization), electrical testing of the part is conducted at datasheet electrical conditions. The goal of electrical testing in stress balancing is to confirm the applicability of electrical parameter trade-offs in the application (electrical conditions and environment), and therefore testing is performed in application conditions. This testing of parts (or sample boards) is different from the assembly-level testing that is required for all uprating processes.

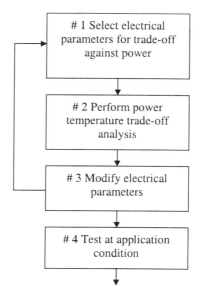

Figure 13.7: Steps in the stress balancing process

Stress balancing can be performed if the part's power dissipation can be decreased by changing some electrical parameter(s). The goal of this step is to assess the acceptability of the electrical parameters to be changed, the magnitude of the change, and the reduction in dissipated power.

Common operating parameters that influence power dissipation include the frequency, operating voltage, output current, fan-out, and the duty cycle of the part. The parameters, which will significantly affect the power dissipation of the part, are selected for modification. For example, the power dissipation has an approximately linear relationship with the effective operating frequency for CMOS parts.

In general, a system will have some minimum performance requirements that are impacted by the part's performance. These requirements affect the minimum possible power dissipation at which the part can operate in the system. For example, in a system that digitizes sensor data, reducing the sampling frequency will affect the power dissipation of the ADC. However, reduction in sampling frequency will be limited by the minimum sampling frequency specification for the system. Thus there may be a range of power dissipation values at which the part can be operated in the given system, the minimum value being the application or part requirement and the maximum value being the manufacturer's specified limit on the part.

Parametric and functional testing of the part serves two purposes:

- It verifies that the part can indeed operate under the new environmental and electrical conditions. Changes made to the electrical parameters to reduce power dissipation may change other electrical parameters of the part. For example, if the supply voltage of an operational amplifier is reduced, its frequency range and output current will change, and the part will saturate at a lower input voltage.
- It checks for the adequacy of margins in the required temperature range on the modified electrical parameters.

The sample size to be tested should be chosen as per the required margins, confidence in testing, and tolerable variances in the parameter values. The tests performed may be parametric go/no-go tests with sufficient temperature margins added.[6] The part is considered successfully uprated for the application if the test results demonstrate that it can operate in the application with sufficient margins. To validate stress balancing, a steady-state temperature needs to be reached for the part and the system to ensure that the junction temperature actually has time to rise due to part power dissipation.

13.6   Reliability issues with uprating

Reliability is the ability of a part to perform within specified performance limits, for a specified period of time, under the life cycle application conditions. Reliability assessment of a part is performed independently of performance assessment (where uprating may be carried out). The recommended operating conditions that are stated in the part datasheet relate only to the electrical parameter limits. Generally, no statement regarding part reliability is made in the part datasheet because the manufacturer does not know the application conditions on which part reliability depends. Exceptions include nonvolatile memory data retention time or read-write cycle endurance, where specific test conditions are quoted.

The part user is responsible for deciding if the part is suitable for a given application, including the reliability requirement. Irrespective of whether there is a mismatch between the part ratings and the application conditions, reliability assessment needs to be performed. In the CALCE electronic part selection and management process, reliability is assessed for a part in its application condition regardless of uprating. For reliability analysis details, the reader is referred to [23]–[25].

It has been observed that the manufacturer's part qualification process is not based on the part's temperature ratings but rather on MIL-STD-883 type testing [26], which has a broader temperature requirement. In other words, the part operating temperature ratings are set for performance reasons rather than reliability reasons. The part temperature ratings are designated in the recommended operating conditions for parts. The limits for reliable operation of parts are designated by the absolute maximum ratings (which are most often wider than the recommended operating conditions). Every part that can be uprated will not necessarily be reliable in its application environment, and some concerns about reliability degradation at higher temperatures are raised by semiconductor manufacturers [27]. There have been reported studies where the reliability of parts were tested beyond their manufacturer-specified temperature limits and the reliability demonstrated [28]–[32].

13.7   Case study: Single inline memory module

Various uprating case studies have been performed at CALCE EPSC. This section presents a case study of an uprating assessment of single inline memory modules (SIMMs) [33]. The SIMMs were rated for an ambient temperature range of 0 to 70°C. Diagnostic tests at temperatures averaging 152°C ambient were conducted on the modules for an extended period of time. The purpose was to determine if the modules were capable of performing outside their temperature specification for an extended period of time.

---

[6] Equipment manufacturers sometimes perform system electrical testing only, without testing individual part parameters. However, this constitutes a risk. Careful design of the tests with sufficient functional coverage is necessary to ensure that the part is uprated.

SIMMs are memory card assemblies often used as extended memory in computers. The SIMM cards can be "plugged" into the memory slots of the computer's motherboard. This provides flexibility to add or remove memory cards. The purpose of the SIMM is to provide additional real estate for the motherboard while increasing the memory capacity of the computer. The SIMM is connected to the motherboard and its edge connector and a SIMM socket. A typical SIMM with surface-mounted memory (typically DRAMs) attached to a printed wiring board (PWB) with edge connectors is shown in Figure 13.8. The edge connector has electrical contacts on both sides of the double-sided PWB; they connect together to form one electrical contact for each pin [34].

The test module was a double-sided PWB SIMM assembled with 12 DRAM parts, 6 per side. There were eight IBM parts and four Micron parts. The parts were plastic small outline J-leaded (SOJ) DRAMs from Micron and IBM (part numbers MT4C1024 and 11E1320PA-70, respectively). There were also some chip capacitors on the SIMMs. Both DRAM parts are rated for 0 to 70°C ambient temperature operation. The modules as a whole are also rated for operation over 0 to 70°C.

Failure of the memory module can lead to program or computer shutdown. Causes of failure depend on such factors as component density, circuit layout, and the manufacturing method, materials and process defects, operational environment extremes, and aging effects [35], [36].

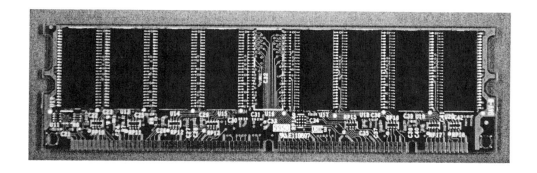

Figure 13.8: A SIMM

For a DRAM to be considered functional, it must be able to perform the following functions with the required voltage and timing performance regardless of the contents of any other cells or previous memory access sequences:

- store a 0 or a 1 in every cell of the memory
- change every cell from 0 to 1 as well as from 1 to 0
- read every cell correctly [35]

Failure occurs only if the memory cell containing a fault is accessed [41]. Categories of faults include stuck-at, transition, coupling, linked, retention, and pattern-sensitive. Faults are classified as permanent if they lead to catastrophic failures. Nonpermanent faults are not expected to cause irreversible damage [38]. Stress on the memory depends on the on-chip current and voltage signals, which in turn depend on the activity the chip is performing.

Memory testing is performed at various stages in the production of the memory chip to identify defects in the chip. Test patterns are developed based on factors such as allocated test time, the degree of access to the internal circuitry, fault coverage, and cost.

For practical purposes, a test to cover all possible faults is impossible. In fact, the development of test patterns over the years has shown that no single pattern can exercise a RAM thoroughly enough to detect all the failure modes [36]. Memories are therefore tested with several patterns. For example, a "walking 1" pattern is used to detect shorts between adjacent memory cells. In the walking 1 pattern, all the memory locations are first initialized to 0. Then, a 1 is written sequentially to each location. That location and all its surrounding locations are read to see if that 1 has "leaked" to the neighboring locations. Then that location is set to 0 and another location is set to 1. The process is continued throughout the entire memory.

The test platform in the case study consisted of a Compaq Presario 924 personal computer (see Figure 13.9). The side cover panel of the computer was removed for accessibility to the SIMMs. Minco thin film resistive heaters were mounted on nonfunctional dummy SIMMs placed in the first and third SIMM slots and were used to heat the test SIMM placed on the second SIMM slot. This arrangement maintained an average temperature of 152°C on the test module during this test.

Type T thermocouples (5 mil, 36 AWG), read with an Omega Type T thermocouple thermometer, were used to monitor the SIMM temperature. Thermocouples were placed on top of various components, underneath components, and on the surface of the SIMM board. A thermocouple was also used to measure the temperature of the computer motherboard. Type T thermocouples have an upper temperature limit of 371°C [41]. The thermocouples give results accurate over their temperature range to within ± 1°C [39].

In situ functional tests were performed using the QAPlus/FE diagnostic software to assess if the voltage and timing parameters met functional requirements during the operation of the computer. QAPlus/FE performs a series of memory tests that access each of the data bits in various patterns to detect faults on the memory chip. The software was continuously executed during the experiment. Each time a test was performed, the temperature readings from the thermocouples were recorded.

The SIMM was operated for 2592 continuous hours (108 days) at an average temperature of 152°C. No faults or failures were detected by any of the in situ functional tests. The results show that the SIMM module can work at temperatures outside those for which it is specified and that it did not fail for an extended period. That is, the module can be uprated for use in a higher operating temperature and can work reliably for some period. Assuming that further modules of this type would produce similar results and if the required application conditions were known, these results could be translated to predict a lifetime for the SIMM in other applications using temperature-dependant physics of failure models [23], [25].

In terms of qualification, Micron performs high-temperature operating life (HTOL) tests on their DRAM parts at 125°C for 1008 hours. This test is also conducted on military temperature range parts and is based on requirements from MIL-STD-883 [26]. In fact, it is common practice to qualify parts to this standard regardless of the operating temperature ratings (in this case, 0 to 70°C).

The testing of parts with extended high-temperature exposure was more severe than the equivalent qualification test conducted by the manufacturer on this part. Results of this test establish both the reliability and electrical functionality of the part at high temperatures.

Figure 13.9: Steady-state temperature test setup

13.8   References

[1]   GIDEP, "GIDEP Alert CM2-P-98-01," April 20, 1998, Analog Devices, <http://www.gidep.org>, accessed June 28, 2002.

[2]   Rockwell Collins, Inc., Standard for the Use of Components Outside Manufacturer's Specified Temperature Range, CPN: 832-6969-001, Cedar Rapids, IA, May21, 2001.

[3]   Syrus T., M. Pecht and D. Humphrey, "Part Assessment Guidelines and Criteria for Parts Selection and Management," *IEEE Transactions on Electronics Packaging Manufacturing*, Vol. 24, No. 4, pp. 339–350, October 2001.

[4]   Syrus T., U. Ramgopal and M. Pecht, "Manufacturer Assessment Procedure and Criteria for Parts Selection and Management," *IEEE Transactions on Electronics Packaging Manufacturing*, Vol. 24, No. 4, pp. 351–358, October 2001.

[5]   Vaughn, W., MEP Cooling Subsystem and Avionics Racks, MEP PDR, December. 16–20, 1991.

[6]   Wright, M., D. Humphrey and P. McCluskey, "Uprating Electronic Components for Use Outside Their Temperature Specification Limits," *IEEE Transactions on Components, Packaging, and Manufacturing Technology,* Part A, Vol. 20, No. 2, pp. 252–256, June 1997.

[7]   IEC/PAS 62240, Use of Semiconductor Devices Outside Manufacturers' Specified Temperature Ranges, Edition 1, 2001-04 (also being developed as GEIA 4900), Zurich, Switzerland, 2001.

[8]   Foucher, B., R. Kennedy, N. Kelkar, Y. Ranade, A. Govind, W. Blake, A. Mathur and R. Solomon, "Why a New Parts Selection and Management Program?" *IEEE Transactions on Components Packaging and Manufacturing Technology* - Part A, Vol. 21, No. 2, pp. 375–378, June 1998.

[9]   Biagini, R., M. Rowland, M. Jackson and M. Pecht, "Tipping the Scales in Your Favor When Uprating," *IEEE Circuits and Devices Magazine,* Vol. 15, No. 4, pp. 15–23, July 1999.

[10] FAA DRAFT Advisory Circular, Compliance Criteria for FAR §33.28, Aircraft Engines, Electrical and Electronic Engine Control Systems, ANE-110 AC No. 33.28 Version 30, December 19, 1997.

[11] National Semiconductor, *Note from the Editor*, National Semiconductor Navigator, Vol. 16, 1999.

[12] Fabula, J., "Up-Screening of Xilinx Products," August 11, 1998, Xilinx Inc., <http://www.xilinx.com/products/qa_data/upscreenletter.html>, accessed July 28, 2002.

[13] Kroeger, R. J. (TI), "Letter to Customer," September 10, 1997, Texas Instruments, <http://www.ti.com/sc/docs/military/cotspem/upscreen.pdf >, accessed July 28, 2002.

[14] Semiconductor Industry Association (SIA), "SIA/GPC Addresses the Importance of Understanding COTS," Robert C. Byrne, National Semiconductor, 1997, <http://www.semichips.org/gpc/cotspape.htm>, accessed May 1999.

[15] Condra, Lloyd, R. Hoad, D. Humphrey, T. Brennom, J. Fink, J. Heebink, C. Wilkinson, D. Marlborough, D.Das, N. Pendse and M. Pecht, "Terminology on Use of Electronic Parts Outside the Manufacturer's Specified Temperature Ranges," IEEE Transactions on Component and Packaging Technology, Vol. 22, No. 3, pp. 355–356, September 1999.

[16] Clifton, S., A. Mathur, R. Solomon, P. Sandborn and M. Pecht, "Equipment Supplier Intervention Techniques," *Future EMS International*, No. 3, pp. 97–100, 2000.

[17] Jackson, M., A. Mathur, M. Pecht and R. Kendall, "Part Manufacturer Assessment Process," *Quality and Reliability Engineering International*, Vol. 15, pp. 457–468, 1999.

[18] Jackson, M., P. Sandborn, M. Pecht, C. Hemens-Davis and P. Audette, "A Risk Informed Methodology for Parts Selection and Management," *Quality and Reliability Engineering International*, Vol. 15, pp. 261–271, 1999.

[19] Ciufo, C., "Sources for Intelligent COTS IC Selection," *COTS Journal*, pp. 12–19, March/April 2000.

[20] Applied Data Systems, "Hot! Hot! Hot! Extending the Temperature Range for 32 Bit Embedded Systems," <http://www.embeddedtechnology.com>, accessed June 11, 2001.

[21] Fairchild Semiconductor, *Design Characterization Package, HC00 Quad 2-input NAND Gate*, Fairchild Semiconductor, South Portland, ME, March, 11, 1998.

[22] Pendse, N. and M. Pecht, "Parameter Re-characterization Case Study: Electrical Performance Comparison of the Military and Commercial Versions of all Octal Buffer," *Future Circuits International*, Vol. 6, pp. 63—67, Technology Publishing Ltd., London, UK, 2000.

[23] Lall, P., M. Pecht and E. Hakim, *Influence of Temperature on Microelectronics and System Reliability: A Physics of Failure Approach*, CRC Press, New York, NY, 1997.

[24] Pecht, M., (ed.), *Integrated Circuit, Hybrid, and Multichip Module Package Design Guidelines,* John Wiley & Sons, New York, NY, 1993.

[25] Pecht, M., R. Radojcic and G. Rao, *Guidebook for Managing Silicon Chip Reliability,* CRC Press, New York, NY, 1999.

[26] United States Department of Defense, MIL-STD-883: Test Method Standards – Microcircuits, 1996.

[27] Biddle, R., "Reliability Implications of Derating High-Complexity Microcircuits," *COTS Journal,* pp. 39–43, February 2001.

[28] Condra, L., G. Kromholt, M. Pecht and E. Hakim, "Using Plastic-Encapsulated Microcircuits in High Reliability Application," *Proceedings of the Annual Reliability and Maintainability Symposium,* pp. 481–493, 1994.

[29] Condra, L., G. Wenzel and M. Pecht, "Reliability Evaluation of Simple Logic Microcircuits in Surface Mount Plastic Packages," *Proceedings of the ASME Winter Annual Meeting,* New Orleans, LA, November 1993.

[30] Emerson, D., et al., "Plastic-Encapsulated Microcircuit Reliability and Cost Effectiveness Study," *IEEE Transactions on Reliability,* Vol. 45, No. 1, pp. 19–22, 1996.

[31] Pecht, M., et al., "Plastic Packaged Microcircuits: Quality, Reliability and Cost Issues," *IEEE Transactions on Reliability,* Vol. 42, No. 4, pp. 513–516, 1993.

[32] Tam, S., "Demonstrated Reliability of Plastic-Encapsulated Microcircuits for Missile Applications," *IEEE Transactions on Reliability,* Vol. 44, No. 1, pp. 8–12, 1995.

[33] Pendse N., D. Thomas, D. Das and M. Pecht, "Uprating of a Single Inline Memory Module," *IEEE Transactions on Components and Packaging Technologies,* Vol. 25, No. 2, June 2002.

[34] SMART Modular Technologies, 1998, "Memory Industry Terms and Acronyms," <www.smartm.com/knowledge/html/memoryterms.html>, accessed May 1999.

[35] Rayapati, V. N., "VLSI Semiconductor Random Access Memory Functional Testing," *Microelectronics and Reliability,* Vol. 30, No. 5, pp. 877–889, 1990.

[36] Sharma, A. K., *Semiconductor Memories: Technology, Testing, and Reliability,* IEEE Press, Piscataway, NJ, 1997.

[37] Sommerville, I., *Software Engineering,* 5th ed., Addison-Wesley, Reading, MA, 1995.

[38] Van de Goor, A. J., *Testing Semiconductor Memories: Theory and Practice,* John Wiley & Sons, New York, NY, 1991.

[39] Svab, A., "Thermocouples, Thermistors and RTDs," July 23, 1998, Sensor Scientific, Inc., <http://www.sensorsci.com/thermocouples.htm>, accessed July 28, 2002.

[40] Brennom, T. and N. Anderson, "Uprating Experience by an Avionics OEM," *Proceedings of Commercialization of Military and Space Electronics,* pp. 317–323, Los Angeles, CA, February 12–15, 2001.

[41] American Society of Testing and Materials (ASTM), *Manual on the Use of Thermocouples in Temperature Measurement*, p. 19, American Society for Testing and Materials, Philadelphia, PA, 1974.

Chapter 14

Reliability Assessment

Michael Pecht and Jingsong Xie

Reliability is the part's ability to function within given performance limits under specified conditions over a specified length of time in a product. This chapter presents a process to assess the reliability of electronic parts during the part selection and management process prior to inclusion of the part in a design.[1]

14.1 Candidate part

Reliability assessment is by definition an application-specific process; that is, the assessment results must be developed for the given application. If the part's application-specific reliability requirements are met, it is accepted for the given application; if not, the part is rejected for the given application due to unreliability.

The assessments of manufacturer quality, part family quality and integrity, and distributor quality processes (see Chapters 6–8) have all been conducted prior to reliability assessment. These assessments enable one to assume an acceptable level of quality, consistency, and control in the part's materials, design, and construction. Thus, a part with inconsistent materials, design, and manufacturing processes will generally be rejected prior to reliability assessment.

Figure 14.1 shows how reliability assessment may be conducted using the part manufacturer's qualification data, virtual reliability assessment, and accelerated reliability tests. The sections that follow discuss the reliability assessment process and the use of each method to assess part reliability.

14.2 Life cycle loads

The reliability of the part depends upon the loads experienced over the entire life cycle. Life cycle loads constitute all the environmental and operational loads, including those encountered during part manufacture, assembly, storage, shipping, and use in field, to which a part is subjected. Life cycle loads include temperature, humidity, chemical contaminants, vibration, mechanical shock, and radiation. The magnitudes of the loads, together with net change, rate of changes, and the duration of exposure, contribute to the stresses induced in a part. That is why reliability assessment is application specific.

---

[1] Traditionally, reliability has been predicted using MIL-HDBK-217, "Reliability Prediction of Electronic Equipment," and its progeny, including PRISM, Telcordia (previously Bellcore) SR-332, Siemens SN 29500, British Telecom HRD 5, and CNET RDF 93. Such reliability assessment methods are flawed for several reasons, and assessment using IEEE Standard 1413 and the IEEE 1413.1 guidebook show that these are poor prediction methods. For example, they assume constant part failure rates, consider the steady-state temperature and voltage to be the only operational loads, ignore actual temperature cycling and vibration, and do not take into consideration the advances in packaging technology. In fact, the U.S. military has labeled the reliability prediction methods in the MIL-HDBK-217 as "unreliable . . . erroneous and misleading" [7], and stated that it is not intended to predict field reliability and, in general, does not do a very good job [21].

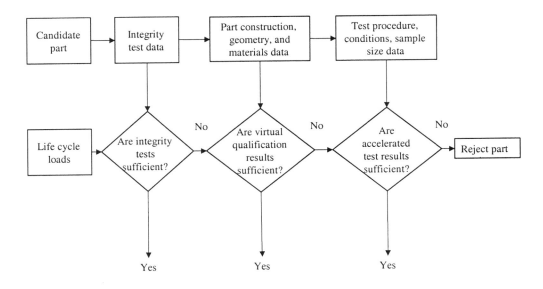

Figure 14.1: Reliability assessment process

14.3   Using tests to assess reliability

Most part manufacturers conduct periodic tests to validate the ability of a part family[2,3] to meet the range of specifications that they are willing to guarantee and to baseline the effects of new materials, processes, and design on the part family.   The tests are also performed in order to transition a part or part family from the engineering phase to the manufacturing phase, or are performed when the part manufacturer considers a change made to the part or part family significant.[4] Part manufacturers often refer to the tests as "qualification" or "reliability" tests. These tests only measure the ability of the part to survive some standard load conditions. They do not assess reliability because the part manufacturer does not generally know the application conditions. These tests are independent of the part's recommended operating conditions.[5] Examples of part manufacturers' tests are presented in Chapter 7.

---

[2] JEDEC Standard No. 26 B, "General Specification for Plastic Encapsulated Microcircuits for Use in Commercial and Rugged Applications," defines a part family as a grouping of part types utilizing similar materials, construction, and fabrication techniques, as determined by the manufacturer.   Manufacturers use varying nomenclature to denote a part family. For example, Motorola notes that "process families" consist of parts with the same semiconductor die fabrication process, and "package families" consist of parts packaged using the same packaging technology. From this definition, parts from the same process family may belong to different package families, and vice versa.

[3] The part manufacturer tests are generally conducted on part families rather than individual parts, but the test results are said to apply to all parts in the family. It is therefore important to know the manufacturer's grouping of parts in a part family to ensure that the results are truly applicable. Chapter 7 discusses part families in more detail.

[4] Integrity tests are not designed to precipitate defects, which may be introduced during the manufacturing process. Defect precipitation is the goal of screens such as burn-in, which are conducted on 100% of the population.

[5] Consider, for example, Maxim's MAX3261 laser diode driver. This part is offered in the commercial (0 to 85°C) and industrial (−40 to 85°C) temperature ranges. Both temperature range parts are subjected to the same test profiles.

Part manufacturers are free to choose the types of tests to which they subject their parts and the conditions and duration of those tests. Due to economic and time constraints, the tests are rarely conducted at each customer's application conditions, but rather at manufacturer-established elevated load levels which are designed to accelerate failure mechanisms.[6] In some cases, part manufacturers use test types and conditions prescribed by standards,[7] such as STACK [33], MIL-STD-883 [34], and AEC-Q-100 [2]. The specified requirements include test types, test conditions and duration, pass/fail criteria, sample sizes, and number of lots. If a part manufacturer's part family tests and test results meet all the criteria set forth, the part family is viewed more favorably than a part family which does not meet all the criteria.

14.3.1 Applicability of tests to reliability assessment

Once the application life cycle loads and test data have been identified, the environmental conditions of the test(s) are compared with those of the part life cycle application. If the test loads exceed the application loads in both magnitude and duration, the tests may be used to validate part reliability. If the part is not subjected to a load (or combination of loads) conditions during testing greater than the part is expected to experience during its life cycle, virtual reliability assessment and/or additional accelerated tests must be used.

14.3.2 Example application

Consider the International Rectifier IRLBA3803P Super 220 HEXFET, a three-leaded plastic package, to be required for use in an application where temperature cycling is expected to occur 1000 times between 15°C and 90°C. The application environment is shock, vibration, and radiation free and is humidity controlled. The part manufacturer, as part of its internal qualification testing procedures, subjects this part to temperature cycling between –55°C and 150°C for 2000 cycles, temperature and humidity bias at 85°C/85% RH for 2000 hours, high-temperature reverse bias at 80% of rated voltage (24 V) at 175°C for 2000 hours, and accelerated moisture resistance at 100% RH, 121°C, 15 lb psig for 96 hours. Both the load and duration to which the manufacturer subjected the part during temperature cycle testing are more extreme than the actual life cycle conditions the part is expected to experience. If the sample size and test results are acceptable, part reliability is said to have been verified through part manufacturing testing.

14.4 Using virtual reliability assessment to assess reliability

Virtual reliability assessment is a simulation-based methodology used to identify the dominant failure mechanisms associated with the part under the life cycle loads and to determine the corresponding time-to-failure distributions. The models used in virtual reliability assessment can incorporate "curve fitting" to describe failure data or can be based on stress-damage models which relate the fundamental mechanical, thermal, electrical, radiation, and chemical failure processes arising from the environmental and operational life cycle loads to the time to failure of the part.

---

[6] Equipment suppliers who purchase an economically significant volume of parts from a part manufacturer may have influence over the types of tests and test conditions and duration to which various part families are subjected.

[7] Some of the tests specified by standards are widely considered non-value-added by part manufacturers. For example, the Rome Laboratories of the U.S. Air Force states that the MIL-STD-883 "end-of-line screens provide a standard series of reliability tests for the industry. Although manufacturers continue to use these screens today, most of the screens are impractical or need modifications for new technologies, and add little or no value for mature technologies" [9].

Virtual reliability assessment uses constructional details, dimensions, and material properties of the part to assess reliability. Based on the part information and the life cycle loads, the potential failure mechanisms are identified, often with the help of failure modes and effects analysis, fishbone diagrams, and databases [25], [31].

With the use of computer simulation, it is possible to list most if not all of the failure mechanism as conditions and assess them all. For example, JEDEC lists the major known and relevant failure mechanisms for semiconductor devices [13]. The University of Maryland CALCE EPSC provides software to assess the times to failure for these models [38]–[40]. Table 4.1 overviews various types of part information, failure mechanisms, failure models, and load conditions.

14.5 Using accelerated testing to assess reliability

Accelerated testing is based on the concept that the part will exhibit the same failure mechanism and mode in a short time under high load conditions as it would in a longer time under actual (real-life) conditions. The purpose of accelerated testing is to decrease the total time and cost required to obtain reliability information about the part under consideration.

Successful implementation of accelerated test strategies requires that (1) failure mechanisms generated in accelerated testing must be the same as those observed and identified in the preliminary physics-of-failure (PoF) assessment and (2) results from accelerated tests must be extrapolated to field life conditions using acceleration transforms as reliability predictors to enable proactive product design. The failure mechanisms must be identified before accelerated life tests are conducted, because the amount of test-time compression achieved in an accelerated test must be determined quantitatively based on the physics of the relevant failure mechanisms [6], [16], [28]–[30], [34].

Failure mechanisms can be categorized into overstress failures and wearout failures, as presented in Figure 14.2. Catastrophic sudden failures due to a single occurrence of a *stress* event that exceeds the intrinsic *strength* of a material are termed "overstress failures." Failures due to the accumulation of incremental *damage* in excess of the material *endurance* limit are termed "wearout failures." Accelerated "stress" tests generally refer to tests that accelerate wearout failure mechanisms.

Overstress failure mechanisms and performance problems (e.g., software and tolerance problems) are not due to incremental damage accumulation and hence cannot be accelerated. This categorization into overstress and wearout failures is critical in understanding the extent to which load conditions can be magnified during accelerated testing.

Load limits are identified in Figure 14.3. "Destruct limits" are irreversible overstress limits that exceed the intrinsic material strength and cause catastrophic failures in a product. Load limits within which a product performs its intended functions satisfactorily are termed "operating limits." These are often termed "reversible overstress limits." When wearout mechanisms are expected during the life cycle of a product, the designer selects "design limits" that allow adequate reliability throughout the intended life cycle.

Table 14.1: Example of part information, failure mechanisms, models and loads

Candidate part information	Failure mechanisms	Failure models	Life cycle loads
Entire package	Fatigue cracking	Attach fracture/brittle model Attach fatigue/brittle model Attach tensile fatigue/ductile model Attach shear fatigue/ductile model Suhir's vertical crack (die) model [11] Suhir's horizontal crack (die) model [11] Westergaard Bolger model [1], [4], [24], [37] Stress intensity vertical crack model [12] Stress intensity horizontal crack model [12] Bond pad fatigue model [11] Wire shear fatigue model [11] Chip cratering model [11] Pecht and Lall model [15] Hu, Pecht, Dasgupta model [11]	Mechanical loads, temperature cycling, vibration
Internal interconnection (i.e., flip chip, wire bond, tab)	Creep	Flip chip interconnect creep model [8], [31]	Mechanical load, temperature
Wire bonds	Diffusion	Kidson's model [14]	Temperature, concentration
Package case	Corrosion	Hermetic metallization corrosion model [1] Plastic metallization corrosion model [1]	Relative humidity, temperature, voltage and contaminants
Package case	Popcorning	Nguyen popcorn model [22]	Temperature, relative humidity
Chip	Electromigration	Black's model [3] Shatzke's and Llyod's model [32]	Current density, temperature, temperature gradient
Chip	Dendritic growth	Dendritic growth modified model [22]	Differential voltage, temperature, contaminants
Chip	Stress-driven diffusion voiding	Okabayashi model : n ≠ 1 [23] Okabayashi model : n = 1 [23]	Residual mechanical loads, temperature
Chip	Slow trapping, time-dependent dielectric breakdown and surface charge spreading	Positive gate voltage model Negative gate voltage model Fowler Nordheim tunnel model [16], [19], [24]	Temperature, gate voltage

A designer often includes safety margins to account for the tolerances of the part's constituents and manufacturing processes, and therefore the obtained "specification limits" are usually less than the part's design limits. The load magnitudes for accelerated wearout testing must exceed the specification limits but be within the overstress limits of the part. For functional accelerated wearout testing, the overstress limits are the operating limits, while for nonfunctional testing the overstress limits are the destruct limits.

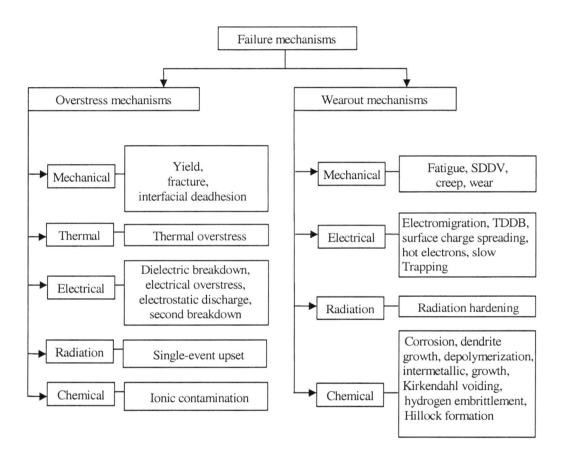

Figure 14.2: Overstress and wearout failure mechanisms

14.5.1   Determine environmental loads to be used for accelerated tests

Based on the dominant failure mechanisms and the critical failure sites, environmental loads are selected that will precipitate the mechanisms. A list of example failure mechanisms and the respective loads that will precipitate failure is presented in Table 14.2. The types of environmental loads used for the accelerated test will dictate the test equipment that is required.

Table 14.2: Examples of failure mechanisms and their respective acceleration stresses [10], [18], [27], [36]

Wearout failure mechanism	Acceleration stresses
Fatigue crack initiation	Mechanical stress/strain range, cyclic temperature range
Fatigue crack propagation	Mechanical stress range, cyclic temperature range
Creep	Mechanical stress, temperature
Wear	Contact force, relative sliding velocity
Corrosion	Temperature, relative humidity
Electromigration	Current density, temperature, temperature gradient
Cyclic fatigue	Cyclic temperature range, mechanical stress range
Delamination	Cyclic temperature range, mechanical stress range, relative humidity
Radiation damage	Intensity of radiation, total dose of radiation
Stress corrosion	Mechanical stress, temperature, relative humidity
Dendritic growth	Voltage differential, relative humidity

14.5.2   Determine test vehicle, test fixture, and test facilities

A test vehicle must be chosen for the accelerated testing. Depending on the failure mechanisms identified and the objectives of the testing, the part or an assembly may be required to serve as the test vehicle to assess design, manufacturing, or reliability issues or to meet contractual or regulatory requirements.

A test fixture can be designed based on the environmental test loads selected, the facilities available, and the test vehicle selected. The test fixture should not prevent the applied loads from being transmitted to the test specimen. Nor should the test fixture cause failures by producing interactions that will not take place during field applications.

The test facilities are determined based on the equipment needed to perform the testing. The environmental loads needed for accelerated testing will dictate the type of environmental testing equipment needed. The size and complexity of the test vehicle will also contribute to the choice of the equipment needed. Appropriate test facilities are chosen that can bring the test vehicle to the desired loads.

14.5.3   Define failure and develop a failure detection scheme

Failure can be defined when an electrical parameter (e.g., voltage, resistance, current) is beyond specification or when the product ceases to operate (e.g., electrical open). A parameter that is beyond specification may be caused by an intermittent failure, where the product continues to operate but cumulative effects will grow until the product ceases to operate. A failure that causes the product not to operate occurs when the load-damage level exceeds the threshold product strength. An example of threshold strength is the temperature at which a component begins to malfunction.

The failure detection scheme chosen for the testing must be able to detect the defined failure. Four main detection schemes [5] for functional test vehicles are:

- *passive* – an unpowered test vehicle is stressed during testing, and failure detection is performed during scheduled stoppages of the test.
- *static* – the test vehicle is powered but is electrically nonfunctioning during testing, and failure detection is performed during scheduled stoppages of the test.
- *pseudo-dynamic* – the test vehicle is powered and electrically functional during testing, and failure detection is performed incrementally while testing. Failure detection may have narrow fault coverage of the test vehicle. The stress limit for failure monitoring may be the operating limit or the destruct limit of the test vehicle.
- *dynamic* – the test vehicle is powered and electrically functional during testing, and failure detection is performed concurrently during testing. Failure detection has wide fault coverage of the test vehicle. The stress limit for failure monitoring may be the operating limit or the destruct limit of the test vehicle.

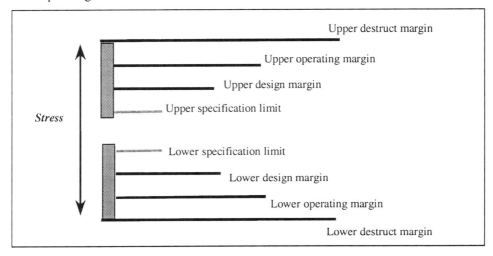

Figure 14.3: Stress limits and margin

Intermittent failures tend to comprise a significant number of product failures [5]; therefore, dynamic failure detection is generally recommended. Intermittent failures that may occur when only a certain load level is applied may not be detected with passive or static failure detection schemes. A pseudo-dynamic scheme may detect the failure if the failure detection scheme is scheduled to run during the time the critical stress level occurs. Dynamic failure detection, which continuously monitors for failures, can be formulated to detect intermittent failures.

The functional test used for failure detection contributes to the fault coverage of the failure detection scheme. Complete fault coverage for test software may be hard to achieve for parts that require extended test intervals. Two main categories for test software are: generic test software developed for in-field use, which usually focuses on exercising many of the software options of the part, and factory test software that focuses on fully exercising the part [5]. When particular failure mechanisms are targeted during testing, test software may be heavily weighted to detect the failure modes of interest.

14.5.4 Determine test load levels to simulate the lifetime

Based on the environmental loads and the overstress limits,[8] the accelerated test load levels are determined. These levels should be defined beyond the specification and design levels but within the operating limits for a functional specimen and between the design and destruct limits for a nonfunctional specimen (see Figure 14.3). Stress-damage models are then used for the relevant failure mechanism to assess the length of a test based on the load level to simulate the part's lifetime.

14.6 Conduct the accelerated tests

The main purpose of accelerated life tests is to evaluate the product's intrinsic vulnerability to applied loads due to wearout failure mechanisms. For successful implementation, failure mechanisms generated must be the same as those observed and identified in the preliminary virtual reliability assessment, and results must be extrapolated to field life conditions using acceleration transforms as reliability predictors to enable proactive product design.

Accelerated life tests can confirm the failure predictions [34]. During accelerated life testing, the anticipated failure mechanism is assessed and the experimental time to failure is recorded. Typically, the test value is somewhat larger than the estimated time to failure due to a certain degree of conservatism built into the failure mechanism models.

14.7 Summary

The results of the reliability assessment process help determine the limits of the part in the particular application conditions. During the assessment, one can also learn the effects of the variation of part strength and the load on a part. The virtual reliability assessment (and, if necessary, the accelerated tests) can be performed at different load levels to determine the point at which a part can perform reliably. These results can be used to modify the part operating conditions. When the conditions warrant, the results of the assessment can also be used to establish if any derating of the parts' operational parameters will be helpful to achieve the reliability goals.

14.8 References

[1]   Ajiki, T., M. Sugumoto, H. Higuchi and S. Kumada. "A New Cyclic Biased T.H.B. Test for Power Dissipating ICs," *Proceedings of the International Reliability Physics Symposium*, pp. 118–126, 1979.

[2]   Automotive Electronics Council, CDF–AEC–Q100 Stress Test Qualification for Automotive-grade Integrated Circuits, Revision E, Detroit, MI, January 31, 2001.

[3]   Black, J. R., "Current Limitation of Thin Film Conductor," *Proceedings of the International Reliability Physics Symposium*, pp. 300–306, 1982.

---

[8] Overstress tests are conducted to determine the stress limits, specifically the operating and destruct limits of the product. The operating and overstress limits are used in designing the accelerated load levels for the accelerated test. The overstress limits may be assigned by the maximum operating limits of a test chamber and not by the intrinsic overstress limits of the product. Failure analysis of the overstress test will verify the dominant failure modes, mechanisms, and sites and the preliminary virtual reliability assessment.

[4]   Bolger, J. C., "Polyimide Adhesive to Reduce Thermal Stresses in LSI Ceramic Packages," *Proceedings of the 14th National SAMPE Technical Conference*, pp. 394–398, 1982.

[5]   Chan, H. A. and P. J. Englert, *Accelerated Stress Testing Handbook: Guide for Achieving Quality Products*, IEEE Press, New York, NY, 2001.

[6]   Dasgupta, A. and M. Pecht, "Material Failure Mechanisms and Damage Models," *IEEE Transactions on Reliability*, Vol. 40, No.1, pp. 531–536, December 1991.

[7]   Decker, F., Assistant Secretary for Research, Development and Acquisition, United States government, 1996.

[8]   Engelmaier, W., "Functional Cycling and Surface Mounting Attachment Reliability," *ISHM Technical Monograph Series 6984-002*, pp. 87–114, 1984.

[9]   Gorniak, M., U.S. Air Force, Rome Labs, "Qualified Manufacturers List (QML) Update," *Reliability Society Newsletter*, Vol. 40, p. 8, 1994.

[10] Hu, J., D. Barker, A. Dasgupta and A. Arora, "Role of Failure Mechanism Identification in Accelerated Testing," *Proceedings of the 1992 Annual Reliability and Maintainability Symposium*, pp. 181–188, January 1992.

[11] Hu, J., M. Pecht and A. Dasgupta, "A Probabilistic Approach for Predicting Thermal Fatigue Life of Wirebonding in Microelectronics," *Transactions of ASME, Journal of Electronic Packaging*, Vol. 113, No. 3, pp. 275–291, September 1991.

[12] Hu, J. M., M. Pecht and A. Dasgupta, "Design of Reliable Die Attach," *The International Journal of Microcircuits and Electronic Packaging*, Vol. 16, No. 1, pp. 1–21, 1993.

[13] JEDEC, "JEP 122-A: Failure Mechanisms and Models for Semiconductor Devices," JEDEC, Arlington, VA, December 2001.

[14] Kidson, G. V., "Some Aspects of the Growth of Diffusion Layers in Binary Systems," *Journal of Nuclear Materials*, Vol. 3, No. 1, pp. 21–29, 1961.

[15] Lall, P., M. Pecht and E. Hakim, *Influence of Temperature on Microelectronics and System Reliability: A Physics of Failure Approach*, CRC Press, Boca Raton, FL, 1996.

[16] Lall, P. and M. Pecht, "An Integrated Physics-of-Failure Approach to Reliability Assessment," *Advances in Electronic Packaging*, ASME EEP, Vol. 4, No. 1, pp 509–524, 1993.

[17] Lee, J., I. C. Chen and C. Hu, "Modeling and Characterization of Gate Oxide Reliability," *IEEE Transactions on Electron Devices*, Vol. 35, No. 12, pp. 2268–2278, December 1988.

[18] Lewis Research Center, NASA, "NASA Preferred Reliability Practices – Environmental Factors," Practice Number PD-EC-1101, Cleveland, OH, May 1995.

[19] Moazzami, R. and C. Hu, "Projecting Gate Oxide Reliability and Optimizing Reliability Screens," *IEEE Transactions on Electron Devices*, Vol. 37, No. 7, pp. 1643–1650, July 1990.

[20] Moazzami, R., J. Lee and C. Hu, "Temperature Acceleration of Time-dependent Dielectric Breakdown," *IEEE Transactions on Electron Devices*, Vol. 36, No. 11, Part 1, pp. 2462–2465, November 1989.

[21] Morris, S., "MIL-HDBK-217 Use and Application," *Reliability Review*, Vol. 10, pp. 10–15, 1990.

[22] NASA Requirements to Preclude the Growth of Tin Whiskers, information received from Jack Shaw, Manager, NASA Parts, Projects Office, NASA Goddard Space Flight Center, Greenbelt, MD.

[23] Okabayashi, H., "An Analytical Open-circuit Model for Stress Driven Diffusive Voiding in Al Lines," *Proceedings of the 5th International Conference on Quality in Electronic Components: Failure Prevention, Detection and Analysis*, pp. 171–175, 1991.

[24] Paris, P. C., M. P. Gomez and W. E. Anderson, "A Rational Analytical Theory of Fatigue," *The Trend in Engineering*, University of Washington, Vol. 13, pp. 9–14, 1961.

[25] Pecht, M., R. Radojcic and G. Rao, *Guidebook for Managing Silicon Chip Reliability*, CRC Press, Boca Raton, FL, 1999.

[26] Pecht, M., *Handbook of Electronic Package Design*, Marcel Dekker, New York, NY, 1991.

[27] Pecht, M., L. Nguyen and E. Hakim, *Plastic-Encapsulated Microelectronics*, John Wiley & Sons, New York, NY, 1995.

[28] Pecht, M. and A. Dasgupta, "Physics-of-failure: An Approach to Reliable Product Development," *Proceedings of the Institute of Environmental Sciences*, Chicago, IL, pp. 111–117, August 1995.

[29] Pecht, M., A. Dasgupta and D. Barker, "The Reliability Physics Approach to Failure Prediction Modeling," *Quality and Reliability Engineering International*, Vol 6, pp. 276–273, 1990.

[30] Pecht, M., *Integrated Circuit, Hybrid, and Multichip Module Package Design Guideline*, John Wiley & Sons, New York, NY, 1994.

[31] Puttlitz, J. K., "Preparation, Structure and Fracture Modes of Pb-Sn and Pb-In. Terminated Flip-chips Attached to Gold Capped Microsockets," *IEEE Transactions on Components, Hybrids, and Manufacturing Technology*, Vol. 13, pp. 647–655, 1990.

[32] Shatzkes, M. and J. R. Lloyd, "A Model for Conductor Failure Considering Diffusion Concurrently with Electromigration Resulting in a Current Exponent of 2," *Journal of Applied Physics*, Vol. 59, No. 11, pp. 3890–3893, 1986.

[33] STACK International, Specification 0001 Issue 12.2 at Notice 2: General Requirements for Integrated Circuits, Herts, UK, November 19, 1999.

[34] United States Department of Defense, MIL-STD-883: Test Method Standards – Microcircuits, 1996.

[35] Upadhyayula, K. and A. Dasgupta, "Physics of Failure Guidelines for Accelerated Qualification of Electronic Systems," *Quality and Reliability Engineering International*, Vol. 14, No. 6, pp. 433–447, 1998.

[36] Viswanadham, P. and P. Singh, *Failure Modes and Mechanisms in Electronic Packages*, Chapman & Hall, New York, NY, 1998.

[37] Westergaard, H. M., "Bearing Pressure and Cracks," *Journal of Applied Mechanics*, Vol. 61, pp. A-49–A-53, 1939.

[38] Osterman, M. and T. Stadterman, "Failure-Assessment Software for Circuit-Card Assemblies," *Proceedings of the Annual Reliability and Maintainability Symposium*, pp. 269–276, January 1999.

[39] Osterman, M., A. Dasgupta and T. Stadterman, *Simulation Guide Testing in Product Qualification and Development in Case Studies in Reliability and Maintenance*, Wallace R. Blischke and D. N. Prabhakar Murthy (eds.), John Wiley & Sons, New York, NY, December 2002.

[40] Osterman, M., *Reliability and Performance of Advanced PWB Assemblies*, Charles Harper (ed.), Chapter 9, "High Performance Printed Circuit Boards," McGraw-Hill, New York, NY, 1999.

Chapter 15

Assembly Requirements and Constraints Assessment

Peter Sandborn and Michael Pecht

A part that has acceptable quality, reliability, availability, and cost may still be unacceptable from an assembly viewpoint if (1) it is incompatible with the assembly equipment or process, (2) it is impossible or impractical to connect or rout the part into the system,[1] or (3) the part cannot be acceptably tested or reworked. Assembly compatibility addresses whether a system that contains the part can be manufactured (assembled). Routing compatibility assesses if the candidate part can be properly interconnected (wired into the system). Test and rework acceptability assesses whether the candidate part can be adequately and economically tested and reworked during assembly.

The three compatibility requirements can be addressed during the "verification" phase of product development after physical design of the system is completed. However, design problems detected during verification may be costly to fix and can have disastrous effects for time-to-market- or time-to-deployment-driven products. Therefore, if the verification activities can be performed concurrently with part selection, the assembly risks associated with the resulting system can be minimized. The process of performing assembly verification activities at the front of the design process is called "virtual verification" [1].

This chapter describes how portions of the verification-level activities can be performed early in the design process, i.e., during parts selection. Figure 15.1 shows the process for assessing process, routing, and test and rework compatibility.

15.1 Assembly compatibility

Assembly process compatibility involves comparing the part's size, shape, and mounting method to the assembly process that will be used to assemble systems containing the part. Figure 15.2 schematically describes the process of evaluating the assembly compatibility of a candidate part.

In Figure 15.2, the spatial compatibility check (see Section 15.1.1) and proximity check (see Section 15.1.2) can be performed only if the sizes of other parts in the system are known. The sizes of the other parts will be known if a part is being selected to replace or upgrade an existing part in an existing system. The spatial compatibility and proximity checks are also useful if assembly compatibility is assessed for a candidate bill of materials (list of all parts in the system). If neither of these situations exists, the spatial compatibility and proximity checks are omitted from the assessment.

15.1.1 Spatial compatibility check

Spatial compatibility is an assessment of whether the candidate part (along with the other parts in the system) can be assembled within a constrained system area. If a maximum board area is specified by the product requirements, a spatial compatibility check is used to verify that an acceptable layout of the parts on the system is possible.

---

[1] In this chapter, "system" refers to a circuit board or substrate with multiple parts assembled on it.

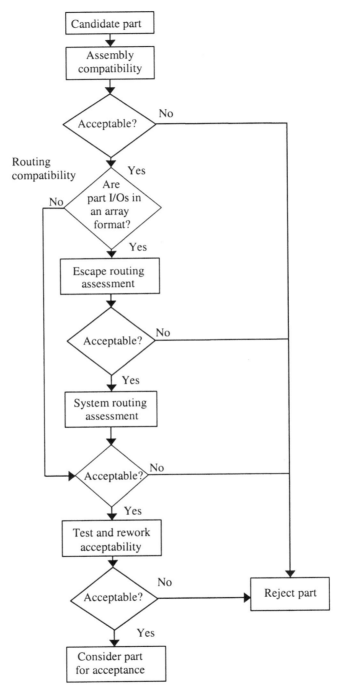

Figure 15.1: Method for assessing process, routing, and test and rework compatibility

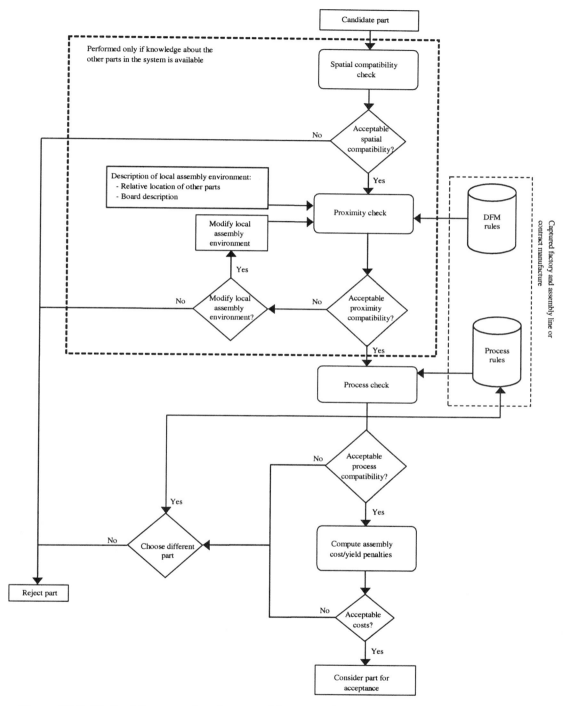

Figure 15.2: Method for assessing the assembly compatibility of a part during parts selection and management (artwork verification not included)

To assess spatial compatibility, we perform a simple accumulation of the area occupied by the parts to determine if the layout is practical. Spatial compatibility is likely if the following relation is satisfied:

$$\text{Maximum system area constraint} \geq 1.7 \text{ maximum } [A_{1\text{-Cumm}} + A_1, A_{2\text{-Cumm}} + A_2] \qquad (15.1)$$

where $A_{i\text{-Cumm}}$ is the area occupied by all other parts in the system on side i of the board or substrate and A_1 and A_2 represent the effective area the candidate part occupies on sides 1 and 2, respectively. A_1 and A_2 are determined using the algorithm shown in Figure 15.3, where Quantity is the number of instances of the candidate part in the system and A_c is the area necessary to layout the part on the board given by eqn. (15.17).

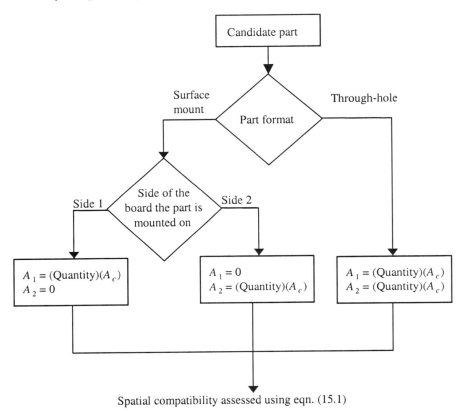

Figure 15.3: Process for performing spatial compatibility checking

The process assumes that the accumulated area of all the other parts in the system is known. If it is not, the process should be performed for each part and the results accumulated in $A_{i\text{-Cumm}}$. The spatial compatibility is acceptable if 1.7 times the accumulated areas in Figure 15.3 is less than or equal to the board size constraint. The factor of 1.7 was determined from parametric studies of part layouts on printed circuit boards and accounts for less than 100% packaging density (i.e., the parts cannot be perfectly tied together on the layout).

15.1.2 Proximity check

Proximity checking involves assessing the location of the candidate part relative to other parts assembled on the board, special board features (i.e., holes, pins, etc.), and the edge of the board. As with spatial compatibility, proximity checking can be performed only if the entire bill of materials and a preliminary placement of the parts on the board are known. Due to these restrictions, proximity checking is viable for inclusion in the parts selection and management process only when the process is being used to select replacement parts for an existing system or during the redesign of an existing system. The process for proximity checking is shown in Figure 15.4.

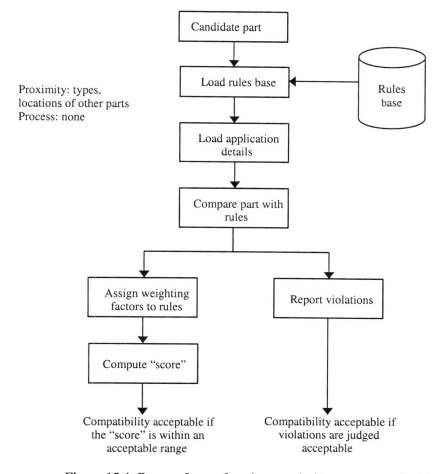

Figure 15.4: Process for performing proximity or process checking

Proximity checking is based on a set of design for manufacturing (DFM) rules associated with the assembly process. The DFM rules are available from many sources: company proprietary guidelines, supplied by contract manufacturers, available in generic form from industry organizations, and included within verification-level DFM software tools, e.g., [2]. All of these sources provide versions and values for the general proximity checking criteria outlined in Table 15.1. Example rules for surface mount (SMT) assembly are given in Table 15.2.

Table 15.1: General proximity checking rules

Proximity checking criteria	Notes
Edge clearance	Minimum distance from board edge to part
Hand clearance	Minimum keepout area around a part that must be manually mounted
Mixed: dual/single inline package (DIP/SIP) to SMT clearance	Mixed = through-hole (TH) and SMT on the same board
Mixed: axial to SMT clearance	Axial = axial leaded device
SMT: chip-to-chip clearance	Chip = SMT parts with two pads
SMT: outline-to-outline clearance	Outline = SMT parts with I/O on two sides
SMT: outline to chip	
SMT: carrier to any SMT	Carrier = SMT parts with I/O on all four sides
TH: DIP to DIP	
TH: DIP to axial	
TH: DIP to SIP	
TH: axial to axial	
TH: axial to SIP	
TH: SIP to SIP	End-to-end, side-to-side, and end-to-side
TH: group SIP	Two SIPs can be placed as close as 100 mils (pin to pin); however a third SIP requires additional spacing from the second
Orientation	Multiple instances of the same part should be oriented in the same direction
Acceptable part rotations	Rectilinear rotations are allowed (0, 90, 180, 270 degrees)

Candidate parts are evaluated against all the other parts in the system using the minimum clearance rules (e.g., Table 15.2) that correspond to the process that will be used to assemble the board. Violations of the rules are reported so that they can be evaluated by manufacturing engineers and the layout of the board changed if necessary.

There are three possible outcomes of proximity checking: (1) cannot be assembled, (2) can be assembled with a corresponding cost and/or yield penalty, or (3) can be assembled with no cost or yield penalties. If proximity compatibility is not acceptable, the local "assembly environment" may be changed. The assembly environment consists of the location of all the other parts in the system (these locations could be modified), the board or substrate technology used (a different board could be chosen), and the assembly process itself (a different assembly line could be chosen). If a change is made to the assembly environment, the other portions of the parts selection and management process are notified and the proximity check is reevaluated.

Table 15.2: Example SMT process side-to-side and end-to-end clearances (all dimensions are in mils)

	Discrete	SMT p > 25 mils	SMT p < 25 mils	SMT connector	Axials	DIP, SIP	PGA	TH connector	BGA
Discrete	18	50	150	100	100	100	100	200	150
SMT p > 25 mils		50	150	100	100	100	200	200	150
SMT p < 25 mils			150	150	150	150	200	200	150
SMT connector				100	100	100	200	200	150
Axials					100	100	275	275	150
DIP, SIP						100	275	275	150
PGA							275	275	150
TH connector								275	150
BGA									150
p = center-to-center pitch of package I/O									

15.1.3 Process check

The process check involves comparing the part's size, shape, and mounting method to the assembly process that will be used to assemble boards containing the part. Checking process compatibility is identical to performing a proximity check (see Figure 15.4). Table 15.3 provides general process compatibility rules.

Unlike the spatial compatibility and proximity checks, the process check can be assessed without knowledge of the other parts in the system. Like proximity checking, process checking is based on a set of DFM rules associated with the assembly process. Process compatibility information is usually contained in the same documents that provide proximity check information.

Table 15.3: General process compatibility checking rules

Process compatibility rule	Explanation
Maximum part height	If the part does not fit in process equipment, manual placement is required at additional cost.
Maximum part dimension	Height<300–400 mils (SMT)
Maximum part aspect ratio	
Minimum part dimensions	
Maximum part weight	In double-sided assembly, the solder paste surface tension must be sufficient to hold underside parts.
Reflow: reflow temperature profile	If reflow is used, all parts must withstand a common reflow profile. Note: lead-free parts may require higher temperatures in the reflow profile.
Wave solder: minimum lead pitch	50 mil minimum
Wave solder: capacitor type	Tantalum capacitors should not be wave soldered due to cracking.
Wave solder: minimum part weight	Lightweight parts may lift during wave soldering.
Automatic TH: maximum I/O count	
Underfill compatibility (BGA and flip chip only)	Is the part compatible with the underfill strategy applied to the rest of the board or is the rest of the board (and parts) compatible with the underfill requirements for this part?
Conformal coat compatibility	Is the part compatible with the conformal coating requirements for the board? All parts on the board must be compatible with the same conformation coating process and materials.
Mixed SMT and TH	Is the part compatible with a mixed SMT/TH assembly strategy for the board (if applicable)?
Moisture control	If the part is moisture sensitive, will the assembler track time out of bag?
Surface cleanliness limits	Does the part have surface cleanliness requirements (for assembly) that are compatible with board cleaning processes in place at the assembly plant?

The maximum weight for parts depends on the type of package the part is in, the number of package I/Os and the format of the package. For BGAs:

$$\text{Weight} \ < \ \frac{\text{Number of I/Os}}{A} \tag{15.2}$$

For peripheral leads:

$$\text{Maximum Weight} < 1.2 \ (\text{Number of I/Os}) \ (b + AL) \tag{15.3}$$

where L is the lead foot length, b is the total lead length, and A is given in Table 15.4.

Table 15.4: Values of A

Type of package	A
Ceramic BGA	35
Plastic BGA, pads $\geq$ 24 mils	25
Plastic BGA, pads < 24 mils	12
Gull wing leads	1.0
J leads	1.35

Candidate parts are evaluated against the rules that correspond to the process that will be used to assemble the board. Violations of the rules are reported so that they can be evaluated by manufacturing engineers. A "score" may also be computed by accumulating rule violations and multiplying the violations by weighting factors that differentiate the relative importance of the rules [3]. A candidate part need not always satisfy all the rules, but if a rule is not satisfied, additional assembly costs (or yield penalties) are likely to result.

The outcomes of the process check are the same as for the proximity check: (1) cannot be assembled, (2) can be assembled, with a corresponding cost and/or yield penalty, or (3) can be assembled with no cost or yield penalties. If process compatibility is not acceptable, a different assembly line can be chosen and the part reevaluated or a different part can be considered.

15.1.4 Artwork verification

Board and assembly manufacturability tools used during verification perform one check that cannot be done during part selection: artwork verification. Artwork verification involves checking the board layout for the correct orientation and location of fiducials, alignment holes, and other structures necessary to facilitate assembly.

15.2 Routing compatibility

Routing compatibility pertains to the layout and routing of a board. If the selection of a particular part causes significant layout and/or routing problems at the board level, the part may be rejected. Rejection of a part is usually based on its use of the routing resources associated with the board. Two routing issues must be considered: how much board area is required to connect (wire) the part to the rest of the parts and off-board connections (or alternatively, how many layers) and how many layers of the board are required to "escape route" the part.[2]

The methods discussed in this section are estimation methods intended to inform the user if the part is likely to cause problems or if the board is compatible with the part. These methods do not result in actual routing of the assembly.

15.2.1 Global routing

A part is always routable into the system given a sufficient number of layers. If the parts associated with the rest of the system are known, routing estimation techniques can be used to determine the effective routing-limited footprint of a part under the constraints posed by a particular set of interconnect design rules (lines, spaces, via/hole capture pad diameter) and layer count, as shown in Figure 15.5. If a candidate part exceeds the fraction of board wiring resources budgeted to it based on board size and cost constraints, it may be rejected.

The first step in global routing estimation is to choose an algorithm that predicts the wire length required by the bill of materials (BOM). The following algorithms can be considered for global routing estimation (see [4]–[6] for more details on the routing algorithms):

- *section-crossings* [7], [8] – Based on estimating the number of wires that cross any section of the substrate. Once the number of crossings is known, the width of the section can be determined from the number of wiring layers and the design rules for each layer (wire width and spacing). This algorithm assumes random placement of parts and therefore is usually useful for establishing an upper bound for wiring requirements.
- *comparison of requirements and resources* – An interconnect capacity-limited part footprint is formulated by setting the total wiring length required to connect all the chips equal to the wiring length available in the substrate. The critical element is the computation of the average interconnection length in the substrate. This length can be computed using three different algorithms:
  - Rent's Rule [9], [10] – Heuristic that relates the number of I/Os a chip has to the number of gates in the chips for different chip types. The heuristic has been hierarchically extended to relate the number of I/Os a board has to the number of chips on the board for different applications.
  - Geometric [11] – Geometric approach that approximates all interconnects as either global (visiting every chip) or nearest neighbor (connecting to only the nearest chip).
  - Hannemann [12] – Heuristic that relates a dimensionless constant (dependent on board dimensions, number of chips, number of I/Os in the system, and design rules) to the number of wiring layers in the substrate. Derived from actual results for DEC and IBM high-density machines of the early 1980s.
- *statistical wiring distribution* [13] – Derived from observations of wire length distributions in medium-scale integration systems implemented on printed wiring boards.

Any of the algorithms listed above can be used on a system once the BOM is known. However, each of the algorithms was formulated under a different set of assumptions and contains different calibration parameters. The accuracy of the algorithms for predicting the wire length required by an application depends on matching the algorithm to the application.

---

[2] Escape routing is applicable only if the part will be surface-mounted with an area array format connection to the board, i.e., a flip chip or BGA package.

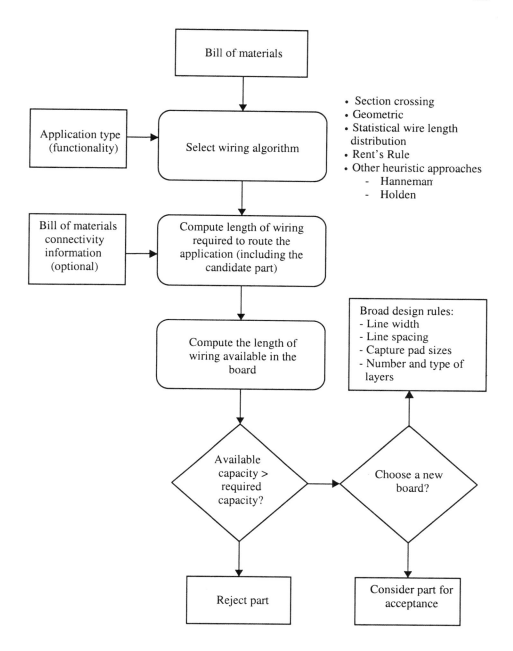

Figure 15.5: Process for performing a global routing assessment

By studying the predictions from the various algorithms for actual applications, a guide for algorithm usage and an estimation of the errors that can be expected can be constructed, as shown in Table 15.5.

In Table 15.5, "parallel" refers to the degree to which the interconnections between chips are organized and "flow" in a regular direction (e.g., a crossbar switch is highly parallel). The opposite of "parallel" is "random." Random wiring is highly disorganized; the interconnection of one I/O on a chip bears little resemblance to the next. Boards containing a microprocessor and associated secondary cache memory usually represent a balanced mixture of random and parallel interconnection.

Table 15.5: Global routing algorithm compatibility [4]

System type	Suggested algorithm	Expected error
Unknown	Rent's Rule	< 25%
Balanced mixture of random and parallel interconnection	Rent's Rule, geometric	< 7%
Highly parallel structures on high-density interconnection	Geometric	< 18%
Medium I/O count chips on PWB (I/O count < 100)	Statistical wiring distribution	< 5%
Low interconnect capacity per layer, many layers	Hannemann	< 10%

In most cases, calibration coefficients associated with each algorithm measure the degree of parallelism in the design.[3] Instead of providing an advisory on the calibration coefficients, we prefer to observe the wire length predictions as a function of the calibration coefficient and judge the stability of the solution from that result. To demonstrate this, consider an example case with 15 chips and 23 discrete parts. Using the chip I/O counts and system block diagram, the required routing for the system can be predicted using the algorithms discussed above. The application is a mix of random and parallel interconnections and contains relatively small (low I/O count) chips. These attributes suggest that we consider using the Rent's Rule algorithm or the statistical wiring length distribution approach. Figures 15.6 and 15.7 show the results from the use of these algorithms. In these figures, the range indicated contains the range of wire length values predicted by the Rent's Rule algorithm. In the example case, testing the two algorithms tells us that the Rent's Rule prediction is that the wire length is between 2040 and 2300 inches. Choosing the center of this range gives us an error of approximately ±6.5%.

15.2.2   Escape routing

If a part's I/Os are in an area array format (as opposed to a peripheral format), the part cannot be wired into the system until all of its I/Os are routed from the part.[4] This situation is encountered for flip chip bonded die and die in various array format chip scale packages such as BGAs.

The process of liberating I/Os from an array is called "escape routing." If the array is small enough (small I/O count) or the line width and line spacing in the board are fine

---

[3] See [4] or [5] for a detailed explanation of the calibration coefficients associated with the algorithms.
[4] Strictly speaking, not all the I/Os on a chip need to be routed. Some I/Os have no electrical connection (they are present only for thermal or mechanical reasons, and some I/Os are redundant and can be connected together).

enough, it may be possible to escape all the I/Os on the top layer of the board. However, more often, some I/Os must use vias to drop to other layers to be completely escape routed.

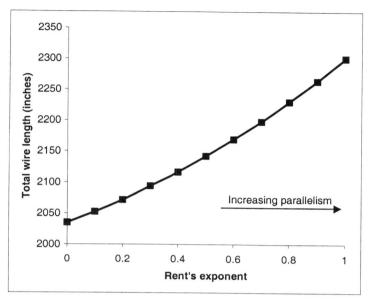

Figure 15.6: Rent's Rule wire length prediction as a function of Rent's exponent

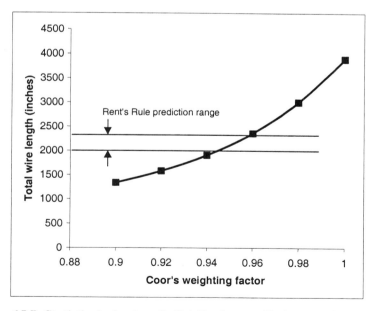

Figure 15.7: Statistical wire length distribution prediction as a function of Coor's weighting factor [13]

If additional layers must be used, the combination of the part I/O pitch and the board capture pad size must allow the placement of a via or hole capture pad on the board within the part I/O array; otherwise, it is impossible to drop unescaped I/Os to deeper layers. In addition, many board technologies, especially microvia printed wiring boards (PWBs), have practical limitations on the number of layers that can be fabricated. A part that requires too many layers to escape route may have to be rejected. A process for assessing the escape routing requirements for individual parts is shown in Figure 15.8. The strategy is to start on the layer where the candidate part is mounted and escape as many I/Os as possible away from the array. The I/Os that remain are dropped to the next layer, and the escape process is repeated until all the I/Os are exhausted or all the layers are exhausted, whichever comes first.

The simplest analytical model of escape routing treats only the number of I/Os that could escape through the periphery of the array [5]. This model geometrically determines the number of I/Os that can escape through the periphery of a full grid square array and assumes that every layer in the board has the same design rules.[5] There are three problems with the simple model. First, it does not consider the space needed to translate unescaped I/Os to deeper layers (i.e., the model does not account for space within the array for capture pads). Second, the algorithm was developed with the assumption that every layer in the board is identical (homogeneous line width, spacing, via/hole capture pad size). The third problem is that the model works only for full grid square arrays.[6]

To model the via/hole capture pad limitations, two capture pad issues have to be considered: (1) sufficient space has to be provided within the array to support capture pads, and (2) the maximum number of I/Os that can escape "inward" in a center-hollow array is limited by the layer with the lowest capture pad density (if via-in-pad technology is used, the capture pad limitations are not relevant).

Key elements of an escape routing model – The first step in any escape routing calculation is to determine the number of lines that can fit between bonds (top layer) or via/hole capture pads (buried layers), as shown in Figure 15.9. The number of lines that can pass between bond or capture pads on layer i is

$$n_i = \left\lfloor \frac{p_i - d_i - s_i}{s_i + w_i} \right\rfloor \tag{15.4}$$

where p_i is the center to center pitch of the I/O pads on the ith layer; d_i is the diameter of the I/O pads on the ith layer; s_i is the minimum space between metal features on the ith layer; and w_i is the line width on the ith layer. In the simplest case (a full square array, Figure 15.10), the number of I/Os that can be escaped on the ith layer is given by[7]

$$N_{i\text{out}} = \sum_{j=0}^{n_i} \left(N_{0i} - 8j \right) \tag{15.5}$$

---

[5] This model was later modified to account for congestion at the corners [14]; that is, if tracks cannot reach the periphery of the array, they cannot escape through the periphery even if sufficient space exists on the periphery.

[6] Several of the problems with the model in [5] have been addressed in [15]. This model accounts for design rules that vary by layer and includes a treatment of the capture pad limited part footprint.

[7] Equation (15.5) assumes that the number of spaces for escapes through the periphery of the array is less than the total number of I/Os left in the array (excluding the outermost row); that is, to use this relation, the following must

be true: $n_i N_{0i} \leq \left(\dfrac{N_{0i}}{4} - 1 \right)^2$.

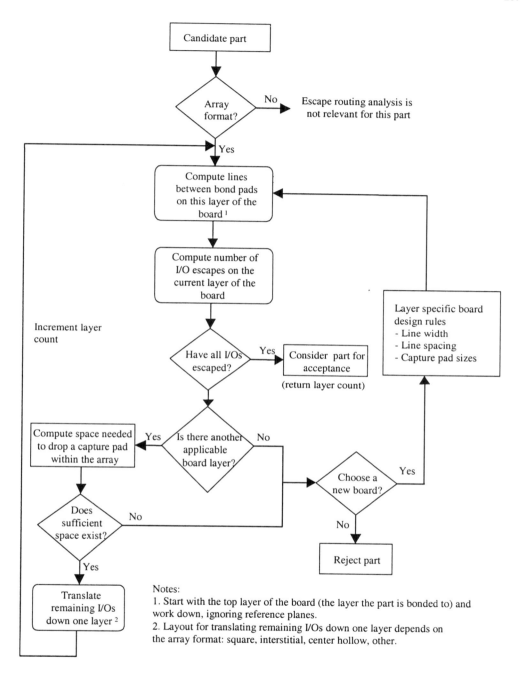

Figure 15.8: Process for escape routing assessment

where N_{0_i} is the number of I/Os in the outermost row on the ith layer and n_i is number of lines between bonds or capture pads on the ith layer given by eqn. (15.4). The estimate in eqn. (15.5) is conservative. Four additional I/Os can be escape routed in the $n_i = 1$ case than predicted by eqn. (15.5).

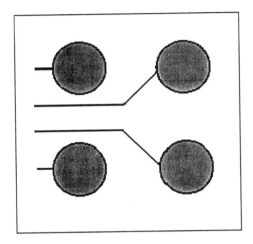

Figure 15.9: Two lines escaping between two capture pads or bonds out of the array to the left

Figure 15.10: Full grid square array examples [16]

The amount of space needed to translate unescaped I/O to the next layer is evaluated by determining if a via or hole capture pad can fit within the array. Figure 15.11 shows the layout that must be satisfied in order to translate an I/O to the next layer.[8]

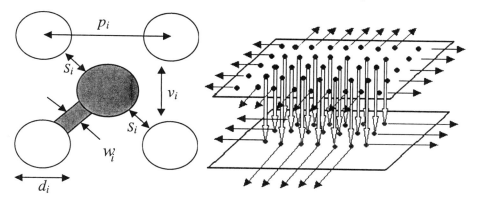

Figure 15.11: Translation of unescaped I/Os to the next layer of the substrate

Unescaped I/Os can be translated to the next layer in the substrate if the following inequality is satisfied:

$$\sqrt{2}\,p_i \geq \left(v_i + 2s_i + d_i\right) \tag{15.6}$$

where v_i is the via capture pad diameter. The last element of the analysis for the square array is to figure out the array size after the unescaped I/Os have been translated to the next layer in the substrate (i.e., the N_0 for the next layer).

$$N_{0_{i+1}} = N_{0_i} - 8(n_i + 1) \tag{15.7}$$

This relation is valid if escape routing was not completed on the ith layer, and it will never predict an array size of 1, i.e., if N_0 for the next layer is predicted to be exactly 0 by eqn. (15.7), one I/O may be present. The escape routing process is completed when the total number of I/Os escape routed is greater than or equal to the total number of I/Os on the part.[9]

To illustrate the concepts above, consider the following example of a full grid square array: 388 mil square package with 256 I/Os in a 16 x 16 grid, 13 mil ball diameter, 20 mil pitch (Tessera microBGA). In this case we wish to determine if this part is compatible with the Dyconex microvia board shown in Figure 15.12.

The first step in the process is to determine the number of lines between bond pads on the top layer of the board in Figure 15.12. Solving eqn. (15.4) for n using $p_i = 20$ mils 0, $d_i =$

---

[8] Figure 15.11 shows the most common situation ("dogbone" layout). Some substrate technologies allow "via in pad," which means that the via capture pad can be directly under a bond pad. Similarly, some substrate technologies allow buried via stacking. In both of these cases, the limitation shown in Figure 15.11 is conservative.

[9] In some cases extra I/Os are included on array parts for reliability and/or mechanical integrity purposes. These extra I/Os may not be electrically active and need not be escape routed. In general, this has a significant effect on the escape routing analysis only if the extra I/Os are located in central portion of the array.

13 mils, $s_i = 3$ mils, and $W_i = 3$ mils, we get $n_i = 0$. Using $N_{o_i} = 60$ and $n_i = 0$ in eqn. (15.5), the number of I/Os than can escape on the first layer is $N_i = 60$. Since there are 196 more I/Os to be escape routed we must drop the remaining I/Os to the next layer of the board using vias. Using $v_i = 7$ mils we find that the inequality in eqn. (15.6) is satisfied, so the remaining I/Os can be translated to the second layer of the board. Following the identical process for layers 2 and 3 yields the results shown in Table 15.6.

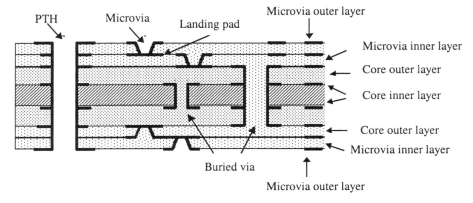

Figure 15.12: Microvia board for escape routing example [17]

From the last column in Table 15.6, if the capture pad size for THs in the core of the board is less than or equal to 16.3 mils, there is space to place the four necessary capture pads within the array and translate to the next layer; however, if the capture pad size for the THs in the core of the board is greater than 16.3 mils (it is 18 mils in this example), the simple analysis performed here concludes that it will be impossible to escape route this part on this board.

So what does the result for the example shown in Table 15.6 mean? It means that there might be a problem and it is necessary to take a more careful look at the part and the substrate. The analysis outlined in this section is intentionally conservative and intended only to alert persons performing the analysis (considering a candidate part) that there could be a problem. For the example considered, completion of the escape routing is "close" (i.e., we are not off by 30 I/O, only 4), and several factors could mitigate a problem of this magnitude:

1. Maybe not all the I/Os need to be escaped (some could be no connects (NCs) used to simply fill out the array or for thermomechanical purposes). Depending on where in the array the NCs are, they may ease the escape routing problem (note: NCs in the outermost row are obviously not going to help the escape routing problem).
2. Some of the I/Os may be power and ground that could be connected together and may not need to be escape routed individually.

In many cases the array of I/Os on the part is arranged on a square grid, but with the center of the grid empty (see Figure 15.13). In this case we must consider not only the I/Os that escape outward from the array, but also those that can escape inward, as seen in Figure 15.14.

Table 15.6: Example escape routing results: the box indicates the final outcome of the escape routing analysis for this example

Parameter	Layer 1 – Microvia outer layer	Layer 2 Microvia inner layer	Layer 3 – Core outer layer
p (mils) – center-to-center pitch	20	20	20
d (mils) - I/O pad diameter	13	7	7
v (mils) - via capture pad dia.	7	7	18
w (mils) - wire width	3	2.5	2.5
s (mils) – space	3	2.5	2.5
n – lines between pads (eqn.15.4)	0	2	2
N_0 – I/O in outermost row (eqn. 15.5)	60	52	28
Total I/Os escaped on this layer (15.5)	60	132	60
I/Os remaining to be escaped	196	64	4
I/Os dropped to next layer	196	64	0

Figure 15.13: Center hollow or peripheral square array [18]

The number of I/Os that can be escaped toward the center ("in") is given by:

$$N_{in} = N_I + nN_I \qquad (15.8)$$

where N_I is the number of I/Os in the innermost row and n is given by eqn. (15.4). Note that eqn. (15.8) is not conservative; it is the best case. The number of I/Os in the original innermost row of a center-hollow array is given by:

$$N_I = N_0 - 8(R-1) \qquad (15.7)$$

where R is the number of rows (e.g., $R = 6$ in Figure 15.13).

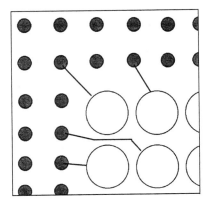

Figure 15.14: I/Os escaping inward in a center hollow array

Equation (15.8) cannot be blindly applied to the array because I/Os routed into the hollow area must go somewhere. A critical limitation that must be imposed on N_{in} stems from the fact that any I/Os that escape inward must find vertical connections (vias or holes) within the area of the center of the array. A vertical connection must exist for each of the I/Os that have escaped inwardly on the current layer and all previous layers within the original center hollow area of the part in all remaining layers of the substrate between the current layer and the last layer needed to finish the escape routing. Therefore the layer with the largest via/hole capture pad design rule determines the quantity that can be escaped inward. Routing into the array of vias/holes on the limiting layer (another escape routing problem, only in reverse) is also necessary. This limitation is relevant for microvia PWBs that have high-density wiring layers are fabricated on coarse PWB cores that have large THs (see Figure 15.15).

To quantify the inward escape routing limitation, we start with $N_{in_{TOTAL}}$, the total number of I/Os that can be routed to the center. The total number of I/Os in the outermost row of the array that must be created in the hollow area is given by:

$$N_{I_O} = 4\left(\sqrt{N_{in_{TOTAL}}} - 1\right) \qquad (15.10)$$

where we have assumed that $N_{in_{TOTAL}}$ is a perfect square (i.e., there is no remainder when we take the square root).

The maximum number of I/Os that can be removed by this new array is given by:

$$N_{\text{in}} = 4(n+1)\left(\sqrt{N_{\text{in}_{\text{TOTAL}}}} - 1\right) \tag{15.11}$$

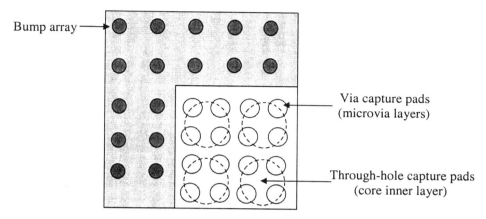

Figure 15.15: Inward escape routing limit

To solve for the largest square array that we can route into in one layer, we must construct the following inequality out of eqn. (15.11):

$$N_{\text{in}_{\text{TOTAL}}} \leq 4(n+1)\left(\sqrt{N_{\text{in}_{\text{TOTAL}}}} - 1\right) \tag{15.12}$$

where n is given by eqn. (15.4) for the new array. Solving eqn. (15.12) for $N_{\text{in}_{\text{TOTAL}}}$ we obtain:

$$N_{\text{in}_{\text{TOTAL}}} = \text{Largest Perfect Square}\left\{\left[2(n+1)+2\sqrt{n^2+n}\right]^2\right\} \tag{15.13}$$

A secondary limitation on the new center array defined by eqn. (15.13) is that it must fit within the center-hollow area, or

$$\frac{A_{\text{Center Hollow}}}{A_{\text{One Via/Hole}}} = \text{Maximum number of vias/holes that will fit} \tag{15.14}$$

The effective area occupied by a single via/hole is given by:

$$A_{\text{One Via/Hole}} = [d_{\text{v/h}} + s + n(w+s)]^2 \tag{15.15}$$

where $d_{\text{v/h}}$ is the diameter of the via/hole capture pad; n is the number of lines between capture pads; s is the minimum space between metal features; and w is the minimum line width.

From eqn. (15.14),

$$N_{\text{in}_{\text{TOTAL}}} = \text{Largest Perfect Square} \left\{ \frac{A_{\text{Center Hollow}}}{[d_{\text{v/h}} + s + n(w + s)]^2} \right\} \qquad (15.16)$$

Equations (15.13) and (15.16) are the two limitations that must be imposed when routing I/Os to the center of an array. The solution strategy is to compute the largest value of $N_{\text{in}_{\text{TOTAL}}}$ from eqn. (15.16) that results in an $N_{\text{in}_{\text{TOTAL}}}$ from (15.13) that is equal or larger.

As an example, consider a center hollow square array shown in Figure 15.15: 220 I/O in a 16 x 16 grid (with 12 NCs in the innermost row), $R = 5$, 20 mil pitch. In this case we wish to determine if this part is compatible with the board shown in Table 15.7. Table 15.7 shows that the problem cannot be escape routed only outward (only 192 I/Os out of 220 − 12 = 208 have escaped when we finish with the third layer).

Table 15.7: Escape routing example results (all dimensions in mils)

Parameter	Layer 1 – Microvia outer layer	Layer 2 – Microvia inner layer	Layer 3 – Core outer layer
p (microns)	20	20	20
d (entering)	13	7	7
v (leaving)	7	7	16
W	3	3	4
S	3	3	5
n	0	1	0
N_O	60	52	36
I/Os escaped outward on this layer	60	96	36
Total I/Os escaped outward	60	156	192

To assess the I/Os escaping into the center hollow, there are two cases to consider:

1. Escape routing cannot be finished before the PWB core (Layer 3) is reached.
2. Escape routing can be finished before the PWB core (Layer 3) is reached.

In Case 1, whatever is sent to the center hollow has to get through the core (Layer 3). Construct Table 15.8 assuming that $A_{\text{Center Hollow}} = [6(20 \text{ mils})]^2 = 14,400 \text{ mils}^2$. We are constrained by the minimum nearest perfect square for a given n, so the best case is $n = 1$, meaning that the total number of I/Os that can be escape routed inward is 16. This total plus those that can be escaped outward from Table 15.7 make 208, which complete the solution.

In Case 2, one finishes before layer 3. This cannot be done in this example, but in any case, let's construct Table 15.7 again (Table 15.9), assuming that $A_{\text{Center Hollow}} = [6(20 \text{ mils})]^2$

= 14,400 mils^2. We are constrained by the minimum nearest perfect square for a given n, so the best case is $n = 1$, meaning that the total number of I/Os that can be escape routed inward is 36. This total plus those that can be escaped outward from Table 15.7 in the second layer make $156 + 36 = 192$, which is not enough to find a solution. As a result, the third layer is necessary and Case 1 is the relevant case.

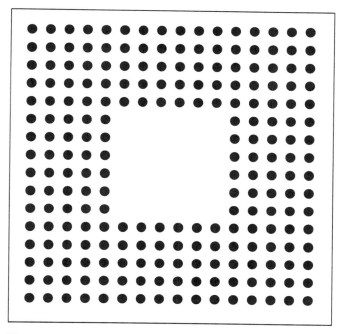

Figure 15.16: Microvia board for escape routing example

Table 15.8: Inward escape routing table (Case 1): best case is shown

N	$N_{in_{TOTAL}}$ from (eqn. 15.13)	Nearest perfect square	$N_{in_{TOTAL}}$ from (eqn. 15.16)	Nearest perfect square
0	4	4	32.65	25
1	46.63	36	16	16
2	118.79	100	9.47	9
3	222.85	196	6.25	4
4	358.89	324	4.43	4

Table 15.9: Inward escape routing table (Case 2): best case is shown

N	$N_{\text{in}_{\text{TOTAL}}}$ from (eqn. 15.13)	Nearest perfect square	$N_{\text{in}_{\text{TOTAL}}}$ from (eqn. 15.16)	Nearest perfect square
0	4	4	144	144
1	46.63	36	56.25	49
2	118.79	100	29.75	25
3	222.85	196	18.37	16
4	358.89	324	12.45	9

Footprint Analysis – One by-product of the escape routing analysis that has been developed in [15] is footprint analysis on the top layer of the substrate. The objective of footprint analysis is to determine the footprint area occupied by the part on the topmost layer of the board (Figure 15.17).

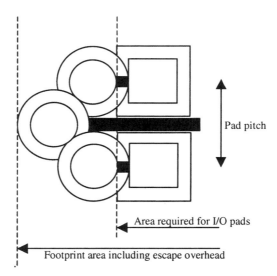

Figure 15.17: Overhead for single line escape [15]

This analysis results in a value for A_c described in the spatial compatibility section (Figure 15.3). The footprint of the package is given by:

$$A_c = \text{Area}_1 + \text{Max}(\text{Area}_2, \text{Area}_3) \tag{15.17}$$

$$\text{Area}_2 = \left(\sqrt{\text{Area}_1} + s\right)^2$$

where Area$_1$ is the part area, s is the minimum assembly spacing between adjacent parts, and Area$_3$ is the area on the top layer needed to escape route the part.

For area array format parts, software tools that perform the escape routing are plentiful. The difficulty is that these commercial tools cannot be used prior to the completion of a detailed layout, an activity that does not take place until long after parts selection has been completed. Therefore, the existing tools are not useful for obtaining early feedback on the routability of candidate parts concurrent with the parts selection process.

15.3 Test and rework acceptability

In some cases, parts must be tested prior to assembly. The cost of testing a part is related to the level of testing performed by the part manufacturer, whether the part is a package or bare die, the function that the part performs, the number of gates and/or bits contained by the part, and the test equipment. If the part does not come from the manufacturer fully tested (e.g., a bare die), then test costs include the cost of creating the test patterns (automatic test pattern generation plus manual test pattern generation) and the cost of applying the test to the part [19]. Predicting testing costs is of little value unless the corresponding fault coverage (fraction of defects detected by the test) is also predicted.

Devices may be tested at the assembly level, but then the cost of rework must be included. The cost of removing a defective part is a function of how the part is mounted to the board, the size of the part, and its proximity to other parts.

In the test/diagnosis/rework process of Figure 15.18, Trichy et al. [20] show that all of the product coming from upstream production is tested. The product that does not pass the test can be either scrapped (disposed of), salvaged (all or part of the product is recovered for reuse in the same or another product), or recycled (broken down to its constituent materials), or the nonpassing product can be reworked to repair the defect.

The first activity that takes place after a product is failed by a test activity is to determine why it failed. This activity is called "diagnosis." Once diagnosis is completed, a decision can be made as to whether a particular product should be reworked, i.e., repaired and sent back into the test, or scrapped. However, diagnosis and rework are not perfect. They can introduce defects, make misdiagnoses, and fail to rework defective products correctly; therefore, a product may go through test, diagnosis, and rework multiple times.

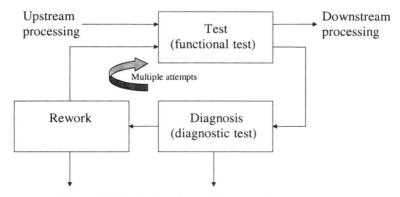

Figure 15.18: A simple test/diagnosis/rework process

15.4 References

[1]   Lindell, P., P. Stoaks, P. Sandborn and D. Carey, "Virtual Prototyping of Electronic Systems," *IEEE Transactions on Components, Packaging, and Manufacturing Technology,* Part A, Vol. 21, Issue 4, pp. 610–616, December 1998. g

[2]   Prasad, R. P., *Surface Mount Technology: Principles and Practice*, 2nd. ed., Kluwer Academic Publishers, New York, NY, 1997.

[3]   Hume, H., R. Komm and T. Garrison, "Design Report Card: A Method for Measuring Design for Manufacturability," *Proceedings of the Surface Mount International Conference*, pp. 986–991, September 1992.

[4]   Sandborn, P. and P. Spletter, "A Comparison of Routing Estimation Methods for Microelectronic Modules," *Microelectronics International*, Vol. 17, No 1, pp. 36–41, December 1999.

[5]   Sandborn, P. and H. Moreno, *Conceptual Design of Multichip Modules and Systems*, Kluwer Academic Publishers, Boston, MA, 1994.

[6]   Pecht, M., *Placement and Routing of Electronic Modules*, Marcel Dekker, New York, NY, 1993.

[7]   Sutherland, I. E. and D. Oestreicher, "How Big Should a Printed Circuit Board Be?" *IEEE Transactions on Computers*, Vol. C-22, pp. 537–542, May 1973.

[8]   Schmidt, D. C., "Electronic System Topology and Design," *Physical Architecture of VLSI Systems*, John Wiley & Sons, Inc., pp. 411–460, New York, NY, 1994.

[9]   Landman, B. S. and R. L. Russo, "On a Pin versus Block Relationship for Partitions of Logic Graphs," *IEEE Transactions on Computers*, Vol. C-20, pp. 1469–1479, December 1971.

[10]   Bakoglu, H., *Circuits, Interconnections and Packaging for VLSI*, Addison-Wesley, Boston, MA, 1990.

[11]   Moresco, L., "Electronic System Packaging: The Search for Manufacturing the Optimum in a Sea of Constraints," *IEEE Tranactions on Components, Hybrids, and Manufacturing Technology*, Vol. 11, pp. 494–508, September 1990.

[12]   Hannemann, R., "Physical Technology for VLSI Systems," *Proceedings of the ICCD*, pp. 48–53, 1986.

[13]   Coors, G., P. Anderson and L. Seward, "A Statistical Approach to Wiring Requirements," *Proceedings of the IEPS*, pp. 774–783, 1990.

[14]   Darnauer, J. and W. Dai, "Fast Pad Redistribution from Periphery-IO to Area-IO," *Proceedings of the IEEE Multi-Chip Module Conference*, pp. 38–43, March 1994.

[15]   Hirt, E., and G. Tröster, "Early footprint comparison for Area I/O Packages and first level interconnect," *Proceedings Electronic Components and Technology Conference (ECTC'99)*, pp. 1210–1216, June 1–4, 1999.

[16]   IBM, Inc., <http://www.chips.ibm.com/products/interconnect/products/bga.html>, accessed September 20, 2001.

[17] Dyconex, "Design Guidelines Dycostrate-C," <http://www.dyconex.ch/html/technologies_products/design_d.htm>, accessed July 1, 2001.

[18] Maxtek, Inc., Technologies, Packaging Technologies-Ball Grid Array, <http://www.maxtek.com/technologies/packaging-ballgrid.shtml>, accessed September 20, 2001.

[19] Dislis, C., J. H. Dick, I. D. Dear and A. P. Ambler, *Test Economics and Design for Testability*, Ellis Horwood, New York, NY, 1995.

[20] Trichy, T., P. Sandborn, R. Raghavan and S. Sahasrabudhe, "A New Test/Diagnosis/Rework Model for Use in Technical Cost Modeling of Electronic Systems Assembly," *Proceedings of the International Test Conference*, pp. 1108–1117, November 2001.

Chapter 16

Obsolescence Prediction and Management

Michael Pecht, Rajeev Solomon, Peter Sandborn, Chris Wilkinson, and Diganta Das

The rapid growth of the electronics industry has spurred a dramatic change in the manufacture of the products and systems we buy. New parts are constantly being introduced with increased speed, reduced feature size, and supply voltage and leading-edge interconnection and packaging technologies. As a result, products and systems that use these parts are exploiting these capabilities to provide product differentiation and to increase demand and sales. Unfortunately, for some product sectors that require a long useful life, the electronic parts that compose the product have a life cycle that is significantly shorter than the life cycle of the product. There are a growing segment of products in which, there is a life cycle mismatch between the parts and the product [1], [2].

A private passenger car has a typical life of 10 years, and thus replacement parts are required to be available for this period. The engine controller manufacturer depends on the same suppliers as the high-volume industries such as the personal computer industry, where the life cycle is typically 2 to 3 years and the underlying technology of the parts used has a similar life cycle. Part technology moves on during this period, preparing for the next high-volume product, and parts manufacturers transition their manufacturing facilities to service the higher-volume products. This leaves the equipment without a manufacturing base, thus rendering it obsolete. As an example, consider the engine control module used in a car. Many similar examples can be found in other industries such as defense electronics, avionics, and industrial controls.

The life-cycle mismatch problem requires that engineers be cognizant of which parts will be available and obsolete when the product is to be manufactured and supported. This is especially important for engineers in the automotive, avionics, and defense industries or other industries where there is a long-term warranty obligation. These systems may encounter obsolescence problems before being fielded and invariably experience obsolescence problems during field life [3]. Next-generation parts with improved performance characteristics must be anticipated in the design to ensure that circuit timing, noise margins, and EMI nonconformances do not suddenly arise at the product level. Parts substituted into the design later may require assembly processes to be upgraded. If the product requires a long application life, then an open architecture or a parts obsolescence strategy, such as preventive redesign, lifetime buy, aftermarket purchases, or part substitution may be required.

If a product requires a long service life, then an open architecture and a parts obsolescence management strategy may be beneficial. Many obsolescence mitigation approaches have been proposed and are being used. These include lifetime or last-time buys (buying and storing enough parts to meet the system's forecasted lifetime requirements or requirements until a redesign is possible), part substitution (using a different part with identical or similar form, fit, and function), and redesign (upgrading the system to make use of substitute parts) [4]. Several other mitigation approaches are also practical in some situations. For example, "aftermarket sources" (third parties that continue to provide the part after its manufacturer has made it obsolete), "emulation" (using parts with identical form, fit, and function that are fabricated using ASIC-type technologies), "reclaim" (using parts salvaged from other products), and "uprating" may be used.

Uprating is a process to assess the ability of a part to meet the functionality and performance requirements of the application in which the part is used beyond the manufacturer's specifications, usually at a wider temperature range than rated by the manufacturer [5]. Uprating is a mitigation approach because the obsolete part is often the "MIL-SPEC" part while the commercial version of the part continues to be available. In some cases, the best obsolescence mitigation approach for manufacturers who need a broader environmental range part (often automotive, avionics and military) is to uprate the commercial version of the part.

Other works have concentrated on understanding the product life cycle in terms of factors including product life-cycle stages, product life, extension of product life, and product marketing issues [6]. The factor of obsolescence is not dwelt upon, but in the case of products, obsolescence may not be an issue, depending on the definition of a product. For example, if a company's product is a subassembly, obsolescence of a critical part, may affect the end product's life. Part selection needs more critical attention, as the root cause of obsolescence at any product level is the obsolescence of a part.

This book reviews life-cycle stages and then presents a methodology for forecasting the years to obsolescence for electronic parts. The prediction of obsolescence enables engineers to manage more effectively the introduction and ongoing use of long-field-life products based on the projected life cycle of the parts. The obsolescence prediction methodology is a critical element of risk-informed parts selection and management processes. Engineers must be aware of part life cycles; otherwise, they can end up with a product whose parts are not available, which cannot perform as intended, cannot be assembled, and cannot be maintained without high life-cycle costs. While technological advances continue to fuel product development, engineering decisions regarding when and how a new part will be used and the associated risks traded off with a new part and technology differentiate winning from losing products.

16.1 Life-cycle stages

Most electronic parts pass through several life cycle stages corresponding to changes in part sales. Figure 16.1 is a representative life-cycle curve of units shipped per time, which depicts the six common part life-cycle stages: introduction, growth, maturity, decline, phaseout, and obsolescence [8].[1] Table 16.1 shows typical life cycle characteristics of the six phases.

16.1.1 Introduction stage

Electronic parts introduced into the market may be either revolutionary or evolutionary. A part is considered revolutionary if it carries out a new function or if it uses a new semiconductor or packaging technology. Generally, a part evolutionary is considered if it is solely an improved version of some existing part, perhaps with added features or enhancements. The first 64Kbit dynamic random access memory (DRAM) from IBM Microelectronics was an example of a revolutionary part. The current 64Mbit DRAM is an example of an evolutionary part.

---

[1] Several additional phases have been proposed including introduction, pending (prior to introduction), and splitting the obsolescence stage into last shipment and discontinued or purged. The word "discontinuance" is sometimes used as an equivalent of "obsolescence." The differences between the two are explained in Section 16.1.6.

The introduction stage in the part life cycle is usually characterized by high production costs because of low yield, frequent modifications, low production volumes, and lack of specialized production equipment. Marketing costs at this stage may also be high. Customers who buy a part in its introductory stage tend to value performance over price.

16.1.2   Growth stage

The growth stage is characterized by the part's market acceptance. Increased sales during this stage may justify the use of specialized equipment for production, which in turn improves economies of scale of production. Mass production, mass distribution, and mass marketing often bring about price reductions. This stage often involves the largest number of competitors, as opportunity-seeking firms are attracted by the part's profit potential.

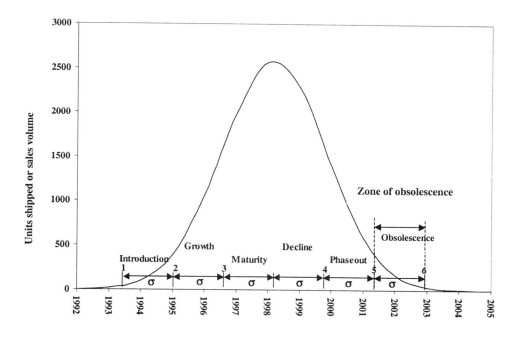

Figure 16.1: A standardized life cycle curve [7]

16.1.3   Maturity stage

The maturity stage of the part life cycle is characterized by high-volume sales. Competitors with a lower cost of production may enter the market, or domestic competitors may shift production facilities to less expensive locations to enable them to lower manufacturing costs. The Pentium processors are examples of mature parts. There are several other microprocessors with similar architecture and functionality available from other manufacturers, which are generally sold at lower prices.

16.1.4   Decline stage

Decreasing demand and generally decreasing profit margins characterize the decline stage. Towards the end of the decline stage, only a few specialized manufacturers generally remain in the market. TTL logic ICs are examples of parts that have continued to be available in this stage due to continued sales in the black-and-white television market.

Table 16.1: Typical life cycle characteristics of the six generic stages of part life cycle [7]

Characteristic	Life cycle stage					
	Introduction	Growth	Maturity	Decline	Phaseout	Obsolescence
Sales	Slow but increasing	Increasing rapidly	High	Decreasing	Lifetime buys may be offered	Sales only from aftermarket sources, if at all
Price	Highest	Declining	Low	Lowest	Low	Not applicable or very high if available from aftermarket sources
Usage	Low	Increasing	High	Decreasing	Decreasing	Low
Part modification	Periodic die shrinks and possible mask changes	Periodic die shrinks	Periodic die shrinks	Few or none	None	None
Competitors	Few	High	High	Declining	Declining	Few
Manufacturer's profit	Low	Increasing	High	Decreasing	Decreasing	Decreasing

16.1.5   Phaseout stage

Phaseout occurs when the manufacturer decides on a date to stop production of the part. Generally, the manufacturer issues a discontinuance notice to customers, provides a last time buy date, and suggests alternative parts or aftermarket manufacturers. As an example, on September 2, 1999, Texas Instruments, Standard Linear, and Logic Group announced the discontinuance of ULN2803A, a Darlington transistor array. Texas Instruments stated that the product would be discontinued on September 2, 2000 [9], with the last (and noncancelable) order date being March 2, 2000.

16.1.6   Discontinuance and obsolescence

Discontinuance occurs when the manufacturer stops production of the part. The part may still be available in the market if the production line or part stocks were bought by an aftermarket source.

A part is obsolete when the technology that defines the part is no longer implemented. Thus, obsolescence occurs at a technology level, while discontinuance occurs at a part number or manufacturer specific level. Diode transistor logic (DTL) and resistor transistor logic (RTL) parts are examples of obsolete part technologies. The National Semiconductor Military Quad SPST JFET analog switch in a ceramic DIP package is a discontinued part. The last-time buy date for this part was December 7, 1999 [10]. A nonmilitary part with the same functionality and technology remains available from National Semiconductor. In this case, the military part is discontinued but the technology is not obsolete.

16.1.7   Special cases of the life-cycle curve

Not all parts conform to the six life-cycle stages presented here. Some parts undergo a false start and die out or may be associated with a niche market. Some parts may also be revitalized after the decline stage. Other possibilities can also arise due to various economic, social, and environmental conditions.

A false start typically suggests that a part starts out with a period of strong growth, only to stall because of one or more of the following factors:

- introduction of a superior competing part
- improvement of a competing part
- identification of a problem associated with the part
- failure to reach the critical mass that allows economies of scale to be realized
- loss of a unique and compelling application for the part

Niche parts generally have some unique applications and thus have a relatively low but steady sales level. An example is GaAs ICs, which have found a niche market in telecommunications, military, and space applications.

Decline can often be delayed or reversed by revitalizing the original part. Defining new market segments, defining new applications, or creating a new image for the part and thereby increasing the demand can cause revitalization.

16.2   Life-cycle forecasting methodology

Traditional methods of life-cycle forecasting include the "scorecard" approach and an availability factor method. In the scorecard approach, the life-cycle stage of the part is determined from an array of technological attributes, and each attribute is given a life-cycle code ranging from 1 through 6 and a corresponding weight. The overall stage for the part is determined by computing a simple weighted average of the life-cycle codes for the attributes. The disadvantage of this approach is that it does not capture market trends accurately because it commonly relies on unquantifiable, technological attributes such as technology complexity and soft market attributes such as usage. This approach also makes the erroneous assumptions that all ICs follow the same life-cycle curve[2] and that all life-cycle stages are of the same length.

---

[2] For example, consider the vastly different life-cycle curves of an op-amp and a DRAM. A DRAM may be in a mature stage that may have only two years until end of life, compared to a mature op-amp that may have ten years until end of life.

The "availability factor" method projects a safe usage window for a part. This approach uses market and technology factors to predict the obsolescence of parts with similar technology and market characteristics. This approach does not use the "life-cycle curve" for the part and cannot determine the life-cycle stage of the part.

This chapter describes a new approach to predicting life-cycle stage and years to obsolescence that solves the problem inherent in the other methods. This approach is based on modeling the life-cycle curve for both parts and technology attributes. Figure 16.2 outlines the methodology for life-cycle forecasting.

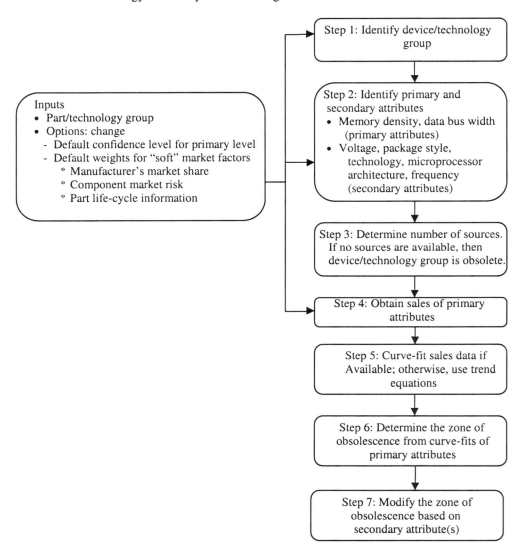

Figure 16.2: The life-cycle forecasting methodology [7]

16.2.1   Step 1: Identify part/technology group

The first step involves the identification of the part/technology group. A part/technology group is a family of parts that share common technological and functional characteristics but may be produced by different manufacturers. For example, a part/technology group is the set of all 16 Mb, 5 V, SOP, EDO (technology characteristics) DRAMs (part functional characteristic) regardless of its manufacturer (e.g., Samsung, Micron, or Hyundai).

16.2.2   Step 2: Identify part primary and secondary attributes

The second step involves identification of the primary and secondary attributes. A primary attribute is a characteristic that defines a part/technology group. For example, the primary attribute of a memory part is memory density. A secondary attribute is a characteristic of the part/technology group that can modify the range for years to obsolescence and/or the life-cycle stage of a part/technology group. For example, secondary attributes for a memory part include package style and supply voltage. The interested reader is referred to [1] of tables on the primary and secondary attributes of major part classes.

16.2.3   Step 3: Determine the number of sources

This step involves determining the number of sources for the part. If no sources can be found, the part is either already obsolete or has not yet been manufactured.

16.2.4   Step 4: Obtain sales data on the primary attribute

In this step, sales data on the primary attribute are collected. Sales data is a direct indicator of the life cycle of the part/technology group. The sales data may be in the form of number of units shipped. If unit sales data are not available, sales in market dollars or percentage market share may be used if the total market does not increase appreciably over time. These sales data are available from market research organizations.

16.2.5   Step 5: Construct a profile and determine parameters

The life-cycle profile is constructed by fitting the sales data of the primary attribute, a characteristic that is unique to, or defines, the part/technology group, to a life-cycle forecasting distribution. Each life-cycle phase is defined in terms of its distance from the mean (μ) measured in standard deviations (σ) when the life-cycle curve is fit with a Gaussian form. The zone of obsolescence is defined as the ordered pair: ($\mu + 2.5\sigma - p$, $\mu + 3.5\sigma - p$), where p is the date of assessment. This ordered pair gives the number of years from the present to the beginning and end of the window of obsolescence.

Gaussian distributions have been used by the Electronic Industries Association (EIA) as their standardized product life-cycle (PLC) curve, and hence are well known and familiar to the equipment manufacturers [11]. The equation of the life-cycle curve is:

$$f(x) = ke^{-\frac{(x-\mu)^2}{2\sigma^2}} \tag{16.1}$$

The function $f(x)$ is a trend equation that gives the relationship of the life-cycle profile parameters (μ and σ for Gaussian distributions) to the primary attribute. Function $f(x)$ is defined by the mean μ, the point in time of the sales peak of the curve, the standard deviation σ, and the year x. The factor k is the sales peak, the number of units shipped, or the percentage demand.

16.2.6   Step 6: Determine the zone of obsolescence

The zone of obsolescence is a period of time in which the part/technology group has a high probability of becoming obsolete. The ordered pair (μ + 2.5σ − p, μ + 3.5σ - p) gives the zone of obsolescence, where p is the date of assessment.   Life-cycle stages are determined by dividing the life-cycle curve for the primary attribute into introduction (μ − 3σ, μ − 2σ), growth (μ − 2σ, μ − σ), maturity (μ − σ, μ + σ), decline (μ + σ, μ + 2σ), and phaseout (μ + 2σ, μ + 3σ).

16.2.7   Step 7: Modify the zone of obsolescence

The life-cycle profile of the part/technology group may require modification by secondary attributes. If the years to obsolescence for any of the secondary attributes fall within the life span ($\pm$ 3σ years) of the main attribute, the years to obsolescence for the generic part will be modified.

16.3   Obsolescence management tactics and strategies

This section discusses the tactics and strategies that can be employed to combat and reduce the adverse effects of obsolescence. Strategies are longer-term efforts to try and reduce the time and cost required to resolve an obsolescence problem. Tactics are the short-term solutions adopted in a particular case. Depending on factors such as system life, period of production, and volume of production, the equipment manufacturer must decide which strategy to use to best handle part obsolescence issues [12]−[19].

Component manufacturing companies issue a product discontinuance notice when they decide to cease production of a part or replace it with a new part that is sufficiently different to warrant a new part number. The timing and content of this notice are of vital importance to the equipment manufacturer, since in many cases the window of opportunity that the equipment manufacturer has to take corrective action is defined by the contents of this notice. The key contents of such notices are the part numbers affected, the effective date of the notice, the last date on which orders for the part will be accepted, and the last date on which shipments of parts will be made. Sometimes part manufacturers will provide suggestions for replacement parts, but these are often not direct form, fit, functionality (FFF) replacement parts and may require some reengineering to incorporate. At the very least, the bill of materials, production drawings, and maintenance manuals have to be updated. The notice may also indicate if the company has entered into any agreements with "sunset suppliers" to take the remaining parts stock or transfer manufacturing information and tooling for continued manufacture.

Many different solutions are available in reacting to an obsolescence problem. These range in complexity from a simple part substitution to a major redesign of the product. The selection of the most appropriate solution is a complex task driven by a large number of factors such as the time available, the expected future production and support lifetime, and the expected occurrence of future product developments. The various strategies available and the driving factors are discussed in the following sections.

16.3.1   Part substitution

Part substitution is the process of selecting a part which is a direct pin-for-pin replacement of an obsolete part. The substitute part may or may not be physically different. Substitution includes FFF replacements of the obsolete part and repackaging a modern solid-state component in the same or equivalent package type as the obsolete component [20].

One method of reducing the obsolescence risk is to select parts in the product design phase that have readily available FFF substitutes. Often manufacturers release lists of FFF replacements for parts they plan to discontinue. A cross-reference and comparison of original versus substitute part characteristics must be conducted, and a waiver is generally required to support the change since it may require relaxing part specifications or performance parameters.

A part characteristic profile should be developed through collection and analysis of engineering, logistics, and technical information during step 1 of the obsolete part analysis. Additional information should be compiled using the manufacturer's documentation. Identification of all critical parameters and characteristics should be carried out to allow comprehensive evaluation of substitute candidates. The steps in selecting the part substitution process are as follows:

1. Component databases and manufacturer databooks and websites should be researched to develop a list of potential substitutes for the obsolete part. A list of parts that most closely resemble the obsolete part should be compiled.
2. For each substitute candidate, an analysis should be conducted to determine whether the substitute part matches required functionality specifications and is a reasonable form and fit replacement for the obsolete part. Characteristic data for each substitute candidate part should be recorded and compared with the obsolete part baseline profiles. It should be determined whether electrical performance parameters remain unchanged at the circuit interface where the substitute part is being used.
3. When researching potential substitutes, attempts should be made to cross-reference obsolete part manufacturing and functional technology to state-of-the-art product specifications in order to minimize the obsolescence life-cycle risk of any substitute selected. Life cycles of the substitute parts, which indicate the life-cycle stage and the number of years the part/technology group is expected to remain commercially available, should be evaluated.
4. A cost-benefit trade-off analysis should be conducted to determine whether redesigning a board with reconfigurable technologies is better than using substitute parts. The cost of support for each substitute candidate and the time frames for part review, approval, and availability should be projected. Cost factors should include:
    -   engineering analysis
    -   source qualification requirements (as necessary)
    -   price of substitute items (including testing)
    -   associated engineering and logistics documentation changes
5. The approval of the substitute part should be based on:
    -   FFF requirements
    -   part family part testing and/or screening requirements and part manufacturer evaluation
    -   module and end product associated integration testing requirements
6. Development and approval of associated engineering change, deviation, or waiver notifications should be ensured.

16.3.2 Negotiation with the manufacturer

A manufacturer's end-of-life (EOL) decision is rarely stopped or even delayed. However, during the period between part phaseout and discontinuance, an equipment supplier may be able to persuade the manufacturer to keep the part from being discontinued by teaming with other end users (distributors) to provide demand for the part and/or by accepting a price increase.

The manufacturer's decision may be dependent on its profitability and manufacturability objectives; if sufficient volume orders can be guaranteed in the near and long-term future, the manufacturer may be willing to continue production of the part. Several equipment suppliers may band together to share information, develop *common* or *standardized* part lists, and provide avenues for communication with component manufacturers. Some manufacturers have responded positively when approached by a group since enough demand is assured to justify continued production of the part. In March 1997, Boeing encouraged suppliers to act cooperatively while dealing with semiconductor manufacturers, and some major suppliers are devising strategies for doing this. The collective equipment suppliers' strategy should be to appear as one virtual customer to the IC makers [21].

The advantages of negotiating with the manufacturer to keep the part active are that the original part is still available and no other system changes are then required. Suppliers may be hard pressed to garner enough industry support and, if successful, may be forced to accept a price increase from the manufacturer. Negotiation with manufacturers occurs in any form of business where a customer may want the part manufacturer to continue manufacturing a part that the part manufacturer wants to discontinue. The part manufacturer (e.g., Texas Instruments), may continue manufacturing the part, may direct the customer to an aftermarket source (Rochester Electronics), or may indicate a substitute part manufactured by Texas Instruments or another part manufacturer [22].

16.3.3 Lifetime buys

In a lifetime (also called "life-of-type," or LOT buy), the equipment supplier buys sufficient parts to meet the system's lifetime needs from the original part manufacturer before the anticipated close of production. This is one of the cheapest solutions because it requires neither reengineering nor requalification nor redesign and is one of the first options considered to mitigate obsolescence. Part manufacturers also promote this option when they plan to exit the market for a specific product or line.

The equipment supplier should determine the last dates for processing minimum order quantity or value and the latest delivery dates for the part from the manufacturer's product discontinuance notice (PDN).

Lifetime buy costs should be calculated as a baseline for comparison with other options, and the lifetime buy execution time frame should be determined and recorded.

- Total material costs should be calculated by requesting a quote from the manufacturing sources based on the total quantity required. Item testing requirements should also be considered as part of the total cost.
- Overhead costs should be calculated. They should include storage and handling, inspection, testing, repackaging, shipping, and the interest costs of tied-up capital. Where special storage conditions are required, such as dry nitrogen storage, this cost should also be factored in.
- Total lifetime costs are calculated by summing the material and overhead costs.

Die banking and repackaging houses result in some suppliers reducing the risk and financial commitment of lifetime buys by purchasing die instead of complete parts. The purchase price of die is less than the cost of a finished part, allowing the purchase of buffer stocks. Controlled storage of die can be done at nominal cost at some assembly houses. Even if the original die manufacturer no longer supports a military product line, as long as the die is available from a source that is already qualified for use by the customer or acquisition agencies, suppliers have an economical solution to the obsolescence problem. Assembly and packaging houses can provide final processing services. Die banking also provides the option of using advanced package types and technologies. Storage and repackaging facilities must be carefully chosen to avoid damage to the stock of die.

The option of making lifetime buys of components slated for discontinuance is tempting since it allows use of the same part without design modifications. No new sources are needed, nor is requalification required. Another benefit of lifetime buys is that parts bought directly from manufacturers also come with the original manufacturer warranties, which may not always be beneficial for customers with long-term storage requirements.

A lifetime buy is also beneficial for semiconductor manufacturers. They receive large and predictable orders that help them to fully utilize the capacities of product lines before the lines are discontinued. The Semiconductor Industries Association Government Procurement Committee (SIA/GPC) has recommended that government acquisition agencies make a contractual provision that allows special payments for bulk purchases of possible DMS parts [23]. The SIA/GPC also proposes that contracting reforms allow for long-term storage of critical semiconductor components for the entire life of a program and for subsequent spares for repair purposes. SIA moves indicate that the semiconductor manufacturing industry prefers consumers to accept the option of lifetime buys.

Maintaining existing systems: Making a lifetime buy of any significant quantity is not just an engineering decision; it is a complex financial decision that requires dependable financial forecasts at both the organizational and macroeconomic levels. Current interest rates and the anticipated holding time of part inventory play an important role in justifying lifetime purchases. Accurately forecasting the required quantity for such purchases is essential. If the affected program has a "sunset" date and the number of units to be produced is known with certainty, then the number of parts required for manufacturing the units may be predicted. For an active system, when numbers yet to be manufactured are uncertain, it is harder to determine the number of parts required for completion of the project and to take into account part environmental and operating requirements, particularly when the part in question is being used under varying conditions in different products. The replacement part requirements for each system will be different. Common reliability prediction methods used to estimate the number of spare parts needed are often inadequate [21], [24].

Even when part requirements can be predicted with acceptable accuracy and when the cost of maintaining a lifetime inventory is not a problem, a bulk purchase may not be the best engineering decision. In order to sustain a long-term program with a lifetime purchase, the parts must have proven dormant storage integrity, as there is always a possibility that manufacturers' warranties will run out before the part is needed. If there are failures or if parts degrade during storage, a poor forecast of the number of parts can mean that parts will be in short supply once the process lines are closed; the engineering justification for bulk purchase of the part in the first place is thus negated.

Lifetime buys also run counter to "lean" manufacturing practices, such as just-in-time production, which reduce inventory overheads and inventory maintenance costs. The Defense Supply Center spends considerable sums of money in lifetime buys every year and consequently holds millions of dollars' worth of components that may never be used.

The U.S. DoD has to operate in accordance with the Defense Federal Acquisition Regulations (DFARs). In particular, there is a general prohibition on the purchase of quantities of parts beyond immediate need [25]. The text of this regulation is reproduced in Table 16.2. This requirement places severe limits on the ability of the government to use last-time buys as an obsolescence solution by setting a limit on the purchase of spare parts. Even then, the military's extended buys normally cover only requirements for spares and maintenance. Although suppliers can procure parts for new production, lifetime buys for new products must be funded from other sources. Manufacturers are also under no obligation to provide the option of a lifetime buy and often do not due to limited availability. As an example, Philips has not offered this option for the majority of its discontinued parts.

In December 1996, Philips exited the military-grade microcircuit market. Completion of end-of-life buys was scheduled for December 1997. On August 14, 1997, Philips abruptly stopped accepting lifetime buying orders for all military parts. Hence, end-of-life buying is not always a reliable option.

Table 16.2: U.S. Code: Title 10, Chapter 131, Section 2213, Limitation on Acquisition of Excess Supplies

Sec. 2213. Limitation on acquisition of excess supplies		
(a)		Two-Year Supply. - The Secretary of Defense may not incur any obligation against a stock fund of the Department of Defense for the acquisition of any item of supply if that acquisition is likely to result in an on-hand inventory (excluding war reserves) of that item of supply in excess of two years of operating stocks.
(b)		Exceptions - The head of a procuring activity may authorize the acquisition of an item of supply in excess of the limitation contained in subsection (a) if that activity head determines in writing -·
	(1)	that the acquisition is necessary to achieve an economical order quantity and will not result in an on-hand inventory (excluding war reserves) in excess of three years of operating stocks and that the need for the item is unlikely to decline during the period for which the acquisition is made; or
	(2)	that the acquisition is necessary for purposes of maintaining the industrial base or for other reasons of national security.

16.3.4   Aftermarket sources

The aftermarket is the period after the original manufacturer has phased a part out of production. Aftermarket sources cater to the continued demand for discontinued parts. If part specifications and test, acceptance, and related technical data are complete and available, aftermarket manufacturers may support continued production of obsolete items. Some aftermarket sources also buy bare die and repackage the die. Part manufacturers often establish routine transfer agreements with aftermarket activities for discontinued product line support. Aftermarket sources are gaining prominence in various consortiums and among manufacturers.

Industry and trade groups such as the SIA endorse using parts from aftermarket sources. In its position section on DMS, SIA states, "...increasingly, major semiconductor suppliers are incorporating qualified and authorized aftermarket sources into their customer support programs." In many cases, these qualified companies can be a system manufacturer's first and most cost-effective option for resolving its obsolescence problem. The emergence of dedicated aftermarket sources that have formal relationships with the original IC manufacturers represents a viable means to support continuing requirements beyond the last time buy cycle. Two major aftermarket sources, Lansdale Semiconductor and Rochester Electronics, have representatives on the SIAs Government Procurement Committee (GPC).

A formal relationship between major part manufacturers and aftermarket sources has flourished in recent years. Product discontinuance notices and press releases include the names of authorized aftermarket sources. This practice can reassure suppliers that the major manufacturers have confidence in these sources. For example, Texas Instruments designates the aftermarket manufacturer Rochester Electronics as its authorized distributor of all obsolete parts. This authorization includes the transfer of responsibility for product support [22].

For equipment suppliers whose products need certification from military regulatory agencies, it is important to have access to qualified parts. Some aftermarket sources have received qualified manufacturer list (QML) approval. A section for aftermarket sources of obsolete parts has been added to the QML by the Defense Supply Center, Columbus (DSCC). Some aftermarket companies listed in or pursuing QML are White Microelectronics, Pantronix, Lansdale Semiconductor, Sarnoff Corporation,[3] Rochester Electronics, OMNIREL, and Austin Semiconductor.

The advantages of receiving recognition from different sources are impressive but do not necessarily guarantee that aftermarket parts have the same reliability and quality as original parts. Confidence of equipment suppliers in aftermarket sources depends on whether these market sources maintain a continuous supply of parts after the original manufacturers have discontinued them. When considering this alternative, the obsolete parts management team should evaluate manufacturer production capabilities and test programs to ensure their ability to meet original item specification requirements. Acceptance tests and qualification requirements for the obsolete part should be identified. The adequacy and accuracy of technical information from the part datasheet should be validated.

Aftermarket firms should be identified and their production capabilities should be validated. Some aftermarket sources distribute catalogs, which identify the products and technologies they support and provide detailed cross references to original part numbers. The original part manufacturers may be contacted to assist in locating suitable aftermarket sources. Often, original part manufacturers distribute lists of cross-referenced parts together with suitable aftermarket sources or even other manufacturers/competitors who make similar parts. Once identified, the supplier's capabilities should be evaluated to meet original item production and qualification requirements, and production plans for the expected duration of obsolete part procurements should be assessed. To accomplish this objective, part technical data and part test and qualification requirements need to be provided to aftermarket sources to assess the ability to manufacture FFF replacements for the part in question. If aftermarket firms are not qualified to the level of the original part specification, an engineering analysis may be appropriate to determine the potential for waiving selected requirements to allow approval of the aftermarket part. Alternately, the feasibility and cost of qualifying supplier facilities to the desired level should be validated.

Bids are requested from aftermarket sources for designated quantities of the obsolete items, and part production and delivery schedules should be determined. Total cost of the resolution should include:

- technical data procurement/development
- qualification of obsolete parts
- manufacturer (aftermarket) assessment
- cost of the obsolete part (including testing)
- documentation changes if engineering change proposals (ECPs)/waivers are required
- material handling and storage if volume procurements are foreseen

The time frame for implementation of this alternative should be determined from in-house procurement lead time through item qualification and material delivery.

---

[3] Sarnoff Laboratories supplies emulated devices. Emulation and reverse engineering are discussed later in this section.

The companies that provide aftermarket services vary greatly in size, capabilities, and experience. It is expected that these variations are reflected in the cost, the reliability, and, most importantly, the quality of their products. This lack of consistency is one of the reasons why many suppliers are reluctant to rely on these companies for purchasing obsolete parts. As a relatively new and seemingly lucrative business opportunity, the aftermarket has attracted many organizations, and while many of them are competent and capable, others are not. Some suppliers have had less than satisfactory encounters with some aftermarket companies and have found that aftermarket parts do not always match the performance of original parts [21]. The availability and lead times to receive aftermarket parts vary significantly. Some aftermarket companies have national (and even international) distribution networks, while others operate from a single source.

Capabilities for quality monitoring and testing vary widely among aftermarket companies. Another difficulty is that, unlike major manufacturers, the availability of detailed quality data on these firms is scarce. While some issues are more relevant for choosing aftermarket sources than for choosing regular suppliers, the manufacturer part family and distributor assessment criteria should be similarly used to assess aftermarket sources. When the aftermarket source does the design, manufacturing, or packaging of discontinued parts, the manufacturer part family assessment [26], [27] criteria should be used. When the aftermarket source does inventory management for original manufacturers, distributor assessment criteria should be used. Due to small production volumes, it might be hard to determine the composition of aftermarket part families. Just as when buying from original manufacturers, equipment suppliers must be knowledgeable about the particular strengths and weaknesses of aftermarket sources.

The number of aftermarket companies is increasing as more organizations identify the aftermarket as a business opportunity. Aftermarket companies offer four main manufacturing services of discontinued parts, each requiring varying levels of technological expertise:

Trading:

1. *Serve as a clearinghouse for excess inventory:* Electronic data transfer has made clearinghouse activities a viable process. Clearinghouse-type aftermarket companies act as intermediaries between organizations holding excess inventory and those with a need for discontinued parts. TACTech's Lo-K-Tor service is one example; Chip Seeker of Second Source is another. With many government and military agencies (e.g., Defense Reutilization and Marketing Service, DTC Parts Trader and Naval Inventory Control Point) offering similar services free to their suppliers, it is difficult for private firms to compete in this market sector.
2. *Operate as a sourcing agent to locate inventory:* Some service companies act as buying agents for organizations facing the obsolescence problem. Qualified Products Laboratory offers a locating service, along with testing and qualification. Second Source also acts as a sourcing agent. Contract assemblers are also well qualified to conduct a search for obsolete parts, as they have developed a wide range of contacts and expertise.

Distributing:

1. *Inventory discontinued parts:* This practice is different from the trading business, as it involves carrying the risk of holding inventory. It also requires proper storage and handling facilities for managing high-reliability parts. Firms' levels of involvement in this process varies substantially. Purchasing inventory can be a collaborative process

between original manufacturers and aftermarket sources or it can be a unilateral effort on the part of aftermarket sources.

Companies such as White Microelectronics monitor technology trends and buying patterns, along with providing roadmaps of individual manufacturers and industry associations; based on these data, they acquire an inventory of parts likely to become obsolete and they stock the parts for future sale. Other companies team with original manufacturers and become authorized distributors of the manufacturer's discontinued parts. This practice is becoming more common, and many companies now include the names of authorized distributors in their PDNs. Some of these agreements are for individual products; the trend, however, is towards maintaining ongoing, long-term contracts, which give the suppliers some assurance of stability. Rochester Electronics has concentrated on developing a niche for itself in this market segment. After securing an agreement with Fairchild in 1987, it is now the authorized distributor of discontinued semiconductor products from companies such as AMD, Analog Devices, Altera, Harris, Intel, National Semiconductor, and Texas Instruments.

Some traditional distributors and wholesalers of electronic parts are also venturing into the business of distributing obsolete parts. Because of their expertise in customer interface, they often take on the complete business aspect of the EOL process. During its departure from the hybrid market, Elantec contracted its entire EOL process to Zeus, a distributor that concentrates on serving the military and aerospace sectors. This work will involve customer notification, inventory handling, order processing, and lifetime-buy fulfillment [28].

2. *Inventory die and wafers:* This business is similar to purchasing and reselling inventory; however, managing die and wafers requires special handling capacities. Stocking die and wafers is important in this transitional period between wafer-size standards for many fabricators.

3. Most of the companies in this segment were originally die and wafer processors that have expanded their businesses to cover purchasing and stocking in response to the obsolescence problem. Some companies stock die as part of their agreements with manufacturers to handle all discontinued inventories. Other companies concentrate on this area alone. Chip Supply Inc. stocks and does value-added processing of die and wafers, such as wafer cutting, backlapping, FPGA programming, bare die burn-in, die bond pad bumping, and functional testing. Chip Supply has entered into an agreement with Analog Devices to distribute bare die.

Assembly and testing:

1. *Repackage base die to meet extended specifications:* Some repackaging companies supply products from their stock of die and wafers bought from original manufacturers or authorized distributors, while others take individual orders for die packaging. Some companies offer subsequent testing and qualification services for parts slated for use in harsher environments.

One major player in this segment is White Microelectronics. White was established in 1967 as a custom hybrid company, but since the beginning of the 1990s, it has focused on standard memory products for extended-temperature, high-reliability use. White Microelectronics specializes in SRAM, flash, EEPROM, and DRAM ceramic monolithics and modules as well as plastic upscreened monolithic memory products. White Microelectronics supplies parts in all standard temperature ranges, as well as in special temperature ranges up to 200°C and the military packaged versions of some products available only in industrial or commercial specifications from the original manufacturers.

Many companies offer spare assembly and packaging capabilities for special packaging needs. Consequently, there are many seasonal and part-time entrants in this business. Some companies such as Team Pacific Corporation, Multi-Products Research & Development, Inc., Chip Supply, High Rel Group, Austin Semiconductor, Inc., SST/Burlington Micro, and VLSI Packaging Corporation specifically serve the DMS area. The package styles and ranges offered by these companies vary from simple TO cans to large area array packages.

2. *Test and certify parts to meet extended specifications:*[4] Many companies established in testing and qualification venture into the market for extending the temperature-range certification of electronic parts. However, they are relatively few. One reason is that many part manufacturers strongly discourage this practice. Another is that the practice of using parts beyond manufacturers' specifications may entail legal liabilities in the event of system failure [29].

    The Plastic Plus program of White Microelectronics tests and certifies industrial temperature-range plastic memory products for use in the full military temperature range. Austin Semiconductor, Inc., performs upscreening, in which "defects" are precipitated lot by lot as parts are used beyond manufacturer-specified limits. Second Source participates in similar activities, with plans to specialize in radiation-hardened products in collaboration with Mission Research Corporation.

As a design and manufacturing house:

1. *Operate discontinued product lines:* The practice of buying and operating entire discontinued product lines that still have potential for mid- to long-term sales is gaining momentum. Often, product lines that are only marginally profitable to the original manufacturer can still be a good business opportunity for smaller and niche-market operators. Such manufacturers purchase and gain rights to fabrication lines, designs, mask sets, testing software, and equipment from the original manufacturer and continue production. This trend is present for both military- and nonmilitary-grade older technology components.

    Lansdale Semiconductor, one of the larger companies in this market segment, has focused on older technology from its inception, when it purchased the germanium small-signal transistor line from Philco-Ford in 1964. Because it focuses on older technology, its major product line is silicon bipolar ICs. It offers products originally manufactured by AMD, Fairchild, Harris, Motorola, Texas Instruments, Intel, Raytheon, Signetics, National Semiconductor, and Western Digital.

    Rochester Electronics has purchased the rights to all of Intel's Fab4 products, which include 2.0-μm NMOS and 1.5-μm CMOS products based on 4-inch-diameter wafers. It also obtained similar rights to International Rectifier's HEXFET power MOSFET parts in metal cans.

    Numerous other companies are involved in aftermarket manufacturing: Austin Semiconductor manufactures Micron Semiconductor products; Multi-Products Research & Development, Inc., makes Raytheon and Siliconix products; TSI Microelectronics Corporation produces hybrid products of National Semiconductor, Fairchild, and Analog Devices; and Quality Technologies Corporation manufactures optical components of Fairchild, General Instruments, Harris, Monsanto, and Philips. More companies are expressing interest in this segment. Semicoa Semiconductors

---

[4] Some suppliers perform such procedures themselves to mitigate the risk of using parts beyond their manufacturer-designated limits through uprating and performance derating. Any strategy to handle obsolescence must consider developing and using these techniques.

caters to the market with product lines of General Semiconductor and Harris, although it is reluctant to apply the term "aftermarket" to its business. Pantronix is also considering obtaining older 3- to 4-μm CMOS technology lines. With the removal of "onshore" clauses from the U.S. military qualification guides, some Asian manufacturers are also exploring this business opportunity.

2. *Design and manufacture equivalent parts:*[5] Some companies design and provide alternative parts for discontinued technologies. Usually these companies concentrate only on their established areas of expertise. PowerTech provides replacement power transistors discontinued by companies like Solitron, PowereX, and RCA. Unless intellectual rights are obtained from the original manufacturers, this method is realistically applicable only to simple digital parts.

Suppliers have raised a number of concerns about purchases from aftermarket companies. Major concerns are cost and quality. Many equipment suppliers tend to trust the product quality of high-volume manufacturers over that of aftermarket sources. High-volume manufacturers generally have superior processes to control yield, quality, and reproducibility compared to parts manufactured by relatively small aftermarket sources.

A qualified defense of aftermarket sources has been made by the U.S. Navy's DMS Technology Center: "aftermarket firms typically have less than 75 employees and vary widely in their structure and focus. It is true that in some instances, the aftermarket vendor may take on the appearance of a brokerage house. But the differing management styles and business techniques employed in this niche market are truly astonishing. Fierce competition in the industry forces the aftermarket firm to choose a product wisely, generally selecting a product with known return on investment" [30].

Aftermarket companies claim that their parts are priced at about the same levels as original parts, but many buyers complain that prices of parts from aftermarket sources are much higher than those of the originals and that minimum order quantities are required. This is not surprising, since demand and production levels are both lower than when the part was originally produced. Some price differential reflects the fact that original manufacturers sold the products close to or below cost when they were in the market. The prices of parts also tend to escalate with time. The price differential depends on the method by which the parts are supplied.

Aftermarket companies that depend on the inventory of original manufacturers will inevitably run out of stock at some point and another part shortage will ensue. Although aftermarket companies can accept a smaller production volume than original manufacturers can, they need some minimum volume for their product lines to be viable. There is no guarantee that in the future enough of a market will remain for them to be able to continue to produce the parts economically. When aftermarket product lines are closed, new obsolescence problems occur. Lansdale Semiconductor discarded its flagship small-signal germanium transistor line when demand for its parts was not high enough.

Repackaging companies that buy commercial and industrial die from original manufacturers may not be able to supply parts with FFF matches if the original manufacturers make major design changes. It is expected that every part will undergo a fabrication facility move, wafer diameter change, electrical design change, die shrink, or combination of these within three years [29]. Without proper analysis and testing of the altered capabilities of the changed die, only special packaging may deliver extended

---

[5] Not included in this category are specialized reverse engineering and emulation companies discussed separately in Section 16.3.7.

environment functionality. More stability occurs when aftermarket sources continue to manufacture the products.

Moreover, aftermarket and original manufacturers enter into exclusive agreements, a practice that can allow critical parts to become single-sourced and at greater risk of becoming unavailable or high priced in the future. Even though the aftermarket has become a parts source in recent years, it has simultaneously hindered the insertion of newer technologies (mostly in the military and avionics marketplaces). Policymakers and regulators should reevaluate their parts procurement strategy so that these products do not remain stuck at "the trailing edge of technology." Rochester Electronics is an aftermarket source for parts discontinued by Texas Instruments. A quote from a lifetime buy notice of Texas Instruments is, "Rochester Electronics Inc. is a Texas Instruments authorized supplier/distributor for discontinued products. Rochester has extensive inventories and can manufacture military parts for continuing supply." In 1994, Boeing reported that of four systems in which buying from an aftermarket source was resorted to – the 767 AWACS interface control unit, the 707 AWACS control power supply, the ARINC 429 dual transmitter, and the ARINC 429 receiver – the obsolete parts were bought from Lansdale Semiconductor and Rochester Electronics.

16.3.5 Redesign

Redesign involves designing-out obsolete parts via engineering changes at various system indenture levels with a view to enhancing system performance, resolving the problem of part obsolescence, and improving reliability and maintainability. There are two approaches for redesign: one is to copy the primitive logic gates in the circuit and map them to the new technology; this approach captures the logic of microprocessors, microcontrollers, and arithmetic logic units as faithfully as possible, with 1 new microcircuit replacing 10 or more older ones. This type of redesign is also known as "technology insertion" and entails replacements for aging electronics technologies primarily at the part and board levels. The second method is to reengineer the board, starting from its intended function and making it interface-compatible with the system.

Redesigns at the component or line replaceable unit (LRU) level may involve significant risk if the item in question has multiple different applications and if extensive system integration testing and qualification may be required. Moreover, depending on the scope and level of the redesign effort, substantial nonrecurring engineering (NRE) and life-cycle logistics costs may accrue. Given the relative magnitude of engineering and logistics cost factors, redesigns may be most appropriate when a large percentage of obsolete parts are resident within a particular system or end item. LRU redesign should be explored only after evaluation of the following alternative solutions:

- selection of an interchangeable replacement component
- alternate source for component (requires requalification)
- use of upratable parts, or alternate components suitable following selective screening to meet higher performance requirements
- die banking, which involves use of a third-party source to hold the die and package the component at a future point in time
- part emulation using newer-technology components that are packaged in the same form factor as the original part, which will emulate the component's functionality
- use of ASICs for replacement of the component or component group

Usually, redesign is the last option, as it is perceived as expensive, but there can be situations when it is viable, both economically and technically, depending on two major factors: the usage pattern(s) of the component(s) in question and the time available for redesign. Component usage patterns determine how many redesigns must be done and who is responsible for them. Usage patterns also determine the requalification process, if any, that the new boards (or modules) will have to undergo. The time available for redesign depends on when the components are projected to become unavailable. In considering redesign, it should also be seen as a marketing opportunity for functional additions, cost reductions, and reliability improvements. This may be the only viable option once the last-time buy parts have been consumed.

Ultimately, all long-life products will require redesign, since the other options are all temporary solutions designed to solve an immediate problem. It should always be considered that long-life systems will be in a constant state of redesign throughout their lives. The cost of redesigning to replace an obsolete component depends on all the factors indicated in the previous paragraph taken together. For example, if the board or module affected by obsolescence of a part has other parts that are obsolete or soon to be obsolete, then redesign might be more cost-effective than other options. If any organization or program has already identified an economically viable and technically acceptable equivalent part or alternative source for the problem part, then redesign is not a priority alternative. Any redesign process undertaken to solve an obsolescence problem must address future availability concerns to ensure that the new design has a longer useful and maintainable life. Another consideration during a redesign process is ensuring that the best currently available and easily upgradeable design and manufacturing technology is used in the new designs. The goal of redesign should reach beyond solving the immediate problem of nonavailability of parts. The steps in the redesign process are as follows:

1. The status of current and planned redesign initiatives for cognizant systems/products should be assessed and associated technical and cost proposals requested.
2. Redesign engineering and cost proposals should be reviewed and approved; feasibility studies, design and cost trade-off analyses, and any other information necessary to validate schedule, technical, and financial aspects of the redesign approach should be gathered.
3. With regard to the first objective, analysis of both existing and potential obsolescence problems on a total system, products, or end-item basis may be necessary to determine the most efficient indenture level for redesign efforts. Design changes should be incorporated on a sufficiently broad scale to minimize continued piece part obsolescence impacts (modular designs). At the same time, parts used in the final engineering change package should also be screened to determine the out-year obsolescence risk.
4. Cost projections for this alternative should be developed, which include:
    -   Up-front system engineering, testing, procurement, and installation, as well as technical data development and modification
    -   Life-cycle logistics support costs (both manpower and material) associated with the redesigned item. A number of logistics cost-estimating models, which provide life-cycle cost factors for products engineering changes and alterations, are available to assist in this analysis.
5. Execution time frames required for completion of the redesign process should be determined.

There are some advantages to using redesign as a solution to component obsolescence. A redesigned system using newer technology and design tools may be less expensive to manufacture and maintain and may achieve better tolerance on electrical parameters than the design it is replacing. Using an application specific integrated circuit (ASIC) built for a particular purpose, as opposed to using generic parts, can reduce the total number of parts required, which can result in reducing the inventory part count. Redesign also provides a possibility for technology insertion and other design improvements such as cost reductions, reliability improvements, or additional functions.

The chief limitation of redesigning a system is that the original design team has often been disbanded, so that the postdesign team has to rely on the drawings and documentation. This point emphasizes the importance of documentation and strict drawing configuration control in enabling the sustainment of a system over a long period. For a redesign, requalification of the system may be necessary and compatibility with the rest of the system may be an issue.

The U.S. Army Missile Command reported success in reducing the cost of obsolescence through redesign of its fuse setter circuit card assembly (CCA) contained in the fire control system of the Multiple Launch Rocket System (MLRS). To preempt an obsolescence problem, the CCA was redesigned using ASICs. The new design has only 3 microcircuits, whereas the old design had about 100.

16.3.6 Emulation

Emulation is the process of creating unavailable electronic components from their slash sheets, datasheets, test vectors, and other information by using macro libraries and VHSIC (very high speed integrated circuit) hardware description language (VHDL) models. Emulated products are sometimes categorized as aftermarket products.

The best-known military-sponsored project for developing and supplying obsolete microcircuits is GEM (generalized emulation of microcircuits).[6] GEM provides a source of microcircuits originally produced with RTL, DTL, TTL, Schottky, ECL, NMOS, PMOS, and CMOS technologies. The GEM BiCMOS process [31], GEM Design System and Library, specification procedures, packaging, interface, and testing procedures were developed under the GEM R&D program. Started as a tri-service manufacturing science and technology R&D initiative sponsored by the Office of the Secretary of Defense, it is managed by the Defense Logistics Agency (DLA). Sarnoff Corporation, a subsidiary of David Sarnoff Research Laboratory, does the actual design and production.

Along similar lines, in 1993 the Office of Naval Research (ONR), in conjunction with the Undersecretary of the Navy for Research, Development and Acquisition, established a program known as Technology for Affordability. That program is executed in many of the Navy's weapon systems procurements and is specifically resident in two ONR programs: Manufacturing Technology and Small Business Innovation Research (SBIR). The technology was developed under the auspices of the SBIR program and is known as Rapid Retargeting. It is executed through a subsidiary company of Titan Systems, VisiCom Laboratories Inc. [32]. Rapid Retargeting rehosts existing logic into new hardware technology. This rehosting process begins by extracting the functionality of the target hardware and capturing it in VHSIC VHDL. The resulting software models are simulated and compared with the original hardware for verification. Once verified, the models are ported to a new hardware design. Subsequently, parts obsolescence is no longer an issue; software models can be rehosted whenever new technology becomes available.

The GEM process re-creates unavailable electronic components from their slash sheets, datasheets, test vectors, and other information by using macro libraries and VHSIC VHDL models. These designs are produced using 1.5-μm BiCMOS wafers. Quality levels are maintained by utilizing a modern single BiCMOS fabrication line that allows continuous

---

[6] A program called AME (advanced microcircuits emulation), was initiated in 1996 by DLA. It is a stand-alone independent program with broader technical scope than GEM.

manufacturing of microcircuits in a quality controlled environment. The customization process occurs at the last stage of creating interconnects on the GEM wafers to create the die. Figure 16.3 schematically shows the process flow of IC emulation.

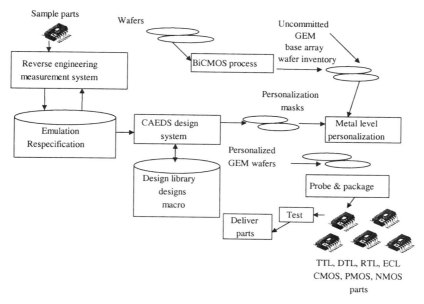

Figure 16.3: The GEM emulation process [31]

Emulation is possible only if the microcircuit in question meets specific physical or functional characteristics with regard to:

- technology type (e.g., RTL, DTL, TTL, NMOS, PMOS, etc.)
- operating voltage
- temperature range
- pin and gate count

Contact with prospective emulation activities will generally quickly determine whether the part is an emulation candidate. The steps in selecting the emulation process are as follows.

Availability and adequacy of part technical data and specifications, including part testing and qualification requirements, to support the emulation process should be assessed. An accurate and complete part specification is essential to ensure that the emulated product matches all of the characteristics of the original part. This specification could be developed from existing government and commercial documentation, as well as from measurements made of existing parts to clarify questionable data. Coordination with original part manufacturers to obtain final acceptance test procedures or other specification requirements may be necessary.

Candidate emulation firms should be identified, and technical data necessary to conduct emulation feasibility analysis (as required) should be provided. Acceptability of the proposed emulation approach and associated material production and delivery schedules should be validated. As in the aftermarket alternative, emulation firms may not maintain item qualification capabilities to the level of the original part specification, and the potential for waiving specifications or funding qualification of the potential source may need to be evaluated.

Bids from selected emulation activities for designated quantities of obsolete parts, and ensuring consideration of varying order quantities may be requested. The total cost of this alternative should be assessed, including:

- technical data/specification development
- qualification of obsolete parts
- manufacturer (aftermarket, emulation agency) assessment
- part test/acceptance costs (including module/end-item drop-ins, and full-system integration testing as required)
- engineering/logistics data revisions
- material handling/storage (if volume procurements are anticipated)

The time required for completion of the emulation process, including in-house lead times and engineering change package analysis and approval, should be estimated and recorded.

There are certain advantages to using GEM parts. They provide FFF[7] replacements of the original parts. GEM microcircuits also have a complete tracking history and meet all the specifications of the original parts, including physical (package type, pin out), electrical (functional DC, dynamic, power), and environmental (ESD, temperature, hermeticity). By their very nature, GEM parts are completely traceable to the wafer level. The program can supply parts in small lots without excessive overhead, since it uses a common wafer for all products. Manufacturing can be done on demand, reducing the need for high inventory and allowing for mistakes in estimating required parts. GEM microcircuits satisfy the original electrical specifications over the full military temperature range (–55 to 125°C) and are packaged and marked to original specifications. The Sarnoff manufacturing center produces QML-(MIL-PRF-38535) standard products. It is also possible to tailor parameters for special applications; for example, in one electronic module of the U.S. Navy's AN/SQR-19, failures could be reduced by using specially tailored GEM products with adjusted timing.

The prototype development time for GEM parts is between 90 and 120 days, not much higher than that for aftermarket sources when premanufacturing activity is required. The program also offers potential stability, as it has only one product line and is not likely to close or discontinue a specific product. GEM products do not have to depend on high-volume production runs for profitability. NRE for each part vary between none and $50,000, depending on whether the same part or similar parts have already been emulated. Unit costs for GEM are attractive compared with other aftermarket options.

Not all obsolescence problems are solvable through GEM technology. The technology range is limited; available GEM technology can only emulate older-generation digital families (e.g., DTL, RTL, TTL, ECL (up to 10K)). Only some borderline analog parts, such as drivers and comparators, have been emulated. Memories are limited to 4K. Other limitations are in pin count (up to 48 pins; 28 pins for older high-voltage systems), gate count (maximum 1120), and propagation delays (minimum 4–5 ns). The price per unit of the emulated microcircuit is extremely sensitive to order quantities.

Because it is not possible to emulate many parts, many suppliers have looked for other solutions to obsolescence that could cover a larger number of parts. This explains why the total number of GEM parts available and under development is still only in the hundreds, even though its capability has been demonstrated in a large number of systems. Some

---

[7] Matches to all datasheet and selected item parameters are obtained, but complete manufacturer test vectors are not guaranteed.

skeptics of the emulation program advise equipment suppliers to rely on it only as a safety net.

The cost structure is reasonable, but this may be due to special agreements with military agencies; once the system is financially independent, its prices are likely to rise. The focus of the GEM program is almost exclusively on military parts. Unless GEM is able to respond to the needs of the consumer marketplace and supply advanced technology parts, its viability is not assured.

GEM parts have been used successfully in programs of the U.S. Air Force (e.g., F-15, F-16, JointStars), Army (e.g., SINCGARS, MLRS), NASA (e.g., Shuttle Orbiter), and Navy (e.g., A-6, Aegis). The U.S. military services claim major cost and time savings from using GEM technology in these programs. In addition, Boeing is piloting the use of GEM technology to replace a microcircuit used in an LRU on the 737 aircraft.

16.3.7   Reverse engineering

Reverse engineering is the process of developing exact replicas of candidate items through review of available technical data and physical disassembly and analysis of item components. Reverse engineering may be appropriate when sufficient technical data or data rights to support reprocurement are unavailable; the goal is to cultivate qualified alternate sources. The reverse engineering process is generally expensive and may be used for cost or technical comparison with redesign or other solutions.

The steps in selecting the reverse engineering process are as follows:

1. Part technical and procurement data availability should be ensured, together with the availability of sample parts for analysis and disassembly during the reverse engineering process.
2. Industrial or government concerns with the potential to reverse engineer the part should be identified; historical contract information and technical data should be provided to the prospective reverse engineering activity. Feasibility analyses, technical/cost proposals, completion schedules, plans for incorporation of product improvement, or redesign initiatives as part of the reverse engineering process should be conducted.
3. Costs for the reverse engineering alternative should be estimated, including:
   - reverse engineering feasibility analyses and studies
   - source setup and tooling
   - prototype production
   - reprocurement data package development
   - review and approval
   - material costs
   - additional life-cycle logistics manpower/material costs resulting from redesign or product improvement initiatives
4. Reverse engineering execution timeframes should be determined and documented.

An emulated IC is not an exact replica of the obsolete part, whereas reverse engineering is an attempt to create an exact replica through review of available technical data and through measurement, testing, and physical disassembly of the part. This is usually expensive. The Defense Microelectronic Activity (DMEA) involves reverse engineering; it has delivered replacements for some obsolete components. In addition to the obvious advantages of being able to manufacture an FFF replica of the original part, newer packaging technologies may be used to meet today's increasing demand for smaller package sizes.

Reverse engineering, like redesign, is an expensive process and requires specialized technical knowledge and infrastructure. For successful emulation and reverse engineering of increasingly complex microelectronic parts, hardware description languages like VHDL should be used when designing electronic products. Issues arise about the intellectual property rights of the original manufacturers. Only if the customer (often government) or supplier has the right to the technology in question can these options be exercised. The U.S. government holds the rights to the products and processes of many technologies that resource military and aerospace suppliers may like to tap.

The U.S. Army reports a success story in which reverse engineering was employed to resolve an obsolescence issue. When the original manufacturer of a hybrid IC used in an Army communications system no longer supported the part, the system manager, to avoid costly redesign, called upon the DMEA engineers to reverse engineer the obsolete IC. From the reverse-engineered IC, a prototype was created which was tested in the communications product. After sufficient proof that the reverse-engineered IC worked exactly like the original obsolete version, the DMEA finalized the design and developed a physical layout that could be cost-effectively manufactured in small volumes. The final parts were tested, qualified, and delivered to the Army for continued support of the communications system. This approach saved the Army millions of dollars and years of wasted time. However, reverse engineering limits the scope of the (reverse engineering) design to outdated technology.

16.3.8 Reclamation

Reclamation is the process of salvaging or "cannibalizing" used parts that have not lost their functionality. The process involves the following steps:

1. The parts that are required for production but are not available must be identified.
2. Possible sources from which the same or equivalent parts can be reclaimed must be identified. The possible sources of reclaimable parts are:
    - systems beyond "economic repair"
    - stocked as spares for discontinued systems
    - excess available stocks

Reclamation is a quick solution for small numbers of obsolete parts. It is also an in-house program with no dependence on outside agencies. It involves major handling and rework efforts, logistics, and disassembly. Quality and reliability assessments of the reclaimed parts should be conducted, though this may prove impossible, as usually these parts have a total lack of history. This poses a risk to the producibility and reliability of the product.

According to TTI Silicon, a wafer distribution and reclamation company, wafer reclamation applications have been increasing, as chip-processing plants want to lower costs. Fabrication facilities are interested in recycling their prime test and monitoring wafers and reuse them for production or testing [33].

The reactive obsolescence mitigation approaches described earlier are methods whereby a sudden or unexpected obsolescence situation can be handled. The methods may fail at any moment and may not provide the parts support required throughout the life of the system. Obsolescence is a certainty; the only unknown is the timing; therefore, it is important to have preventive obsolescence mitigation solutions at the design stage.

16.3.9   Design for obsolescence

Products designers must be knowledgeable about the components industry, the forces acting on it, and the trends in its response to those forces. Strategically, this involves education about the industry and the supply chain upon which it feeds rather than training in specific components and component designs. In the product design process, manufacturers should develop and use processes to consider future expected component changes, including obsolescence, in their internal design approval processes (preliminary design review [PDR], critical design review [CDR] etc.).

Using the available knowledge of the forces acting on them, equipment manufacturers can estimate the amount by which such details as feature size or supply voltages will decrease or speed will increase. These trends can be accommodated in the designs by partitioning the design to ensure that high-risk items are contained on replaceable modules within the products known as "shop discardable units" (SDUs). Replacement SDUs, which are FFF replacements for the original module, can be substituted with minimal affect on the remainder of the LRU. Even though the future cannot be exactly predicted, equipment suppliers can prepare for it.

Equipment suppliers should develop plans for their products with regard to expected component changes over the entire life cycle of the product. Obviously, equipment suppliers cannot predict exactly when or how component designs will change or when they will become obsolete. The picture becomes especially unclear when trying to look beyond the first 5–10 years of product life.

The equipment supplier is responsible for the entire life cycle, and if a life-cycle plan for the product is developed, the equipment supplier can at least begin to understand what the right questions are. Each equipment supplier should develop a general plan for dealing with component obsolescence. In addition, there should be a specific plan, to be reviewed in PDR, CDR, etc., for the product. The plan might include, for example, a list of key components with a severe obsolescence impact, plans for last-time buys of key components, scheduled redesigns, and the impact on and recommendations for modifications to product operation, maintenance, and support should particular contingency plans be implemented.

It is important that circuit designers have easy and direct access to the obsolescence forecasting information relating to the components that they are choosing. Equipment suppliers may also find it beneficial to maintain a preferred parts list. The emphasis here is on preferred or recommended parts since such lists are inherently difficult to keep up-to-date and a mandatory list would not be consistent with the product performance requirements.

Developers of new designs should be required to select components that can pass an internal selection process gate. The process should ensure that components are qualified for the application, are from a vendor in whom the purchaser has confidence and which can be produced with adequate ongoing quality. There is an incentive to use components that have already demonstrated these attributes since the approval costs are lowered, but there should be no absolute requirement to do so. The progress of technology leads to a proliferation of new parts selections and of the corresponding component identification codes that will undoubtedly continue. We might take some steps to restrain this trend by considering the true cost of introducing a new component by considering the approval process and stock-holding cost throughout the product's life cycle.

There is also the loss of consolidation benefits when opportunities for selecting existing approved components are not taken. Addressing this issue has the beneficial effect of reducing component selection costs and at the same time enlarging the market for those components that remain. Taken across the industry, this translates into an enlarged market for the component supplier and reduces the chance of that market being insufficient to sustain the supplier's interest. Part selection and management plans should take into account the market size for each part. However, such a policy is only partly effective since a step technology capability increase can wipe out a market in a very short time as the volume users switch en masse to the newer components.

Only when the service life of an electronic product ends before the end of service life for all its components is there no risk of obsolescence associated with that product. For military/avionics systems, 10 changes/upgrades of parts and 5 upgrades for assembly of the product can be expected during the system's life cycle. The goal of system design teams should include reconfigurability of subsystems (printed wiring boards and higher levels) of assemblies for easy modifications later on in the system's life cycle by using:

- Programmable logic devices such as field programmable gate arrays (FPGA) or programmable logic devices (PLD). Redesigns can benefit from the use of programmable devices.
- Flexible designs using alternative package styles/footprints (to account for transitions from one manufacturing process to another, – e.g., through-hole technology to surface mount technology) occur rapidly, which prevents extensive redesigns even when using mature technology in new designs is unavoidable.
- Modular designs/upgrades at the subsystem level; such designs can make module functional interfaces impervious to module changes, redesigns, substitutions, and upgrades. While the components used in individual modules are likely to become obsolete at the same rate as other components, with proper controls system integrators can preserve functions and interfaces. The modular design approach can minimize system redesign impact, and if modules are used across multiple products, some economies of scale may be obtained. Thus, by using modular designs, products and subassemblies can be made reusable, upgradable, and scalable. In order to install upgraded versions of an existing component or package design, circuit card layout changes are common. Hence, high obsolescence risk components should be placed in subassemblies (line discardable units or LDUs) that should be discarded and replaced with alternative pin-for-pin compatible assemblies even if the individual components change. PC manufacturers with microprocessors and memory cards do this routinely. An acceptable "threshold" cost, below which it is economical to discard a subassembly and therefore consider it an LDU, should be defined.

Typically, the engineer should keep the following in perspective when implementing new designs:

- Has the design been partitioned to simplify solutions to expected component obsolescence problems?
- Have memory growth paths been incorporated in the design layout?
- Do microprocessors have roadmaps defined by the manufacturer?
- Does the design consider plans for future enhancements/upgrades?
- Are preplanned product upgrades identified?
- Are planned design upgrades compatible with those of similar products using similar components?
- Is design information archived in sufficient detail to facilitate future upgrades if the original design team is not available?
- Are high-risk parts placed on subassemblies that are economical to discard?
- Are cost estimates available for scheduled design upgrades?
- Are support guarantees compatible with forecasting component availability?

- Can minor variations in product operating procedures allow the use of parts with a lower risk of obsolescence?

16.3.10 Open systems

The forgoing sections describe the processes that an equipment manufacturer can use to mitigate the effects of obsolescence. The rate at which obsolescence is affecting equipment manufacturers is increasing rapidly as the market life of commercially available components decreases. Thus the burden associated with managing this change is also increasing. To try to minimize this burden, an approach to design that eases this problem is becoming a necessity. One such approach is "open systems." There are many levels at which "open systems" can be defined. From the equipment manufacturer's point of view, open systems is an approach that permits modules to be replaced as an obsolescence fix rather than continuing to address the issue at the component level.

Open architecture and standardized interfaces permit FFF replacement modules to be substituted during repair for those that have become unrepairable due to obsolescence. The existence of standardized interfaces minimizes the consequential effects of changes and sets up change containment boundaries within which change can proceed independently. The common approach to commercial-off-the-shelf (COTS) development is the use of industry standard modules of a particular family (e.g., VME, Compact PCI) that are purchased from a third party or manufactured directly by the equipment manufacturer and assembled into a chassis. These all share the desirable attribute of adhering to public specifications with strictly defined interface characteristics in terms of mechanical, electrical, and protocol considerations. These modules are both electrically and mechanically standard and are available from more than one source, although there may be differences that are not significant to the end application. The modules usually include a number of functions that are not needed for the particular application. They thus carry a cost, size, weight, and power consumption overhead. They also carry a verification overhead due to the necessity of showing that the unused functions cannot cause unintended effects under the reasonably expected operating conditions and scenarios. This approach to open systems is popular, but the above-mentioned drawbacks need to be fully considered when choosing which open systems approach to adopt. For example, a bus-based open architecture based on standards, e.g., IEEE 1394, may utilize custom card formats and designs yet still maintain the essential attribute of change containment.

The opens systems approach is being actively pursued by military avionics manufactures (e.g., [34] and others) and is a draft policy of the U.S. DoD [35], [36]. Open systems promise to radically change the business model of equipment suppliers and rearrange the industry into distinct hardware, software, and systems integrators.

16.3.11 Hardware-software independence

A major concern with hardware changes is the consequential effects on the system through faulty operation of previously correct software. Such faults are very subtle and hard to predict and may show up only under very constrained circumstances which testing has little hope of revealing. In an avionics context, this is the topic addressed by [37]. A solution to this problem may be to use a layered architecture of hardware, hardware services software and applications software, and well-defined and stable specifications for the interfaces. Such architecture has been promulgated in previous work for avionics [40]. Figure 16.4 shows one such avionics implementation where APEX is the interface definition for software services and COEX is the interface definition for hardware services.

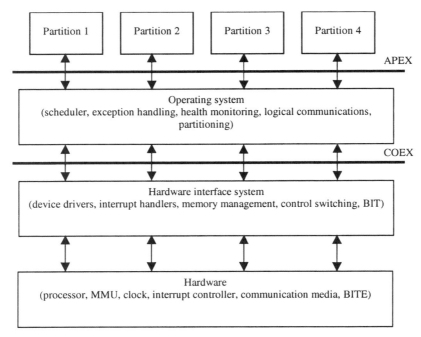

Figure 16.4: ARINC 653 APEX software architecture [39]

16.4 Economics of obsolescence management strategies

A decision on the choice of an obsolescence management option should be based on the life-cycle cost impact of initiating and implementing the decision. Information about the immediate piece part and nonrecurring engineering costs is not enough. This section describes some of the factors that play a role in determining the life-cycle cost of obsolescence management.

The various affected parties view the costs of parts obsolescence differently, and the metrics used to measure the costs vary according to the different interests of the entities involved in the process, i.e., suppliers, customers, and acquisition and regulatory agencies. The cost of managing obsolescence also varies, depending on the range of preventive action taken, from relatively low to very high. The process of handling obsolescence can even result in long-term cost savings. The task of the parts selection and management team is to choose solutions that result in an optimal life-cycle system cost. The various components to be considered when assessing the cost of obsolescence are introduced briefly below.

All solutions to the problem of obsolescence have some associated cost elements in addition to the cost of the solution itself. These may be manifested as increasing costs of ownership such as:

- decreased reparability
- reduced mission effectiveness
- configuration management
- support equipment proliferation
- loss of installation baseline controls

The process of responding to obsolescence has specific associated costs, such as:

- redesign or emulation cost
- time needed to procure replacement
- additional cost or premium paid for replacement
- testing costs (for part, board, box, system, platform)
- qualification costs (for part, board, box, system, platform)
- manufacturing and repair labor costs

Not all these costs will be incurred in all situations. Estimating the costs of obsolescence is a process of comparing alternatives in time and technology. Some of these cost alternatives and assessments are discussed in previous sections where different strategies for obsolescence management are presented. For each alternative, the cost of various options is to be compared against the cost of all other options. When there is a large number of available options, the decision-making process can be very complicated. The importance and difficulty of this process of cost estimation and optimization reemphasizes the importance of the parts selection and management team and process. Determination of the most cost-effective postobsolescence solution, typically a solution involving redesign of the product versus a last-time or lifetime buy, involves both technical and economic trades: technical trades to understand the options available to resolve the obsolescence issue and economic trades to understand the cost-benefit relationships and the most cost-effective time to initiate a redesign activity. Boeing Electronic Products [14] has developed for internal use a simple set of relationships to allow the product design engineering team to assess the cost effectiveness of the solution and to determine the optimum year to redesign for a minimum-cost solution. These net present value (NPV) techniques are used to determine the cost effectiveness of trades between multiple solution options.

Variations in obsolescence management costs come from three major sources: the first two are the part-level and system-level sources, and the third source is non-system-specific economic factors. These three sources are briefly discussed as follows.

16.4.1   Cost variations at the board or component level

In response to obsolescence, board-level changes are often made by different means, including piece part replacement from alternative sources and redesign of the board to use a different part. The impact of the choice made at this level goes beyond the direct cost of replacement or the cost of man-hours doing redesign.

A given substitute part for a given board may not be equally functional in all locations on the board. Some socket locations might require local adjustments such as adjusting digital timing in asynchronous design applications. Such adjustments add to the cost of the substitution process.

On the other hand, careful planning may reduce the overall cost of such changeovers. The design and procurement cost of similar parts can be shared across the platform. Redesign of one board might release enough inventories for life-cycle maintenance of another board. Board-level redesign can reduce the number of part types that must be maintained in inventory, reducing inventory costs.

16.4.2   Cost variations at the system or module level

Some decisions regarding cost estimation for obsolescence management can only be made at the system or higher levels. Cost estimates also should factor in the level of accuracy required in the test vectors for validating the various levels of assemblies while making any type of modification. The options are either to aim for very high fault coverage at each level or to have merely adequate coverage to guarantee performance at the next higher level of assembly. Another set of options has to do with deciding at what level FFF will be maintained. At each level of design, the changes may be made transparent to the next level of assembly and connectors, or some changes can be carried over to the level above.

16.4.3 Cost variations due to economic policy factors

Contributors to the difficulty of accurately estimating the costs of obsolescence management are macroeconomic factors and public and business policies. The cost of supporting a system (with or without modifications) changes with time. This support cost depends on economic factors such as rate of inflation, cost of capital, and interest rate. The level of available funding for different budget segments also influences the decisions made regarding how to handle obsolescence.

Once a discontinued item is identified, the manager carries out a four-step process to learn the extent of the problem and determine a resolution. The first step is to determine where the discontinued item is used and the total future requirements for the part in each application. The second step is to determine whether a quick fix, such as sufficient stock on hand or substitute parts to cover future requirements, or a bridge buy is feasible. If a quick fix is not feasible, the third step is to gather logistical information about the part. This includes demand levels, future requirements, prequalified sources, and availability of descriptive information such as specifications, performance data, and interface descriptions. Finally, the manager decides what resolution or mix of resolutions will solve the obsolescence problem and then implements the solution. A complete analysis of each option is costly and time-consuming. An analysis of each option must consider factors such as cost and execution time frames.

16.5 References

[1]   Pecht, M., R. Solomon, P. Sandborn, C. Wilkinson and D. Das, *Life-Cycle Forecasting, Mitigation Assessment and Obsolescence Strategies*, CALCE EPSC Press, College Park, MD, 2002.

[2]   Dasgupta, A., E. B. Magrab, D. K. Anand, K. Eisinger, J. G. McLeish, M. A. Torres, P. Lall and T. J. Dishongh, "Perspectives to Understand Risks in the Electronic Industry," *IEEE Transactions on Components, Packaging and Manufacturing Technology*, Part A, Vol. 20, pp. 542–547, December 1997.

[3]   Bumbalough, A., "USAF Manufacturing Technology's Initiative on Electronics Parts Obsolescence Management," *Proceedings of the 44th International SAMPE Symposium*, pp. 2044–2051, May 1999.

[4]   Stogdill, R. C., "Dealing with Obsolete Parts," *IEEE Design and Test of Computers*, Vol. 16, pp. 17–25, April 1999.

[5]   Wright, M., D. Humphrey and P. McCluskey, "Uprating Electronic Components for Use Outside Their Temperature Specification Limits," *IEEE Transactions on Components, Packaging and Manufacturing Technology*, Part A, Vol. 20, pp. 252–256, June 1997.

[6]   Levitt, T., "Exploit the Product Life Cycle," *The Harvard Business Review*, Vol. 43, pp. 81–95, November 1965.

[7]   Pecht, M. and D. Das, "The Electronic Part Lifecycle," *IEEE Transactions on Components and Packaging Technologies*, Vol. 23, pp. 190–193, March 2000.

[8]   Solomon, R., P. Sandborn and M. Pecht, "Electronic Part Lifecycle Concepts and Obsolescence Forecasting," *IEEE Transactions on Components and Packaging Technologies*, Vol. 23, pp. 707–717, December 2000.

[9]   Texas Instruments, 1999 Standard Linear and Logic Product Withdrawal Device List, Dallas, TX, September 2, 1999.

[10]   National Semiconductor, "Product Folder, LF11331-4 Normally Open Switches with Disable [Discontinued]," January 14, 2002, <http://www.national.com/pf/LF/LF1133 1.html>, accessed, January 14, 2002.

[11]   Electronic Industries Alliance, Product Life Cycle Data Model, EIA-724, Arlington, VA, September 19, 1997.

[12]   Wilkinson, C., J. Wasson, L. Condra, J. Fink, C. Scheidt, T. Brennom, Z. Porter, J. M. Chopin, R. Hoad, J. Matthews, J. Chapman, D. Humphrey, P. Feussinger and Y. Sagot, "Component Obsolescence Management for Aerospace Electronic Equipment," *Proceedings of the 1999 Avionics Conference: Civil and Military Convergence?*, London, November 1999.

[13]   Condra, L., D. Followell, G. Houchens, J. Jenks, M. Koehler and Z. Porter, "Minimizing the Effects of Electronic Component Obsolescence," *Aero Magazine*, Vol. 10, April 1, 2000, Boeing Company, Seattle, WA, <http://www.boeing.com/ commercial/aeromagazine/aero_10/elect.pdf>, accessed January 14, 2002.

[14]   Porter, Z., "An Economic Method for Evaluating Electronic Component Obsolescence Solutions," Seattle, WA, May 1, 1998, <http://www.geia.org/sstc/G12/ porter_paper.pdf>.

[15]   Condra, L., "Combating Electronic Obsolescence by Using Common Processes," *Proceedings of the 2nd Annual NDIA Systems Engineering and Supportability Conference*, September 1999.

[16]   Wasson, J. and C. Wilkinson, "Component Obsolescence Management for Military and Space Electronics," *Proceedings of the 5th Annual International Conference and Exhibition, Commercialization of Military And Space Electronics*, Los Angeles, CA, February 2001.

[17]   Cavill, P. J., "Whole Program Life COTS: Warding off the Curse of Component Obsolescence," *Military and Aerospace Electronics*, December 1, 2000, PennWell Corporation, <http://mae.pennwellnet.com/Articles/Article_Display.cfm?Section= Archives&Subsection=Display&ARTICLE_ID=88120>, accessed, January 15, 2002.

[18]   Condra, L., A. Anissipour, D. Mayfield and M. Pecht, "Electronic Components Obsolescence," *IEEE Transactions on Components, Packaging, and Manufacturing Technology* – Part A, Vol. 20, pp. 368–371, 1997.

[19]   Condra, L. and J. Matthews, "The Growing Problem of Component Obsolescence," *Avionics Magazine*, pp. 26–31, March 1999.

[20]   Livingston, H., "SSB-1: Guidelines for Using Plastic Encapsulated Microcircuits and Semiconductors in Military, Aerospace and Other Rugged Applications," *Defense*

Microelectronics Activity, December 14, 2000, <http://www.dmea.osd.mil/ssb1_paper.pdf>, accessed, January 15, 2002.

[21] Pecht, M., J. Fink, E. Hakim and J. Wyler, "An Assessment of the Qualified Manufacturer List (QML)," *IEEE Aerospace and Electronic Systems Magazine*, Vol. 12, pp. 39–43, July 1, 1997.

[22] Military Products: TI Military Lifetime Buy Notices, Texas Instruments, Dallas, TX, December 31, 1999.

[23] "White Section: SIA/GPC Perspective and Recommendations on DMS, Semiconductor Industries Association," 1996, http://www.semichips.org/, accessed November 6, 2003.

[24] Cushing, M. J., D. E. Mortin, T. J. Stadterman and A. Malhotra, "Comparison of Electronics-reliability Assessment Approaches," *IEEE Transactions on Reliability*, Vol. 42, pp. 542–546, December 1, 1993.

[25] "U.S. Code: Title 10, Chapter 131, Section 2213, Limitation on Acquisition of Excess Supplies, U.S. Department of Defense," <http://www4.law.cornell.edu/uscode/10/2213.html>.

[26] Syrus, T., U. Ramgopal and M. Pecht, "Manufacturer Assessment Procedure and Criteria for Parts Selection and Management," *IEEE Transactions on Electronics Packaging and Manufacturing*, Vol. 24, pp. 351–358, October 2001.

[27] Syrus, T., M. Pecht and D. Humphrey, "Part Assessment Guidelines and Criteria for Parts Selection and Management," *IEEE Transactions on Electronics Packaging Manufacturing*, Vol. 24, pp. 339–350, October 2001.

[28] Press release: Elantec Semiconductor Announces Military Market Exit, Elantec Semiconductor, Milpitas, CA, July 1, 1997.

[29] Kroeger, R., "Manufacturer Part Assessment: A Pragmatic Approach," *First Workshop on Use of Components Outside Specified Temperature Limits*, CALCE EPSC, University of Maryland, College Park, MD, January 1997.

[30] Easter, R., VIEWPOINT: Steer Clear of Aftermarket Parts Leprechaun, Military and Aerospace Electronics, PennWell Corporation, Tulsa, OK, May 1, 1994.

[31] U.S. Navy, "GEM: The Generalized Emulation of Microcircuits, Navsea Crane DMS Technology Center," <http://dtc-dms.crane.navy.mil/html/gem.htm>, accessed January 15, 2002.

[32] Fitzhugh, G. L., "Rapid Retargeting, A Solution to Electronic Systems Obsolescence," *Proceedings of the IEEE 1998 National Aerospace and Electronics Conference* (NAECON 1998), pp. 169–176.

[33] "With Wafer Reclamation Growing, Company Renames Itself TTI Silicon," *Semiconductor Business News*, January 19, 2000, CMP Media, <http://www.siliconstrategies.com/story/OEG20000119S0036>, accessed January 15, 2002.

[34] Walden, D., K. Blanshan and G. Kranz, "Technology Roll Lessons Learned on Embedded Avionics Platforms," *IEEE Aerospace and Electronics Systems Magazine*, Vol. 15, pp. 19–23, May 1, 2000.

[35] Mandatory Procedures for MDAPs and MAIS Acquisition Programs, Draft DoD 5000.2-R, U.S. Department of Defense, April 1, 2000.

[36] Joint Technical Architecture V3.0, U.S. Department of Defense, <http://www-jta.itsi.disa.mil/jta/jtav3-final-19991115/finalv3.html>, accessed, November 6, 2003.

[37] CAST position paper P32, Federal Aviation Authority, Washington, DC, May 1, 1999.

[38] Design Guide for Integrated Avionics, ARINC Report 651, Aeronautical Radio Inc., Annapolis, MD, November 1, 1991.

[39] Avionics Application Software Standard Interface, ARINC Specification 653, Aeronautical Radio Inc., Annapolis, MD, January 1, 1997.

[40] Wilkinson, C., M. Pecht, J. Wasson and L. Condra, Recommendations for Future Avionics Hardware Development, 2002-01-2983, World Aviation Congress, Phoenix, AZ, November 2002.

Chapter 17

Part Acceptance and Risk Management

Peter Sandborn

The part selection and management process involves "risk-informed" decision making and thereby a prediction of the impact (adverse consequences) of critical decisions. It is likely that the level of effort expended on risk analysis will vary from part to part. Risk assessment during part selection and management may be as simple as a single engineer deciding that the candidate part is the only viable choice or a complex formal process with the involvement of many stakeholders. Some parts play a trivial role in the system and are easily substituted for if problems occur. In this case, the consequences associated with problems that may occur are relatively minor; therefore, the part acceptance process is simple. In other cases, a formal process is necessary, not only to reduce selection risk but also to document the decision making process.

17.1 Risk assessment process

"Risk" is defined as the probability of an unfavorable event occurring. General methodologies for risk analysis (both quantitative and qualitative) have been developed and are widely available, e.g., [1], [2]. The objective of this chapter is to present a process that is specific to the electronic part selection and management process. The process for assessing the risks associated with accepting a part for use in a specific application is shown in Figure 17.1. Although Figure 17.1 appears complex, it is really just a formalization of the selection process that engineers already use. The advantage gained by formalizing the process is that the risk of overlooking an important issue is reduced. The steps in the risk assessment process are as follows:

- *Start with a risk pool* – The risk pool is the list of all known risks, along with general knowledge of how those risks are quantified (if applicable) and possibly mitigated. The information in the risk pool is general and requires the addition of application-specific knowledge in order to form a useful risk catalog for the part under consideration or the product under design.
- *Determine an application-specific risk catalog* – Not all risks in the risk pool will be applicable to every product or part. Using the specific application's properties, risks are selected from the risk pool to form an application-specific risk catalog. The application properties most likely to be used to create the risk catalog include functionality, life cycle environments (manufacturing, shipping and handling, storage, operation, and possibly end-of-life), manufacturing characteristics (schedule, quantity, location, suppliers), sustainment plans and requirements, and operational life requirements.

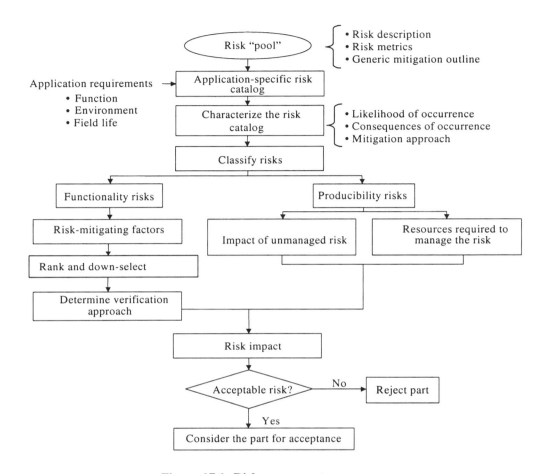

Figure 17.1: Risk assessment process

- *Characterize the risk catalog* – Application-specific details about the likelihood of occurrence, consequences of occurrence, and acceptable mitigation approach(es) for each of the risks in the risk catalog are generated. The likelihood and consequence of occurrence should be independent of each other.

- *Classify risks* – Each risk in the risk catalog is classified into one of two categories: functionality risks and producability risks. Functionality risks are risks for which the consequences of occurrence are loss of equipment, mission, or life. Functionality risks impair the product's ability to operate to the customer's specification. Producability risks are risks for which the consequences of occurrence are financial (reduction in profitability). Producability risks determine the probability of successfully manufacturing the product, which in turn refers to meeting some combination of economics, schedule, manufacturing yield, and quantity targets.

- *Determine risk-mitigating factors (functionality risks)* – Factors may exist that modify the applicable mitigation approach(es) for a particular part, product, or system. These factors include the type or technology of the part under consideration, the quantity and type of manufacturer's data available for the part, the quality and reliability monitors employed by the part manufacturer, and the comprehensiveness of production screening at the assembly level. Besides technical and quality issues ascribed to the part, mitigating factors may include less quantitative experience such as the history with the part supplier and personal relationships.

- *Rank and down-select (functionality risks)* – Not all functionality risks require mitigation. If the likelihood or consequences of occurrence are high, the risk should be addressed. Alternatively, if the likelihood or consequences of occurrence are low, the risk may not need to be addressed. The ranking may be performed using a scoring algorithm that couples likelihood and consequence into a single dimensionless quantity that allows diverse risks to be compared.[1] Once the risks are ranked, those that fall below some threshold in the rankings can be omitted. The threshold level is determined by a review of the functionality risk ranking coupled with the application-specific safety and reliability requirements.

- *Determine the verification approach (functionality risks)* – For the risks that are ranked above the threshold determined in the previous activity, the mitigation approaches defined in the risk catalog are considered. The acceptable combination of mitigation approaches is selected from the mitigation approaches being considered and becomes the required verification approach.

- *Determine the impact of unmanaged risk (producability risks)* – The likelihood of risk occurrence is combined with the consequences of occurrence to predict the resources associated with risks that the product development team chooses not to manage proactively.

- *Determine the resources required to manage the risk (producability risks)* – The resources necessary to manage producability risks are determined. The risks are managed by creating a management plan and performing a prescribed regimen of monitoring the part's field performance, vendor, and assembly/manufacturability as applicable.

- *Determine the risk impact* – The impact of functionality risks is assessed by estimating the resources required to develop and perform the worst-case verification activity allocated over the entire product life cycle (production and sustainment). The value of the product that may be scrapped during the verification testing is included in the impact. For managed producability risks, the resources required are used to estimate the impact. For unmanaged producability risks, the resources predicted in the impact analysis are translated into costs.

- *Decide if the risk is acceptable* – If the impact fits within the overall product's risk threshold and budget, then the part selection can be made with the chosen verification activity (if any). If the impact is not viable, design changes or alternative parts must be considered.

17.2 The risk pool

This section contains some of the risks that must be considered during the part selection process. The risk pool includes risks organized by part selection activity, risk classification – functionality or producability, risk metric(s) – the production, field use, or life cycle

---

[1] Risk "scoring" algorithms are most prevalent in the assessment of financial and investment risk [3]; however, the development and use of scoring algorithms for other types of risk is not uncommon. Limitations of methodologies that involve quantitative risk assessment, such as scaling or scoring, can be found in [4].

measure that will be affected by the risk, and applicable mitigation approach(es) if the OEM chooses to mitigate the risk and what form(s) the mitigation can take. The risks articulated herein are very general and are usually customized for particular organizations. Not all the risks listed in the risk catalog need be applicable to every application.

Availability risks (see Table 17.1) – Availability risks are associated with the delivery of a part to the manufacturing process. The following risks are associated with availability:

- *limited volume* – the part manufacturer or distributor cannot supply the part in the time or at the rate required by the application. Example: production plans for a product that contains one instance of part A are 100,000 units per month. The manufacturer or distributor of part A has the capacity to supply only 50,000 parts per month.
- *late delivery* – in this case, the manufacturer or distributor promises to deliver the required number of parts, but the delivery schedule slips and the parts are delivered to manufacturing late.
- *early delivery* – in this case, the manufacturer or distributor promises to deliver the required number of parts on a specified date, but the delivery arrives early. Early delivery risk is not an issue if small quantities of parts are involved and appropriate part storage facilities already exist.

Table 17.1: Availability risks

Availability risks	Risk classification	Risk metric(s)	Mitigation approach(es)
Limited volume	Producability	Rate (e.g., parts/mo)	Develop a second source
Late delivery	Producability	Production schedule delay	
Early delivery	Producability	Storage cost Handling/storage defects	Create appropriate part storage and handling capabilities

Quality risks (see Table 17.2) – Quality risks focus on poorer than expected quality levels or on quality levels that are erratic or decline over time. The following risks are associated with quality:

- *lower than expected quality* – the parts received for assembly into a system have a poorer quality than the quality quoted or expected by the part manufacturer or distributor.
- *lot-to-lot quality variation* – significant variation in quality is observed between different lots.
- *quality decline over time* – a gradual decline in quality is observed over some period of time.

Table 17.2: Quality risks

Quality risks	Risk classification	Risk metric(s)	Management plan
Lower than expected quality	Producability and functionality	Test scrap rate	Supplier intervention
Lot-to-lot quality variation	Producability and functionality		
Quality decline over time	Producability and functionality		

Supplier intervention can range from notification of the supplier, to audits of the manufacturer's processes, to employing third parts surveillance houses (see Chapter 11). The manufacturer should be informed about the quality problems observed. Possible business implications of not improving the quality function should also be communicated to the manufacturer. Providing assistance to the manufacturer in establishing the quality functions that the manufacturer lacks should be considered. Clearly, such intervention involves a substantial cost and time commitment and should be undertaken only if extended business relations are anticipated.

Qualification risks (see Table 17.3) – Qualification risks are associated with how much effort the OEM must dedicate to the qualification or requalification of a part. The following risks are associated with qualification:

- *OEM must perform all qualification* – the manufacturer of the part does not perform any qualification activities.
- *periodic requalification required* – the part design, fabrication, or packaging changes frequently, requiring requalification of the part.
- *system qualification ramifications of changes to or changing the part* – the part has significant system qualification ramifications (hardware and/or software) associated with its use.

Table 17.3: Qualification risks

Qualification risks	Risk classification	Risk metric(s)	Management plan
OEM must perform all qualification	Producability	Schedule delay Cost of qualification	Supplier intervention
Periodic re-qualification required	Producability	Cost of qualification	Supplier negotiation
System qualification ramifications	Producability	Schedule delay Cost of qualification	Supplier negotiation

Supplier negotiations may involve putting in place an agreement with the supplier not to make changes to the version of the part being used for a specified period of time.

Table 17.4: Reliability risks

Reliability risks	Risk classification	Risk metric(s)	Management plan
Time to failure not known	Functionality		Develop test/modeling plan to determine part reliability Supplier intervention to develop a process for implementing failure-free operating periods for parts
Time to failure much less than product life	Producability	MTBF, FFOP Maintenance cost	Supplier intervention
Consequences of failure	Functionality	Cost Customer confidence	Preventive maintenance plan

Reliability risks (see Table 17.4) – Reliability risks are associated with mismatches between the product (system) life and the part life. The following risks are associated with reliability:

- *time to failure of the part much less than product life* – the failure-free operating period is significantly less than the product life.
- *consequences of failure* – field failure of the part results in a significant loss (financial, property, life).

Assembly risks (see Table 17.5) – Assembly risks are associated with mismatches between the part and the board (or other parts on the board). The following risks are associated with assembly:

- *difficult to assemble* – nonstandard or manual assembly steps required; extra setups or calibrations of equipment required.
- *poor testability* – testing is slow or fault coverage is poor
- *poor reworkability* – difficult (expensive) or impossible-to-rework defective parts during the assembly process.

Table 17.5: Assembly risks

Assembly risks	Risk classification	Risk metric(s)	Management plan
Difficult to assemble (part and board not well matched)	Producability	Assembly cost, quality, and time	Additional operator training Increased equipment monitoring
Poor testability	Producability	Test cost and time Test fault coverage	Refine testing process Additional operator training
Poor reworkability	Producability	Rework cost, time Rework success rate	Refine rework process Additional operator training

Obsolescence risks (see Table 17.6) – Obsolescence risks are associated with mismatches between the part and the product life cycles. The following risk is associated with obsolescence:

- *part life is less than the product life* – the part becomes obsolete before the product becomes obsolete.
- *part obsolescence significantly impacts nonhardware system components* – if the part needs to be replaced with a different part, the impact on nonhardware portions of the system, i.e., system software, needs to be understood.

Table 17.6: Obsolescence risks

Obsolescence risks	Risk classification	Risk metric(s)	Management plan
Part life is less than the product life	Producability	Time to obsolescence or risk of obsolescence	Mitigation approaches: Manufacturer negotiation Lifetime buy Substitute part Aftermarket source Emulation Redesign Reclaim Reverse engineering
Part obsolescence significantly impacts nonhardware system components	Producability	Time to obsolescence or risk of obsolescence	Mitigation approaches: (same hardware mitigation approaches) plus modularization of software

The part/product life cycle mismatch is a problem if the part's time to failure is also less than the product life or if additional units have to be manufactured for reasons other than spare replenishment. If the part's time to failure is longer than the product life and no additional production is planned, no parts should require replacement during the product's life and part obsolescence is irrelevant. A detailed discussion of the part obsolescence risks appears in [5].

Environmental mismatch risks (see Table 17.7) – Environmental mismatch risks are associated with mismatches between the manufacturer-specified operating environment, and the local environment the part will be exposed to in the product. The following risks are associated with environmental mismatches:

- *poor characterization of the local environment* – the local environment that the part is exposed to is not known, or is not known for the entire life cycle of the part in the product.
- *necessary notifications not available* – product change notifications are not issued or are not received by the appropriate people in a timely manner.
- *poor fault coverage during functional test* – functional testing is allowing too many types of faults to pass undetected.
- *part handling and testing-induced defects* – defects are being introduced into parts during their handling or testing.

- *legal liability* – the product manufacturer (OEM) may be held legally responsible for product failures caused by using parts outside of the part manufacturer's specifications.

Table 17.7: Environmental mismatch risks

Environmental mismatch risks	Risk classification	Risk metric(s)	Management plan
Poor characterization of local environment	Functionality	Product failure rate Maintenance cost	Determine life cycle of local environment for the part via modeling or measurement
Necessary notification(s) not available	Functionality	Product failure rate	Request that part supplier (or distributor) provide product change notifications (PCNs) Establish a part-specific PCN routing procedure within your company
Poor fault coverage during functional testing	Functionality	Product failure rate	Improve fault coverage by changing testing software or hardware
Part-handling and testing-induced defects	Producability	Scrap rate (cost of additional parts)	Improve handling and testing procedures
Legal liability	Producability	Cost of lawsuits	Involve legal council in part selection process when parts are used outside of the manufacturer's specifications

Functional risks (see Table 17.8) – functional risks are associated with functionality that is possible within the product but either unknown or known but unnecessary. The following risks are associated with functionality:

- *unintended functionality* – the product functions in an unintended manner due to effects such as sneak circuits, EMI, and switching noise.
- *undeclared additional functionality* – extra functionality that the requirements specification for the product did not include.

Table 17.8: Functional risks

Functional risks	Risk classification	Risk metric(s)	Management plan
Unintended functionality	Functionality	Operational downtime	Document problems
Undeclared additional functionality	Functionality		

Sustainment risks (see Table 17.9) – sustainment risks are associated with long-term resources incurred after the part is fielded. The following risks are associated with sustainment:

- *poor reparability* – the product becomes difficult (expensive) or impossible to repair due to the presence of the part.
- *poor upgradability or reconfigurability* – the product becomes difficult (expensive) or impossible to upgrade or reconfigure due to the presence of the part.
- *poor maintainability* – the product becomes difficult (expensive) or impossible to maintain due to the presence of the part.
- *expensive end-of-life (EOL) requirements* – the product becomes difficult (expensive) to EOL due to the presence of the part. Only applicable if EOL is a requirement for the product.

Table 17.9: Sustainment risks

Sustainment risks	Risk classification	Risk metric(s)	Management plan
Poor reparability	Producability	Repair cost and time	Additional training for repair personnel
Poor upgradability and reconfigurability	Producability	Upgrade cost and time	
Poor maintainability	Producability	Maintenance costs	Create EOL plan including resources to perform EOL operations
Expensive EOL requirements	Producability	EOL costs	

17.3 Managing risks during part selection

The risk assessment process in Figure 17.1 breaks down into two different types of risk:

- *managed risks* – risks that the product development team chooses to manage proactively by creating a management plan and performing a prescribed regimen of monitoring the part's field performance, vendor, and assembly/manufacturability.
- *unmanaged risks* – risks that the product development team chooses not to proactively manage.

If risk management is considered necessary, a plan is prepared. The plan contains details about how the part is monitored (data collection) and how the results of the monitoring feed back into various parts selection and management processes. It is important to estimate the feasibility, effort, and cost involved in management processes prior to making the final decision to select the part.

At the completion of the part selection process, the decision whether to use the part or to reject it is made. The decision is based the part's assembly performance, field performance, and sales history, which may be essential to ascertain the validity of the predictions made during the part selection process. If the feedback calls for changes in selection criteria, such changes are incorporated into the part selection process. Prospective parts in the future are judged based on the altered part selection criteria. Part monitoring is also needed to make changes in parts that are already in use. For example, part monitoring field data might indicate that a change in operating conditions is required for the part to perform satisfactorily.

The existence of one or more of the risks discussed in the last section does not mean the risks have to be managed. The decision to manage risks should be based on a comparison of the resources required to manage the risk and the likely resources required if the risk is left unmanaged. The sections that follow define processes for estimating the resources associated with managed and unmanaged risks. The key metric used to determine whether risks should be managed or not is resources. Resources consist of time, data, or money (all ultimately translate into costs).

17.4 Management plan

During part selection, the resources required to manage the part throughout the application life cycle must be identified and evaluated when the decision is made to accept or reject the part. The purpose of the management plan is to define how to collect information and how to make decisions using this information (see Figure 17.2).

The management plan consists of determining what information needs to be collected during the part's field life, where the information should be collected (the information sources), and how often the information needs to be updated.

Data collection involves monitoring the part to obtain information about it. The data collected depend on the part and the application in which the part is used. Data collection may be active or passive. Active data collection involves directly monitoring the performance of the part. Passive data collection involves waiting for notices of part changes from the manufacturer or reports of changes in product quality from the customer. Evaluating the feasibility of data collection involves knowing where the data come from, how often they will be collected, and when data collection will stop. All these decisions are specific to the application and must be made so that data collection is optimized.

The data collection procedure provides information about the part's performance. Based on this information, certain corrective actions have to be performed. These might involve employing parts management strategies like obsolescence management or they might lead to changes in the part grading data.

A management plan must also address what actions are taken in response to the information gained through the data collected. The questions that must be answered in the management plan are: What data indicate that action is necessary? Who makes the decision that action is necessary? Under what conditions is monitoring terminated? Once a plan is developed, the next step is to estimate the resources required to implement it. If the estimated resources are too large, the part may be rejected or the risks may not be managed.

17.5 Management during part use

For risks associated with a part that will be managed during the product's life cycle, a process like the one in Figure 17.2 should be developed and followed. The process begins with monitoring of the application, followed by a continuous assessment of whether unforeseen problems are arising. If problems do arise, risk management plans are executed. If risk management plans do not exist, then the problem has to be treated on a reactive basis.

After a part is selected for use in a product, it is necessary to monitor its performance and logistics data. Data collected about the part's performance are used to determine if it is necessary to perform management actions on the part and, if so, to determine the management function to be carried out. The collected data are also used to provide feedback to the part-grading process. This ensures that the part's performance is reflected in the decision to select the part in the future.

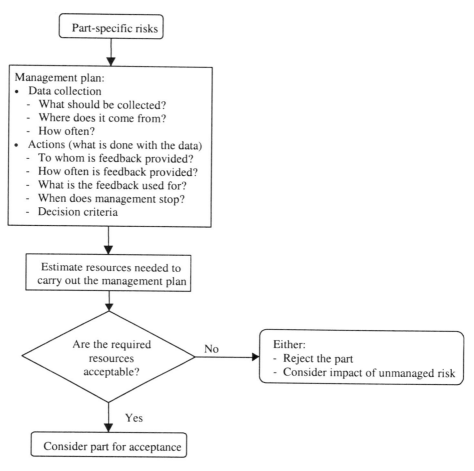

Figure 17.2: Parts selection risk management process

The part is monitored at three different points (vendor, assembly, and field), as shown in Figure 17.3. Data about the part's performance and quality can be obtained from notification from the part source, tests performed on the product after assembly, isolation of parts responsible for failure, and returned products received from the customers.

Notifications from the vendor regarding quality changes or faulty parts that have been shipped in the past are received either directly from the vendor or through independent organizations. Organizations including, the Government Industry Data Exchange Program (GIDEP) [6] obtain notices from the vendors. Notices about faulty parts are usually posted with a date code. If and when such notices are received, it is necessary to check if any parts were received from the vendor in question within the specified time interval and to take suitable corrective action. Corrective action might involve stopping the assembly of these parts or, if the part has already been assembled and shipped to the customer, informing the customer about problems in the part.

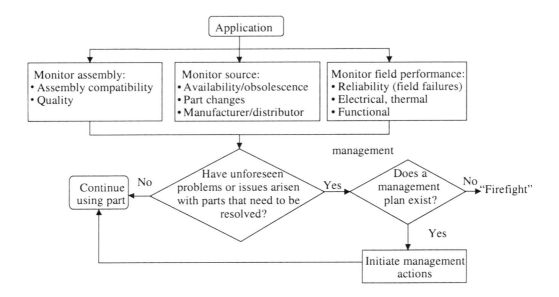

Figure 17.3: Risk management methodology after part selection

Samples from finished products may be tested for mechanical and electrical compliance to requirements and specifications. If a failure is recorded, the product is analyzed further to determine whether the cause of failure can be attributed to a specific part or some other systems attribute, e.g., the assembly operation, the board, etc. If a faulty part is the cause of failure, the data are recorded. Corrective action in the form of changing part-grading data or discontinuance of the part is taken as appropriate.[2] Data collection and analysis procedures at supplier and subcontractor assembly operations help access relevant data when a part failure is documented at any stage, allowing for continuous improvement and cost reduction. Databases containing the part number, manufacturer, lot number, date codes, and, if available, the cause of failure and failure mode are usually maintained for all testing operations. The data are made available to and accessed by all stakeholders associated with product marketing, sales, development, and support to facilitate corrective action and the application of lessons learned to avoid duplication of effort.

Parts should be continuously monitored for life cycle status.[3] In addition to monitoring the manufacturers, distributors, and independent agencies for notices of parts being discontinued, a proactive approach should be followed to determine the life cycle status. This approach should be based on continued awareness of possible changes in technology, part usage in the industry, and the number of manufacturers for specific parts.

---

[2] Note that corrective action may not be warranted for the first failure of a part. Corrective action may be required only when a nonnegligible number of instances (depending on the application) of part failure have been recorded.
[3] Monitoring of the part's life cycle status is relevant only for long-life products, i.e., products that either will be manufactured for a long period of time (with minimal changes) or have to be supported for a long period of time.

17.6 Unmanaged risks

The decision not to manage a risk can be taken in two different stages. One is during the part selection procedure, when the probability of a risk materializing is considered too low to merit constant monitoring or any other management action. The other stage is during part use, when a situation associated with the part occurs and the decision of whether any corrective action has to be taken is considered. The situation is left unmanaged if the cost of management exceeds the cost of ignoring the risk. A process for considering the impact of unmanaged risks is shown in Figure 17.4.

Any proactive approach to parts management involves devising a management plan for the parts. The management plan is specific to the part's application. Carrying out the management plan involves expenditure and work, and it is necessary to first make an estimate of this cost and compare it with the cost (associated with the adverse consequences) of not managing the part. The factors to be considered for this comparison are the probability of the risk materializing and the cost of not managing the risk once it materializes.

When data obtained from part monitoring are compared with the expected performance of the part, and the two do not match, corrective action is usually deemed necessary. At this point, the cost and effort of managing the situation should be evaluated and compared with the cost of leaving the situation unmanaged. In some situations, like obsolescence, there is no option but to take corrective action.

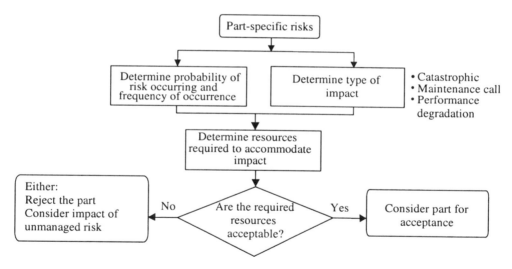

Figure 17.4: Process for considering unmanaged risks

17.7 Summary

The part selection and management process requires risk-informed decision making and the prediction of the impact of critical decisions. Performing a formal risk analysis ensures that the key issues are known and addressed at an appropriate management level.

17.8 References

[1]   Bradley, J., *Elimination of Risk in Systems: Practical Principles for Eliminating and Reducing Risk in Complex Systems*, Tharsis Books, Saanichton, Canada, 2002.

[2]   Conrow, E. H., Project Risk Management: Keys to Successful Implementation, American Institute of Aeronautics and Astronautics, Houston, TX, 2000.

[3]   Bluhm, C., L. Overbeck and C. Wagner, *An Introduction to Credit Risk Modeling*, CRC Press, Boca Raton, FL, 2002.

[4]   Conrow, E. H., "Some Inherent Limitations on Quantitative Cost Risk Assessment Methodologies," *Proceeding of the 29th Annual DoD Cost Analysis Symposium*, February 1996.

[5]   Pecht, M., R. Solomon, P. Sandborn, C. Wilkinson and D. Das, *Life Cycle Forecasting, Mitigation Assessment and Obsolescence Strategies*, CALCE EPSC Press, College Park, MD, 2002.

[6]   Government-Industry Data Exchange Program, <http://www.gidep.org>.

Chapter 18

Environmental and Legislative Issues

Paul Casey and Yuki Fukuda

Electronic products should not contain substances that are harmful to human health and the environment. In order to comply with regulations, part manufacturers make changes to both solders and molding compound formulations. These changes can impact costs associated with part assembly, reliability, and obsolescence. It is imperative that those involved with parts selection and management are aware of legislative and part changes.

18.1 Legislation for eliminating lead in electronics

The hazards of lead in the environment and its harmful effect on humans have been documented for some time. The use of lead for plumbing, paints, and other applications has been banned in developed countries. In the United States, legislation banning such uses of lead was enacted in 1991. As of 2001, virtually all environmental legislation exempted electronic products from regulation. Yet the regulatory environment in Europe and Japan concerning hazardous substance bans and electronic waste recycling has been growing and may be indicative of what is to come in the United States.

The primary mechanism by which lead contamination reaches soil and water is from landfills and disposal sites. Therefore, much of the legislative and pilot project activity has been aimed at evaluating methods to prevent electronics containing lead from ending up in landfills to begin with. Even though the contribution of electronic parts to landfilled lead is slight, the volumes have been increasing rapidly enough to convince lawmakers in some countries that such products ought to contain fewer toxins.

18.1.1 European regulations

European environmental directives concerning electronic equipment include chemical bans and product take-back requirements. The costs are to be absorbed by manufacturers, and this includes the cost of recycling not only products sold after the date of enforcement, but also the costs of properly disposing of past electronic waste.

The most significant European Union (EU) actions are the directives on Waste in Electrical and Electronic Equipment (WEEE) and Reduction of Hazardous Substances (RoHS). The WEEE legislation is primarily concerned with preventing the accumulation of electronics in the waste stream. Further concerns involve the reduction of waste through recovery, recycling, and reuse of electronic equipment so that the environmental impact of such equipment is reduced throughout its life cycle. The important aspects of the WEEE are the producers' responsibility for end-of-life–related costs and the inevitable design and manufacturing costs related with a high percentage by weight of product recovery. Producers are permitted to label the recycling costs of past electronic waste for a limited time, but the costs of recycling current waste may not be labeled on shelved products.

The directive on the use of certain hazardous substances in electrical and electronic equipment (RoHS) supplements the WEEE by making specific mention of materials to be regulated. The European Parliament was concerned that the mitigating effects of implementation of the WEEE directive on hazardous materials in the waste stream would be insufficient and that, in particular, the presence of mercury, cadmium, lead, chromium, polybrominated biphenyls (PBB), and polybrominated diphenyl ether (PBDE) would pose health risks. It was therefore decided that hazardous materials in electronics had to be banned and replaced with benign alternatives [1]. Of the banned substances, lead has the greatest impact on the electronic industry. Lead is commonly used in all electronic assemblies.

On January 27, 2003, the revised version of the RoHS directive was made available and the phaseout date for lead has been finalized as July 1, 2006. While neither the WEEE nor the RoHS directives state what concentrations of hazardous substances will be used as cutoff limits, the draft commission decision on amending annex II of the directive on the end-of-life of vehicles reads: "It is in fact evident that a 'zero level' of heavy metals is in some instances impossible to achieve. It is therefore proposed to insert a paragraph in annex II stating that a concentration of up to 0.1wt.% per homogeneous material of lead, hexavalent chromium and mercury and 0.01wt.% per homogeneous material of cadmium shall be tolerated provided these substances are not intentionally introduced" [2]. Thus, even though no industrywide definition of lead-free currently exists, it appears that the likely European definition of 0.1% by weight of material may win out over the JEDEC definition of 0.2% [3], which seems to be preferred by U.S. companies.

Additional bans have been proposed within the EU on the national level. In 1998, Denmark proposed a ban on the import, sale, and production of most products containing more than 50 ppm of lead. Lead in electronic products was exempted pending further action. In Sweden, the Ordinance of April 27, 2000, named "Producer Responsibility for Electrical and Electronic Products," put the entire burden of electronic waste disposal squarely on manufacturers. The law requires producers to accept any electronic equipment that serves the same function as the equipment currently manufactured.

The European Committee of Domestic Equipment Manufacturers is at odds with the Swedish law because it violates the three principles of fairness concerning producer responsibilities:

- polluter pays
- nonretroactivity
- cost sharing

The polluter pays principle is based on the belief that producers are only responsible for equipment made or put on the market by them, and cannot be held responsible for wastes accumulated by defunct companies and wastes generated by other companies. The principle of nonretroactivity deems the costs of disposing of "historical" waste, i.e., waste generated prior to the date of enforcement of take-back legislation, as inapplicable to producers. If historical waste is to be the responsibility of producers despite efforts to the contrary, the principle of cost sharing applies. It consists of tagging on the costs of recycling and final disposal to products so that the cost burden is not exclusively the producer's. One of the concerns of lawmakers regarding this principle is that companies or retailers might be prone to tagging on artificially high disposal costs.

18.1.2   U.S. regulations

In the United States, the proposed Lead-based Paint Abatement Act of 1991 considered imposing taxes for the use of lead. Subsequently, the Lead Exposure Act was passed in 1992, but by 1994 the Reid Bill-Lead Exposure Act was passed with the notable exception of solders, save for plumbing applications [4]. Lead and lead compounds were originally cited in the Emergency Planning and Community Right-to-Know Act (EPCRA) of 1986, Section 313, but as of January 19, 1999, the Environmental Protection Agency shifted them to the persistent bioaccumulative and toxic substances (PBT) lists, which changed reporting requirements from 25,000 or 10,000 lb to 100 lb used or consumed annually [5]. It is worthwhile to note that the EPCRA in no way limits the use of lead; it only regulates reporting of lead use and discharge. The United States has yet to draft further legislation that affects lead consumption in electronics.

There has also been some activity on the state level in the form of electronics recycling pilot projects. The State of Minnesota and Sony have collaborated to form Sony product drop-offs that are free of charge. Massachusetts banned the disposal of CRTs as solid waste and is setting up collection sites for electronic wastes. South Carolina, Florida, and Wisconsin are also setting up electronic waste pilot programs or discussing possible future recycling legislation. The cities of New York and San Francisco currently have electronic waste collection and recycling programs aimed at the proper disposal of domestic and small business electronics. These services may be utilized by the commercial sector as well, but not free of charge.

18.1.3 Japanese regulations

Japan's environmental agency was upgraded to the Ministry of the Environment in 2000. Three major laws were enacted in April 2001, namely, the Household Electric Appliances Recycling Law, the Green Purchasing Law, and the Law for Promotion of Utilization of Recyclable Materials. Ever since environmental energy-saving and recycling laws were enacted, the Japanese electronics industry has gradually shifted towards "green" electronics.

The Household Electric Appliances Recycling Law required manufacturers to recycle discarded electronic products. This necessitated a change in the electronic product development cycle, including design, manufacturing, and product support. Elimination of lead used in printed wiring boards was deemed more appealing than the cost burden of separating lead out prior to recycling at product end-of-life.

Measures have been taken in the design and manufacturing stages to control other toxic chemicals, such as cadmium and halogenated flame retardants. These toxic substances, as well as lead and its compounds, are mentioned in phaseout targets. This task is made harder by the fact that more than 300 chemical substances are developed and introduced into the Japanese electronics market each year, and about 19% of them are toxic [6]. For lead, the Ministry of International Trade and Industry (MITI) set fixed amounts of reduction by mass for the years leading to 2005.

Under the Green Purchasing Law, the government encourages electronics manufacturers to make "green" products in accordance with designated standards. Though manufacturers are not obliged to make green products, the development and introduction of green products satisfying the standard can attract consumers and generate additional market share. A survey carried out among a large sample of Japanese companies demonstrated that about 33% of companies reported higher product sales due to additional consumer attention on account of the Green Purchasing Law. In addition, 11% of those surveyed succeeded in reducing manufacturing costs [6]. Even though these surveys are rather subjective, they still demonstrate the desire of Japanese companies to be the world leaders in lead-free electronics.

18.2   Lead-free components

One of the issues raised by the replacement of tin-lead solder is whether or not a more or less global replacement will be made available. Tin-lead solders are very well understood materials that have been used by the electronic industry since its beginnings. They have been used as quasi-universal joining materials, and there seems to be no material versatile enough to replace them.

Of all the candidate lead-free solders, those cited as the most promising for all-purpose use are the tin-silver-copper (Sn-Ag-Cu) family of alloys. Despite the good overall performance of Sn-Ag-Cu alloys, they are not one-to-one substitutes for eutectic tin lead. For this reason, a variety of alternative solder alloys have been developed. For example, if an inexpensive alloy with a low melting point were desired, it could contain Sn, zinc (Zn), and possibly bismuth (Bi) as well. In higher reliability-products, Zn could be replaced by antimony (Sb) to improve wetting characteristics, reduce corrosion, and improve the microstructure. Sn-0.7Cu is commonly cited as the low-cost wave-soldering alloy of choice, but it has a higher melting temperature than some of the Bi-containing wave alloys used in Japan.

Sn-Ag-Cu alloys are composed of available metals, are easy to produce, and are relatively nontoxic due to their small copper and silver content. Within the Sn-Ag-Cu family, Sn-3.9Ag-0.7Cu (National Electronics Manufacturing Initiative [NEMI] recommended), Sn-3.8Ag-0.7Cu (from the Ames Lab/Iowa State research in the 1990s), and Sn-3.0Ag-0.5Cu (popular in Japan since it is relatively less expensive) have been identified as the most common substitutes to tin-lead eutectic solder.

18.2.1   Lead-free alloy intellectual property

Many lead-free alloys are patent protected or may become protected if applications are awarded. For example, there is some indication that the NEMI alloy may be affected by patent applications in the United States and, more importantly, in Europe. As of March 2003, there was one U.S. patent application and three in Europe that were filed by Japanese companies. If awarded, these applications may affect the use of the NEMI alloy.

On the other hand, an encouraging sign is that the licensing umbrella of Sn-Ag-Cu is expanding. By mid-1998, Ames Lab had already licensed their Sn-Ag-Cu alloy to Johnson Manufacturing, Multicore, and Nihon Superior. Subsequently, a dispute erupted because Senju and Matsushita were awarded a patent in Japan in January 2000 for the same alloy that Ames Lab had patented in the United States. In April 2001, rights held by Senju and Matsushita, and those held by the Ames Lab and its licensees, were unified into a single set for licensing to other solder manufacturers [7]. As a result, equipment manufacturers can purchase Sn-Ag-Cu from Senju Metal, Nihon Superior, or any vendors licensed by both companies, and sell equipment in the United States and Japan without fear of infringement.

A variant of Sn-Ag-Cu with an Sb grain refiner dubbed CASTIN, was developed and patented by AIM but has since been licensed by Cookson and other solder manufacturers. Another example of a heavily licensed solder is the Oatey alloy (Sn-Bi-Ag-Cu), which was originally intended for plumbing applications. This is an interesting case because it demonstrated that patents not originally drafted with electronics in mind can be enforced in that domain. In fact, it opens up the possibility that other lead-free solders patented for glass working, antifouling, brazing, and the like can become enforceable for electronics if those with rights to such inventions elect to explore that possibility.

18.2.2 Issues with solder balls and lead finishes

Electroplating is among the more attractive deposition techniques for ball bumping applications [8]. Complex alloys suffer from a lack of uniformity, nullifying some of their supposed benefits. Of the binary lead-free variety, tin may be alloyed with Ag, Cu, or Bi. Bi is a lead by-product, so the long-term use of Sn-Bi is unlikely. Sn-Ag and Sn-Cu have a niche of applications in which they may be successful [9], but their melting points are high enough to cause concern.

Leaded packages have a few plating options in lead-free configurations, with the more promising ones being nickel palladium (Ni-Pd), Ni-Pd-Au, Ni-Au, and tin-based finishes such as matte tin, Ni-Sn, and the Sn-Bi alloy. Reports of tin whiskers with matte tin finishes are very rare and many companies report never having observed excessive whiskering with this finish. Yet use of tin platings, with or without nickel underplating, is usually considered unacceptable in mission environments where reliability is of any consideration. In achieving the desired balance between cost and reliability, it is likely that manufacturers of low-end consumer products will gravitate toward matte tin (especially in the United States), while the others opt for more costly alternatives.

18.2.3 Assembly and rework issues

Many lead-free solders reflow at significantly higher temperatures than the Sn-Pb eutectic solder they are intended to replace. For example, Sn-Ag-Cu lead-free solders have melting temperatures of around 220°C, and thus the reflow temperatures for these solders should typically be in the range of 235 to 260°C.

With reflow temperatures of up to 40°C higher than those of leaded solder, and moisture ingress, package resistance to moisture is essential to prevent popcorning and other quality and reliability problems from arising. Popcorning occurs when moisture accumulates in slightly delaminated interfaces in a package, vaporizes upon reflow, and cracks when the vapor pressure exceeds the strength of the package. Studies have shown that the peak reflow temperature is not necessarily the parameter that most affects internal pressure and therefore popcorning. Some studies show that other reflow parameters, such as preheat and total heat energy transfer, better represent the likelihood of pressure cracking [10].

Rework poses more problems with lead-free solders having higher melting points. These solders require higher rework temperatures, which have the potential to damage materials used in boards and components near the rework sites. As rework processes require at a minimum two reflow temperature excursions, thermal damage may cause "excessive warpage, delamination, and solder mask discoloration. In addition, multiple reflow cycles during rework may result in the degradation of the PCB surface finish and result in the oxidation of the copper pads" [11]. For lead-free solders with high tin content, time spent during rework at elevated temperatures promotes accelerated intermetallic growth. For the Sn-Ag-Cu line of solders, which have good mechanical strength, strain is taken up at the interface between the solder bulk and the intermetallic [12]. With this type of failure, the impact of rework on intermetallics plays a role in the fatigue life of Sn-Ag-Cu solder joints, particularly in high strain applications, such as in flip chips and BGAs.

18.2.4 Reliability implications

Companies like Sony and Matsushita have already manufactured products with environmentally friendly parts. However, there are risks that must be addressed in the parts selection and management process.

Reliability concerns with lead-free alternatives involve understanding the use conditions. Some solders can withstand thermal cycles better than others; some solders can withstand vibration and shock loads better than others. Some solders are apt to cause tin whiskers.

The variety of lead-free solders, leadframe platings, and board platings can also generate compatibility problems that render management of electronic parts increasingly difficult for leaded and lead-free assemblies [13]. Leaded and lead-free solders have different melting and phase transition temperatures, and mixing materials can lead to unintended consequences. For example, when conventional tin-lead eutectic solder on a board is mixed with a lead-free Bi-containing leadframe plating, there is a substantial risk of forming the ternary eutectic alloy 16Sn-52Bi-32Pb, which has a melting point of 96°C. This can lead to severe strength degradation and is the reason why components with Sn-Bi leadframe platings will have a designated marking.

While the issue of tin whisker formation is not new, the introduction of lead-free leadframe plating is making this problem reappear. Eutectic Sn-Pb is resistant to whisker growth, but lead-free platings are prone to develop compressive stresses and surface oxides [14]. Stresses build up until the oxide layer ruptures, permitting the tin whisker to extrude [15]. The effect of external stress on whisker growth is unclear. Some experiments have shown no correlation between external stress and whisker growth [16]; others have shown a seemingly strong correlation between the two [17].

18.2.5   Implementation costs

The selection of lead-free parts can have significant cost implications. Costs that may run beyond normal expectations include the need to track multiple versions of parts and products based on part use. Other cost include those associated with equipment and training to make future changes or repairs to parts and assemblies. Environmentally friendly parts also tend to cost more than the traditional parts due to the need for additional development and testing.

18.3   Environmentally friendly molding compounds

Reliability of the new mold compounds must be taken seriously when selecting parts. Changes to molding compound additives and to resin systems themselves create performance differences between conventional molding compounds and green variants. Some properties that are prone to change and can have an effect on assembly and reliability are viscosity, moisture absorption, glass transition temperature, and coefficient of thermal expansion. Moisture absorption is of critical importance, as it affects hygro-thermal stresses and interfacial adhesion strengths. From a practical perspective, one should be mindful of "floor life" when selecting parts. Surface mount parts with higher moisture sensitivity, as defined by the joint industry standard J-STD-020B, may require additional procedures to ensure that the desired assembly yields are achieved.

In both peripheral and area array packages, the use of green molding compounds may precipitate wire sweep, degradation of wires, and possibly even bond pad cracking. Wire sweep is a problem that seems to have been worsened by green molding compound formulations. With higher typical filler loadings and viscosities, green molding compounds could create a larger pressure differential across bonding wires during the transfer molding process. This pressure, in turn, is manifested in the form of lateral wire displacement, which impedes qualification of a process when it exceeds a few percentage points. Green encapsulants may also accelerate device electrical failures by wire lift-off, pad cracking, or segregation of the ball from the wire at the neck [18].

Another notable reliability issue with green molding compounds involves substitutes for the bromine (Br) and Sb flame retardants. For example, one replacement, red phosphorous

(P) in particulate form, was found to cause device malfunction by introducing stray conductivity [19]. Conventional qualification tests did not catch any problems, but the mold compound significantly reduced the time to part wearout in the field. As a result, the use of red P particulate fire retardants has been discontinued.

The industry is seeking replacements that are nontoxic but also achieve the UL94V-0 flammability rating. Examples of some current alternatives are [20]:

- metal oxide with silicone polymer and coated red P
- high filler loading, magnesium hydroxide, and multilayer coated red P
- no flame retardant additive, but use of a fully aromatic resin
- hardener, high filler content, and Zn borate

Each of these must be properly assessed prior to use in electronic parts.

18.4   Industry groups involved with lead and halogen-free electronics

The idea behind industry groups is to eliminate the duplication of work where possible by utilizing resources across a spectrum of companies to conduct studies of importance to a particular sector. Many such groups exist within the electronics industry, and they undertake activities ranging from lobbying, to conducting tests, to organizing conferences and generating roadmaps.

18.4.1   North America

The CALCE Electronic Products and Systems Center at the University of Maryland has over 50 industrial partners and carries out numerous projects related to green electronics. In 2000, the CALCE Consortium initiated a comprehensive multiyear research effort to investigate the enablers for and the consequences of environmentally friendly electronics. In particular, CALCE has been evaluating the constitutive and durability behavior of popular lead-free alloys, determining the influence of plating systems on intermetallic growth, characterizing lead-free solder contact interfaces, studying creep corrosion on PEMs with noble leadframe platings, assessing green molding compounds, and developing methodologies for managing supply chains.

NEMI has completed a number of lead-free projects. One such project, the lead-free assembly project, resulted in industry standard recommendations for reflow and wave soldering, which were Sn-3.9Ag-0.6Cu, and Sn-0.7Ag, respectively. Other NEMI lead-free projects cover topics such as lead-free rework and accelerated testing for tin whisker growth.

The National Center for Manufacturing Sciences (NCMS) has been involved in the selection and characterization of lead-free solders. They ran two major lead-free alloy evaluation programs, one for replacements to tin-lead eutectic solder and the other to assess candidate replacements in harsh environments with operating temperatures ranging from −55°C to 160°C.

The Institute for Printed Circuits (IPC) has been active in organizing industry discussion and information sharing in the lead-free area. The IPC has hosted joint conferences with JEDEC on an annual basis.

The High Density Packaging User Group (HDPUG) has been running a number of cooperative lead-free projects. They have various lead-free and low temperature lead-free soldering projects underway and have completed a number of halogen-free material evaluations.

18.4.2 Europe

Orgalime represents 32 trade federations consisting of approximately 100,000 companies spread over 21 countries with 1.2 trillion euros of revenue per year. It represents metalworking, mechanical engineering, and electrical and electronic engineering. Orgalime issued a statement in response to the WEEE/RoHS directives and lobbied for changes, citing the potential effects of the legislation on the competitiveness of the EU electronics sector.

Soldertec, which consists of over 40 member companies, has been coordinating lead-free related research projects, and has had lead-free alloy selection and testing projects since 1991. Current projects as of 2003 include assembly, designing reliability into and predicting reliability of products, and characterizing lead-free alloys in detail.

The European Lead-Free Network (ELFNET) is a research organization that consists mostly of European universities but also includes Interuniversity Micro-Electronics Center (IMEC) and Soldertec. Its main purpose is to help resolve technical problems to increase the competitiveness of European companies with respect to lead-free technologies. Another similar organization is EFSOT, which stands for "next generation Environmentally Friendly Soldering Technology." EFSOT has both government and corporate participation, and aims to develop intelligent manufacturing solutions for improving the environmental performance of electronics.

18.4.3 Japan

The Japan Electronics and Information Technology Association (JEITA) was formed on November 1, 2000, as a result of a merge between the Electronic Industries Association of Japan (EIAJ) and Japan Electronic Industry Development Association (JEIDA). The EIAJ and JEIDA were originally involved in developing lead-free roadmaps for the Japanese electronics industry. Yet in the 2002–2003 period, there has been greater international cooperation between JEITA and International Tin Research Institute (ITRI) in coming to terms on an appropriate definition of "lead-free." In January 2003, a joint JEITA-ITRI roadmap stated that the appropriate goal to strive for was 0.1% by weight of lead. The roadmap specified that by 2005, the average manufacturer should have succeeded in completely eliminating lead from products. The organizations also recognized that a mutually acceptable system for labeling was necessary.

The Japan Institute of Electronics Packaging and (JIEP) and the New Energy Development Organization (NEDO) have engaged in evaluations of lead-free candidate solders from the assembly and usage standpoint. The Japan Institute of Electronics Packaging (JIEP) conducted a detailed study of Sn-Ag-Bi alloys that contained 7–25% by weight of Bi. Results from reliability studies showed that alloys containing more than 7% Bi were prone to fillet lifting and exhibited brittle characteristics. In contrast, the Japan Welding Engineering Society was primarily assigned the task of determining the metallurgical characteristics of lead-free alloys.

18.4.4 ISO 14001: Eco-labeling of organizations

ISO 14001 certification can enhance the image of an organization and improve its ability to market products or services. Even though ISO 14001 certification is not necessarily indicative of environmental performance, it does show a willingness to be pro-active in terms of environmental management. It also demonstrates to the local community that the organization is not reacting to regulations and scrambling to meet them, but rather is dedicated to consistently meeting or exceeding pollution requirements.

ISO 14001 can help save money by making companies aware of all the areas in which it can economize and reduce natural resource consumption in a systematic manner. The same is true for companies that are burdened by large energy bills or high disposal costs. These companies most often have the greatest potential for improvement, and therefore benefit the most from resource and energy savings and the marketing possibilities that result from the ability to publicize major improvements.

18.4.5 Component markings

Lead-free marking of components continues to be conducted in package codes, but it is generally agreed that more readily identifiable markings are needed to distinguish between leaded and lead-free packaging materials. A handful of Japanese companies have addressed this issue by implementing their own lead-free component and board markings, two examples of which are shown in Figures 18.1 and 18.2 [21], [22]. The Sharp symbol is used for marking lead-free solder mounted boards, whereas the Fujitsu mark is for components.

Figure 18.1: Fujitsu's mark

Figure 18.2: Sharp's mark

18.5   References

[1]   "Directive of the European Parliament and of the Council on the Restriction of the Use of Certain Hazardous Materials in Electrical and Electronic Equipment," *Official Journal of the European Union*, L 37, pp. 19–23, February 2, 2003.

[2]   "Directive 2000/53/EC of the European Parliament and the Council of 18 September 2000 on End of Life Vehicles," *Official Journal of the European Communities*, October 21, 2000.

[3]   "JEDEC Announces Lead-Free Definition," JEDEC Press Release, November 15, 2001, <http://www.jedec.org/Home/press/press_release/leadfreePR.pdf>, accessed April 14, 2003.

[4]   Senate Record Vote Analysis of the Lead Exposure Act, 103d Congress, 2nd Session, Vote 133, May 25, 1994.

[5]   "Lead and Lead Compounds; Lowering of Reporting Thresholds; Community Right-to-Know Toxic Chemical Release Reporting; Final Rule," *Federal Register*, Vol. 66, pp. 4499–4547, January 17, 2001.

[6]   "Environmental Guidelines; Fiscal Year 2000 Version," Ministry of Environment, Government of Japan, <http://www.env.go.jp/en/eco/o-epi2000.pdf>, accessed April 23, 2003.

[7]   "Pb-Free Solder Patent Problems Finally Resolved," Nikkei Electronics Asia, Hong Kong, HK, April 2001.

[8]   Magill, P. A., et al., "Electroplating Lead-Free Solders for Flip-Chip Packaging," <http://www.unitive.com/techDocumentation/docs/electro.pdf>, accessed March 21, 2003.

[9]   Fukuda, Y., P. Casey and M. Pecht, "Evaluation of Japanese Lead-Free Consumer Electronics," presented at the CALCE Technical Review Meeting, October 2002.

[10]   Gannamani, R. and M. Pecht, "An Experimental Study of Popcorning in Plastic Encapsulated Microcircuits," *IEEE Transactions on Components, Packaging, and Manufacturing Technology*, Part A, Vol. 19, No. 2, pp. 194–201, June 1996.

[11]   Gowda, A., K. Srihari, and A. Primavera, "Lead-Free Rework Processes for Chip Scale Packages," <http://www.uic.com/wcms/WCMS.nsf/index/White_Papers_5 .html>, accessed April 25, 2003.

[12]   Frear, D. R., J. W. Jang, J. K. Lin and C. Zhang, "Pb-Free Solders for Flip-Chip Interconnects," *JOM*, pp. 28–32, June 2001.

[13]   Zheng, Y., C. Hillman and P. McCluskey, "Effect of PWB Plating on the Microstructure and Reliability of SnAgCu Solder Joints," AESF SUR/FIN, Chicago, IL., June 24–27, 2002.

[14]   Choi, W. J., et al., "Structure and Kinetics of Sn Whisker Growth on Pb-free Solder Finish," *Proceedings of the Electronic Components and Technology Conference*, pp. 628–633, 2002.

[15]   Brusse, J. A., G. J. Ewell and J. P. Siplon, "Tin Whiskers: Attributes and Mitigation," *CARTS 2002: 22nd Capacitor and Resistor Technology Symposium*, pp. 221–233, March 2002.

[16]   Dunn, B., "A Laboratory Study of Tin Whisker Growth," *European Space Agency STR-223*, pp. 1–50, September 1987.

[17]   Xu, C., Y. Zhang, C. Fan, and J. Abys, "Understanding Whisker Phenomenon: The Driving Force for Whisker Formation," <http://www.circuitree.com/CDA/ ArticleInformation/features/BNP_Features_Item/0,2133,75917,00.html>, accessed May 13, 2002.

[18]   Lin, T.Y., C. M. Fang, Y. F. Yao and K. H. Chua, "Development of Green Plastic Encapsulation for High Density Wire Bonded Leaded Packages," *Microelectronics Reliability*, Vol. 43, pp. 811–817, 2003.

[19]   Ouishi, M., M. Kawai and T. Shirakura, "Fujitsu HDD Issue," *Nikkei Electronics*, pp. 102–119, Vol. 833, October 21, 2002.

[20]   Pollock, J., "Green Molding Compound Evaluation for CSP," *Proceedings of the Third International Conference on Lead Free Electronic Components and Assemblies*, pp. 265–270, April 23, 2003.

[21]   "Fujitsu Lead-Free Package Plan (Policy)," <http://edevice.fujitsu.com/fj/ CATALOG/AD81/81-00004/4e-2.html>, accessed May 1, 2003.

[22]   "Sharp Environmentally Conscious Products," <http://sharp-world.com/corporate/ eco/ report2002/22.html>, accessed May 1, 2003.

Chapter 19

Legal Liabilities

Ray Biagini and Michael Pecht

From a legal perspective, several alternatives should be considered before a part is selected for use. First, parts that satisfy the electrical requirements and that are specified to meet the actual conditions they are expected to experience throughout their life cycle should be explored. If such a part is not available, it may be necessary to modify the conditions the part will experience so that it satisfies the specifications. For example, fans, cooling rails, and heat sinks may be used to modify an environment that is too warm for the part specifications, and heaters may be used for environments that are too cool. The part may also be moved to a different location within the system to keep it in an environment that meets its specifications. Isolating the subsystem from the outside environment can also modify the environment to which the part is subjected. In that case, a complete subsystem may be kept within a controlled environment. Environmental management options, however, may bring with them associated costs from added size, weight, power, and performance constraints, as well as additional maintenance requirements. All of these options must be considered and documented.

The use of parts outside their manufacturer-specified limits should be considered as an option only when no other feasible alternative can be found. Such use is called "uprating" [32], defined as a process of assessing the capability of a part to meet the performance requirements of the applications in which the part is used outside the manufacturer's specifications.

This chapter is a primer on the legal system and the legal issues that a company should consider when selecting parts. It sets forth basic parameters of a product liability prevention program for companies, emphasizing the acute need for top-to-bottom education and training of company personnel – from program managers to test technicians to hazard analysts.

19.1   Determining the plaintiffs and the defendants

The plaintiff can be anyone who has suffered harm, including financial and emotional harm. Plaintiffs usually sue all companies in the supply chain to ensure recovery from the responsible party. If during the course of the trial it becomes apparent to the plaintiffs that they cannot prove their case against a particular supplier, the plaintiffs often drop their case against that defendant and continue the trial against the remaining defendants. If more than one defendant is found liable in a case, individual defendants may pay only a percentage of the recovery amount for which they are found responsible.

19.2   Determining the court that will hear the case

There are two sets of court systems in the United States: the U.S. federal court system and the U.S. state court system. Each system has its own set of rules and judges. U.S. federal law is usually limited to issues that are delegated to the U.S. Congress by the

Constitution. If a particular issue is not delegated specifically to Congress, such as regulating interstate commerce, then the U.S. state law regulates. Although the laws of individual states regulate most product liability claims, many claims are brought in U.S. federal court.

A claim can be brought in U.S. federal court for only three reasons:

- Disputes concerning issues of U.S. federal law and the U.S. Constitution are brought in U.S. federal court. Such disputes are called "U.S. federal question cases."
- Cases where the U.S. government is a party are also brought in U.S. federal court.
- Disputes between residents or corporations of different U.S. states can be brought in U.S. federal court when the amount being claimed is at least $50,000.

If a case meets any one of the aforementioned criteria, one of the parties will most likely move the case to U.S. federal court. However, if a case does not meet one of these criteria, the case can still be brought in state court. Even if a federal court hears the case, state law is still applied to the issue.

19.2.1   U.S. state courts

Most U.S. states have four levels of state courts. The lowest court in the U.S. state system is the court of limited jurisdiction; it only has the authority to hear specialized cases such as those involving family law or bankruptcy. The next court is the court of general jurisdiction. This court hears cases on all subjects except those delegated to the court of limited jurisdiction. The court of general jurisdiction is the trial court for the state. Above this is the intermediate appellate court, which hears appeals from the courts of general jurisdiction and limited jurisdiction. The highest court in a state is the court of last resort. This court hears appeals from the intermediate court and pronounces the final decision for the state. If a case involves a U.S. constitutional issue, however, a party may appeal from the state's court of last resort to the U.S. Supreme Court. The individual states determine the number of courts they have; this decision often depends on the population and needs of the state.

19.2.2   U.S. federal courts

There are three basic levels of U.S. federal courts. Listed in order from the lowest court to the highest, they are the United States District Court, the United States Circuit Court of Appeals, and the Supreme Court of the United States. Legal claims in the federal system are first heard in the U.S. district court. Appeals from the district court are heard in the circuit court of appeals for the geographic area, which includes the district court [30]. The circuit courts of appeals are divided into three-judge panels to hear cases. Each panel considers cases for two weeks or some other specified time period; then the judges in the circuit are rearranged into different three-judge panels. If two different panels in a circuit reach contradictory conclusions in similar cases, or if a majority of judges in the circuit believe that a particular panel's decision deserves further consideration, a case may be reconsidered en banc, with all judges in the circuit deciding the case [30].

In the case of a conflict of law, the parties' debate about which U.S. states law should apply in their case. The U.S. federal court decides which U.S. state's law will apply by considering the following factors:

- State of business incorporation
- Business and transaction location
- Physical location of the business
- Geographical place of injury

- Plaintiff's domicile
- Fairness (at judge's discretion)

19.3   The plaintiff's claims

In the majority of product liability actions, the injured party (i.e., the "plaintiff") will typically seek recovery from any entity involved in placing the product into the stream of commerce. In a hypothetical case involving the use of electronic parts outside a manufacturer's specification, those entities (identified as "defendants") could range from the original manufacturer of the electronic part to the last party in the supply chain who places the end product in the hands of the user. The parties between those two ends of the supply chain – companies that perform the testing of the electronic parts, companies that integrate those parts into larger subsystems and systems, manufacturers of the end product, distributors, and retailers – will also likely be named as defendants in the product liability case.

At its core, a plaintiff's product liability case will allege that the design of the product at issue is unsafe because of its use of parts outside the manufacturer's specification or that the defendant companies failed to warn the product's users of potential hazards by placing warning labels on the product or in technical manuals. In addition, a plaintiff will often contend that the product contained some unspecified manufacturing defect (as contrasted with a design defect), although product liability cases rarely turn on that issue. This is primarily due to the fact that it is easier for the plaintiff to prove a design defect or failure-to-warn case than a manufacturing defect case.

In most cases, a plaintiff will have the option of filing his or her product liability action in a state or federal court. State courts are usually the preferred forum for plaintiffs based on the belief, although not always true, that state court judges and juries tend to be more sympathetic to plaintiffs. A defendant, however, is not necessarily bound to the plaintiff's selection of a state court to hear the case. Under certain circumstances, a defendant may attempt to "remove" the case to federal court. Based on the facts of the particular case, removal may be advantageous to a defendant who has certain defenses available, particularly the government contractor defense, which will be discussed in this chapter.

In all product liability cases, a plaintiff will seek to be compensated for actual damages. Such compensatory damages may include the cost of medical treatment, lost earnings, lost future wages, and pain and suffering. In many jurisdictions, punitive damages may also be sought when a party engages in conduct that exhibits a conscious disregard for the end user's safety. Other states require a showing that the defendant acted maliciously or wantonly in allowing a defective product to be placed into the stream of commerce. Regardless of the standard adopted by the state, punitive damages often result in damage awards several times greater than the actual harm caused by the defective product.

The practice of uprating causes several questions to come to mind. In the process of uprating parts, are the hazards and risks associated with the use of parts known? If so, what is being done with such information? Are customers informed of the risks and, if so, are the appropriate people notified? In supplying uprated electronic parts to the government, are the parts being certified to conform to specified requirements? If yes, to what standard, guide, or procedure is the certification made? These questions should be addressed to avoid or limit a product liability suit brought against companies that engage in the uprating of electronic parts.

A company that uprates parts may encounter several different legal theories in the event that company becomes a defendant in a product liability case. The principal legal theories under which a company may be sued include strict liability, negligence, and failure to warn.

The following is a brief discussion of these product liability theories and the potential damages that may be recovered.

19.3.1   Strict liability

Strict liability is defined as responsibility without fault and applies without regard to (1) the reasonableness of the individual defendant's conduct or (2) whether or not the injured party bought the product from the particular defendant. Most states have adopted a strict product liability standard under which any party in a product's supply chain will be found liable for injuries caused by a defective condition in the product. Under strict liability, the supplier who causes accidental losses is exposed to liability for all of those losses, regardless of whether reasonable care was exercised, even if everything possible was done to try to prevent the accident [14]. In order to recover damages, a plaintiff need only show that he or she suffered injuries as the result of a design or manufacturing defect in a product and that the defendant played some role in introducing the product into the stream of commerce.

The basic elements of a strict liability claim and the points the plaintiff must prove are:

- the supplier had an absolute duty to make safe products
- the supplier breached that duty
- the supplier's breach caused harm to the plaintiff

In most U.S. jurisdictions, a defendant is strictly liable when the defendant had the duty to make a product absolutely safe. This high standard is applicable only when the defendant deals in an ultrahazardous or abnormally dangerous activity. If the court does not accept the plaintiff's claim that the use of parts at temperatures beyond those for which they are specified is ultrahazardous, the plaintiff will not have a strict liability claim against the defendant.

Each U.S. state defines its own version of strict liability. However, the basic principle remains the same. For example, in Missouri the plaintiff must demonstrate that the product was defective and dangerous when put to a use reasonably anticipated by the supplier and that the plaintiff sustained injury or damage as a direct result of the defect. The plaintiff also must show that the defect existed when the product left the control of the supplier and entered the stream of commerce. In California, however, "the doctrine of products liability does not apply as between parties who: (1) deal in a commercial setting, (2) deal from positions of relatively equal economic strength, (3) bargain the specifications of the product and (4) negotiate concerning the risk of loss from defects in it" [17].

19.3.2   Negligence

Negligence is defined as the failure to use such care as an average or a reasonably prudent and careful person would use under similar circumstances [6]. Under the legal theory of negligence, those involved in the placement of products into the stream of commerce owe a duty to exercise "ordinary and reasonable care" to avoid foreseeable risks of injury caused by the defective nature of a product. This duty of care is owed to any person who can reasonably be expected to use a product. In order to recover damages, a plaintiff must demonstrate the existence and breach of this duty against each defendant and that the plaintiff's injuries were proximately caused by the defendant's conduct. Unlike strict liability, which focuses solely on whether a product was defective, negligence requires an examination of each defendant's conduct. Thus, where a defendant's actions do not contribute to or cause the plaintiff's injuries, a defendant will not be found liable under the claim of negligence. The general standard applicable in most negligence cases is one of reasonable care under the circumstances. The term "reasonable" allows the standard objectivity.

The basic elements of a negligence claim include:

- a duty of care owed from the defendant to the plaintiff
- a breach of that duty
- harm caused to the plaintiff as a result of the defendant's breach
- financial damage to the plaintiff as a result of the breach

The plaintiff must prove each of the four elements of negligence to win a case brought under the negligence claim in court, and will begin by asserting that a duty of care existed between the defendant and the plaintiff. In the U.S. justice system, every person or entity does not have a duty towards every other person or entity. A duty of care is created by a specific law, in which the legislative body explicitly states that there is a duty from one person or entity to another. The persons or entities involved for purposes of that statute would be explicitly defined. The plaintiffs in the fictitious case, however, will have little difficulty proving that the defendants owed them a duty of care because generally courts do find that a duty of care exists between a merchant and a consumer.

The plaintiff's greatest task will be to prove that the suppliers breached their duty of care. This element of negligence is usually more unsettled than the other elements because it is fact sensitive. In order to prove a breach of a duty of care, the plaintiff will present facts that prove that the defendant did something wrong. To do this, the plaintiffs must prove that the defendants did not do what a reasonable person would have done under similar circumstances.

If the plaintiffs prove the first two elements of negligence, the third element of harm and the fourth element of damage will likely be easy to prove. Although part use beyond the manufacturer's temperature specifications may not have been a conclusive reason for the accident, a jury will be told that it *is* the reason for the accident because no other reason has been suggested, satisfying the harm element of negligence. The resulting medical bills will be enough to prove the element of damage.

If the court awards damages to the plaintiff, the court may also award punitive damages. Punitive damages are an amount of money the court believes will adequately punish the defendant for his wrongdoing as well as deter any future breaches of the duty of care. Courts have a large amount of discretion regarding punitive damages, and the plaintiffs in this case will rely heavily on the lives of the passengers that were destroyed as a result of the defendant's wrongdoing.

19.3.3 Failure to warn

In most states, a supplier of a product has a duty to warn its customers of potential dangers about which it knows or should know concerning use of the product. Absent such a warning, typically contained in a user manual or placed on the product itself, the product is considered defective.[1] Most states also impose a continuing duty to warn of hazards that the manufacturer or company discovers or should have discovered following the sale of the product. Thus, a manufacturer or company can be found liable as a defendant in a product liability action several years after placing the product in question into the stream of commerce. A claim for failure to warn can be brought under the standards of negligence or strict liability.

---

[1] Although a product may not contain a physical defect, basic principles of product liability prescribe that a supplier's failure to adequately warn a consumer about the potential dangers of a product is in itself a defect.

Whenever a supplier markets a product, the supplier must provide instructions regarding proper use and warn of hidden risks. If adequate instructions and warnings are not provided to purchasers and users, the supplier may be liable for supplying a defective product even if the product in question is free of flaws and even if the product design is not, judged on its own merits, unreasonably dangerous [14].

19.4   The defendant's rebuttal

Before an equipment supplier is named as a defendant in a product liability action, there are several steps that can be taken to ensure that it is able to raise the significant defenses that are available. Two such defenses are particularly important: the government contractor defense and the contract specification defense. The significance of these "bookend" defenses is that, if successful, they will permit the company to be dismissed from the case prior to a trial on the merits of the plaintiff's claims.

19.4.1   Government contractor defense

The primary product liability defense for those who sell products to the United States is the government contractor defense. In 1988, the U.S. Supreme Court ruled that a company providing products to the government cannot be subject to a product liability action brought by a third party where the company proved that (1) the government approved reasonably precise specifications for the product at issue, (2) the product conformed to those approved specifications, and (3) the company provided any information it possessed concerning dangers associated with the product that were actually known to the company but not to the United States [7]. The Supreme Court gave several practical justifications for the existence of the defense. First, the immunity created by the defense allows companies to develop products at the limits of technology without fear of judicial review and the application of state tort law standards of care. Second, the defense avoids the shifting of liability from suppliers to the government through increased contract prices. Finally, the defense protects those companies serving the national interests in getting the government's work done [7].

Although the Supreme Court's 1988 decision involved a defective design case brought against the prime contractor of a major weapon system, several courts have subsequently expanded the scope of the government contractor defense to cover subcontractors who supply subsystems to the government. For example, the defense has been successfully raised by a subcontractor providing ejection seats for the U.S. Navy F-18 aircraft [24], a supplier of engine bearings for use in U.S. Army helicopters [19], and a subcontractor providing combat computer systems placed aboard U.S. Navy attack submarines [33]. In addition, the vast majority of courts considering the issue have ruled that the government contractor defense applies equally to contracts involving military and nonmilitary governmental agencies [10]. Finally, a number of state courts have concluded that the government contractor defense is available to those who contract with state governmental agencies.

For suppliers of uprated electronic parts to the United States, either directly or through a prime contractor or higher-tier subcontractor, the government contractor defense can be a powerful tool to use against product liability claims. To be effective, however, the company must take various preventive steps *before* the product liability action is filed to enhance the likelihood that the defense will result in dismissal of the case. The overriding theme in all of the steps described below is to keep the government involved at each stage of contract performance. A company that warns of hazards known to it during contract performance, seeks and follows government instruction, and documents this process will maximize its ability to use the defense as a shield. The following are some of the steps that should be taken before and/or during performance of present and future government contracts.

- Provide the government with the written procedures used to substantiate uprating. Ideally, secure the government's meaningful review and approval of these processes, as well as approval of the specific use of electronic parts outside of the manufacturer's specifications. Document the government's approval and memorialize government involvement in any key design decisions. Such evidence will go a long way towards establishing the first element of the defense: government approval of reasonably precise specifications.

- To help demonstrate the second element, conformance, the company must document its test results (or substantiating information) proving that the electronic parts can be used outside of the manufacturer's specifications. If possible, allow the government to witness any conducted tests or, at a minimum, offer the government access to the documented test or substantiating results. Document government acceptance and recognition that the delivered part conforms to the applicable specifications.

- The company must raise potential safety hazards relating to the use of the uprated electronic parts with the government. Such information can be brought to the government's attention through the submittal of government-approved safety and hazard analysis reports. Where the company is acting as a subcontractor, the company should inform not only the prime contractor of any potential dangers, but the procuring governmental agency as well. Once potential hazards have been discovered, the company must obtain closure from its customer or the government on such issues.

- Document efforts to find alternatives to uprating, including the search for parts that satisfy the part datasheet requirements. If such a part is not available, document efforts to modify the environment the part will experience (i.e., fans, cooling rails, and heat sinks) or to move the part to a different location within the system to keep the part in an environment that meets its specifications. If the decision to uprate is made, justify why the alternatives were not feasible. Most importantly, seek customer involvement in and approval of this process and ensure that any hazards are shared with the customer.

19.4.2 Contract specification defense

Companies providing electronic parts to commercial entities outside of the government procurement context may be able to raise a defense akin to the government contractor defense in response to a product liability action. Under the contract specification defense, recognized by several states, a company may avoid liability if it can demonstrate that (1) the product design or features had been reviewed and approved by the customer, (2) the product conformed in all material respects to the approved requirements, and (3) the company warned the customer about hazards of which the company knew or should have known.[2] As with the government contractor defense, successful application of the contract specification defense can result in complete dismissal of a product liability case before the case is heard by a jury.

Although the defenses are similar, two important distinctions must be made between the contract specification defense and the government contractor defense. First, under the former, a company will be held to a "should have known" standard. In other words, the

---

[2] For example, California, New York, Pennsylvania, and Florida all recognize some version of the contract specification defense.

company must inform its customer of any hazards about which the company knows or should have known concerning use of the product that is the subject of a lawsuit. By contrast, under the government contractor defense, a company is only responsible for sharing information about which it actually knows. As a result of these differing knowledge standards, the company raising the contract specification defense will remain subject to a post-lawsuit assessment by the court to determine which hazards the company should have known about.

Second, the nongovernment contractor is subject in most states to a "continuing duty to warn" of dangers about which the contractor becomes aware. Thus, even where a company has completed its performance under a particular contract, the company may still have an obligation to warn its former customers about subsequently discovered hazards. Such an obligation may continue for several years after contract performance and may also require the company to halt further distribution of the product or initiate a product recall. By contrast, some courts have concluded that the government contractor's duty to warn ends with completion of the contract at issue; in other words, the duty to warn does not survive the warranty period of the contract.

As with the government contractor defense, a company can take several steps to maximize the effectiveness of the contract specification defense well before any lawsuit is ever filed. Ensure that the customer understands the electronic part in question, as well as its usage. If the ultimate customer of the electronic part is an entity other than the company's direct contact, the company should make a similar disclosure to that customer. Where alternative parts or methods are examined, ensure customer involvement in and approval of the part selection process.

Secure customer approval of the written selection procedures and hazard analyses utilized by the company, as well as use of the specific electronic part in question. Document test results and make such findings available to the customer. If feasible, the company should provide these results to the manufacturer of the electronic part in an attempt to convince the manufacturer to modify its own specifications. In addition, the company should share the results of any safety and hazard analyses with the customer and bring closure to any issues that may be uncovered.

Provide regular training sessions to engineers and technicians in the procedures for designing and testing systems, documenting tests and analyses, and determining whether parts can be safely used outside of the manufacturer's specifications. Such training will enhance the company's quality assurance program and alert the company to potential new safety hazards about which it must then inform its customers.

Establish protocols for effectively addressing reports of part failures or defects, including responding to failure notices, notifying the original manufacturer, halting further usage of problematic parts, and recalling products if necessary. By developing a "hazard tracking system," a company will be able to monitor the actual performance of the parts.

19.4.3 Other significant defenses

In addition to the government contractor defense and the contract specification defense, other significant product liability defenses may be available to companies. Unlike the defenses detailed above, however, none of these defenses are likely to result in a quick dismissal of a product liability case.

An argument raised frequently in response to a product liability claim is that the product or process complies with the "state of the art" or, similarly, that the defendant conformed its conduct to the applicable "industry standard"[3] [19], [20]. Courts will generally evaluate these facts to determine whether the defendant has satisfied its duty of care owed to the injured parties. In most jurisdictions, therefore, compliance with the state of the art is a factor to be considered in determining whether a defect exists and is not necessarily conclusive on the issue of liability. Although compliance with the state of the art is raised often, plaintiffs frequently contend that the applicable industry standard is lagging behind in its knowledge concerning a particular product or process. Therefore, compliance with the state of the art or industry standard should not be viewed, standing alone, as a means to avoid liability. Nevertheless, if a company could demonstrate that it adhered to widely accepted methodologies in determining the acceptability of using a particular part, the company may be able to prove that it acted reasonably.

In response to a failure-to-warn allegation, a company can satisfy its duty to warn the ultimate end user by providing its customer with information concerning risks associated with use of the product. Under the "sophisticated user doctrine," the duty to warn individual end users is shifted from the company that used the parts to the entity that integrates those parts into a larger system or end product [18]. The sophisticated user defense, although raised most frequently by the makers of prescription drugs, should also be available to companies that provide electronic parts to an intermediate entity.

Where a third party alters or uses a product in an unforeseeable manner, the company will not be liable for any resulting harm. Pursuant to the theory of "alteration or product misuse," where an injury would not have occurred had the product been used in its unmodified condition or intended manner, the company would escape liability [4]. In the context of companies that use uprated electronic parts, this defense may become applicable where the company's customer uses the uprated parts in unintended or unforeseeable applications and such use proximately caused the plaintiff's injuries. For the same reason, electronic part manufacturers who notify companies not to use their parts outside of the manufacturer's specifications will be able to raise a product misuse defense.

Government contracts issued pursuant to the Defense Production Act (DPA) may have a built-in defense to product liability actions. Under the DPA [2], the United States has the authority to issue "rated orders" for items that have been designated a priority for purposes of national defense. The DPA makes the performance of such "rated" orders mandatory and imposes civil and criminal penalties on the contractor for nonperformance [2].

Important for product liability purposes, the DPA provides that "no person shall be held liable for damages" arising out of the performance of a rated contract issued under the act [2]. Such language would therefore encompass the situation where a product liability action is brought against a government contractor acting under a DPA-issued rated order. A word of caution however: the existence of a DPA immunity defense has not been definitively resolved one way or the other by the courts.

A defense raised often by regulated industries is that of "regulatory compliance." Companies in these industries frequently argue that they should not be subject to liability when the product that is the subject of the lawsuit complied with the applicable governmental or regulatory standards [29]. Compliance with such standards, while generally admissible to prove the reasonableness of the defendant's conduct, is not conclusive on the issue of liability. Therefore, as with compliance with the state-of-the-art and industry standards, proof of regulatory compliance is but one factor that will be examined in judging the appropriateness of the defendant's overall conduct.

---

[3] For a U.S. court to consider a standard to be an industry standard, the standard must be followed by a majority of the equipment suppliers, including the largest equipment suppliers in the industry. This shows that the industry agrees with the standard, has faith in the standard, and believes that its method is the best way to complete the task.

The "U.S. state secrets privilege," is a rule that allows the U.S. government to withhold information in a case when disclosure of the information would be inimical to national security [22]. Only the government can raise the U.S. state secrets defense.[4] The government official who has firsthand knowledge of the matter must intervene in the case so that top-secret information is not made public. In some cases, the court will seal files so that the information is not available to the public but is available to the parties in the case. However, if the government asserts the U.S. state secrets defense, the court may decide not to allow the information in court.

Lastly, although technically not a legal defense, the use of an "indemnification clause" on a purchase order form may provide the company that uses uprated parts with some measure of relief. If a company has sufficient leverage with its customers to do so, the company can require its customers to indemnify it against any claims or judgments against the company that arise from the customer's use of a part.

19.5   Unique aspects of international law

This section of the chapter presents information about product liability law in the United Kingdom, France, and Japan. Both the United Kingdom and France are members of the European Economic Community (EEC). The EEC adopted a directive regarding product liability in the member states on July 25, 1985. "Member states have three years to pass national laws that comply with the EEC Directive on Product Liability, but there is a ten year transitional period during which variations in national law are permitted" [11]. As a result of the directive, the law regarding product liability in the United Kingdom and France may change rapidly. The information in this chapter is for guidance purposes only, and counsel should be obtained for the most current status of the law.

19.5.1   United Kingdom

The law in the United Kingdom is very similar to the law in the United States. This is probably because many of the fundamental principles of U.S. law originated in Britain. This section focuses on some of the differences between the law in the United States and the law in the United Kingdom.

19.5.1.1   Negligence

The three elements of negligence are:
*   there is a duty of care – there must be a sufficient relationship between the defendant and plaintiff so that the defendant reasonably contemplates that carelessness on his part may damage the plaintiff
*   the duty of care is breached through carelessness
*   foreseeable damages result (the damages cannot be too remote) [11]

A caveat involves a situation where there are multiple suppliers. Buyers of parts should be particularly aware of the following provision:

> One of the relevant factors in determining whether or not the manufacturer or assembler of the *finished product* has been negligent will be whether he has devised

---

[4] The U.S. supreme court stated, "The privilege belongs to the Government and must be asserted by it; it can neither be claimed nor waived by a private party. It is not to be lightly invoked. There must be a formal claim of privilege, lodged by the head of the department, which has control over the matter, after actual personal consideration by that officer. The court itself must determine whether the circumstances are appropriate for the claim of privilege, and yet do so without forcing a disclosure of the very thing the privilege is designed to protect" [25].

and maintained a screening process which is adequate to detect faults in component parts. In these circumstances a manufacturer owes a duty to take reasonable care to ensure that the component parts can be properly used. If he carelessly uses parts which are themselves unsafe or inappropriate for the purposes he will be in breach of duty. [11]

19.5.1.2   Strict liability

There are three categories of persons who may be held strictly liable under the British Consumer Protection Act of 1989. The three categories are:

- the manufacturer of the product
- any person who has held himself out as a producer by putting his name on the product or using a trademark or other distinguishing mark in relation to it
- any person who has imported the product into a member country of the EEC from outside the EEC in the course of his business to supply it to another [11].

Product labels and printed warnings, with respect to their potential dangers, are likely to be relevant in revealing the degree of safety consumers are entitled to expect from the product. However, Section 9 of the 1989 Consumer Protection Act prohibits any attempt to limit liability arising under the 1989 act, whether the limitation is contained in a contractual term or notice [11]. Thus, an appropriate warning may include language regarding potential dangers. However, a supplier may want to avoid language that specifically limits liability on those dangers.

There are three defenses regarding strict liability that are somewhat unique to the law of the United Kingdom. The first is that a supplier may escape liability simply by naming his own supplier [11]. This allows the plaintiff to work his way back to the original manufacturer and hold him liable. If the supplier to the plaintiff cannot name his own supplier, then he will be liable. The second defense unique to the United Kingdom is that the defect arose because of a national or EEC regulation that had mandatory requirements on the preparation and labeling of certain products [11]. Compliance with such a regulation should exonerate a supplier. The last unique defense is that if a product has been used inappropriately as a component part of another product, the supplier of the component may not be liable [11].

19.5.1.3   Failure to warn

The United Kingdom does not recognize a separate failure-to-warn claim. Instead, the failure-to-warn claim is incorporated into the negligence claim. If a supplier has not adequately warned the buyer of a defect or danger, the failure to warn is a negligent act. The plaintiff must prove the elements of negligence as stated in Section 19.5.1.

19.5.1.4   Defenses in the United Kingdom

Three noteworthy defenses in the United Kingdom are contributory negligence, assumption of the risk, and disclaimers and limitations on liability.

Contributory negligence may be invoked when a negligent act or omission by the plaintiff either contributed to the accident or exacerbated the damage. Contributory negligence is not a complete bar to recovery. Instead, the tribunal reduces the amount owed by the defendant to account for the negligence of the plaintiff [11].

The assumption of the risk defense may be used when the plaintiff agreed to carry the risk of loss or damage arising from a defective product. In order for this defense to be applicable, the plaintiff's knowledge must be complete and he must take this risk willingly. "The defendant will have good prospects of relying on the defense if he has warned the plaintiff of the defect or indicated that the article is not to be used without prior testing. If the plaintiff discovers a defect but continues to use the product he may be regarded has having assumed the risk" [11].

The disclaimers and limitations of liability defense may be used provided that the plaintiff has proper notice of the provision before the contract is made [11]. "A notice or disclaimer which purports to limit or exclude the rights of one party against another will usually have to be incorporated into a contract as a term between them to be effective" [11].

19.5.2 France

France's system of product liability is derived from the judicial interpretation of the French Civil Code. The Civil Code was drafted in 1804. Therefore, there is a lack of specific statutes covering product liability. As a result, the court system has used the Civil Code to create product liability law that protects the consumer. Through this process, France has created a strong precedent for negligence and strict liability.

19.5.2.1 Negligence

Negligence is a fault-based system. Therefore, the plaintiff must prove that the product was defective and that the defect was the cause of damage [12]. The courts generally hold that marketing a defective product per se constitutes fault on the part of the supplier or the distributor" [12].

One of the most important duties of a supplier or distributor is the duty to ensure that warnings or information is adequate. However, 'the distributor's duty is limited to informing the customers of the risks or dangers of which he is aware or could possibly have been aware" [12].

19.5.2.2 Strict liability

Strict liability in France has been created by the judicial system. The French Civil Code states that a person is liable for damage caused by persons or objects in his custody [13]. The courts have stretched this provision to create strict liability.

> The original supplier can be strictly liable when the damage has been caused by the structure of the product while the distributor would be liable when the cause lies in its utilization, although some cases suggest that where the technical expertise of the distributor is equivalent to that of the supplier the latter may be relieved from liability. [12]

19.5.2.3 Defenses in France

Four noteworthy defenses in France are disclaimers, contributory negligence, misuse of the product, and state of the art.

The French legal system uses the term "exclusion to liability" instead of "disclaimer." However, the concepts are fundamentally the same. Of paramount importance in the field of product liability is the implied obligation to advise and/or inform the customer about the particularities of the product regarding use and risks, which the courts have imposed on both the supplier and the distributor.

There are two caveats to this rule:

- The obligation is most stringent for products which are either new to the market or involve high technology or risks, especially when the customers are not purchasing for business purposes.
- The obligation is imposed more severely on the supplier than on the distributor because it is believed that only the supplier is in a position to know the properties of this product, while the distributor usually sells according to the supplier's

instructions. However, in some instances distributors are as liable as they would be if they were "professional" distributors who shared with the supplier part of the technical knowledge relating to the product" [12].

In France, the duty to inform covers not only precautions, but also proper instructions for use [12]. However, exclusions to liability are usually held invalid because of the strict liability doctrine [12]. The exception to this rule is when the purchaser is not only another businessperson, but also a specialist in the same field. Thus, the rule is "if the defect was such that a specialist should normally be able to detect it, an exclusion of liability is valid; conversely, if the defect is such that it could not be detected, even by a specialist, then the exclusion of liability is invalid" [12].

The defendants may claim contributory negligence when the plaintiff's negligence substantially contributes to the cause of damage. In France, the determining factor is often the nature of the defect. If the defect was apparent or the purchaser was technologically competent in the area, then the court may rule for the defendant [12]. However, if the defect was latent and the purchaser was not technologically competent in that area, then the court may rule that the defendant breached his duty to give adequate instructions [12].

The misuse of product defense is used when the defendant believes that the plaintiff used the product incorrectly and thus caused the harm. Sometimes a product is unfit because of a cause other than a defect, i.e., use of the product in a way not suited to its purpose. In that case, the courts often decide on how much the defendant fulfilled the duty to inform, advise, or warn the plaintiff. If the instructions are deemed conspicuous, complete, and accurate enough, then the courts may hold the plaintiff responsible [12].

The state of the art defense is used when the defendant could not have known about the defect because of the level of pertinent scientific and technical knowledge at the time of manufacture [6]. In France, the seller of a defective product remains liable even if he proves that the state of scientific and technical knowledge at the time he put the product into circulation did not enable the existence of a defect to be discovered [12].

19.5.3   Japan

The product liability law in Japan took effect on July 1, 1995. Under the law, suppliers, processors, and importers are liable for damages caused by defects in any product that they have manufactured, processed, or imported [28].

In Japan, the plaintiff must prove that the product was defective when it was placed for sale. This level of proof may be difficult for a plaintiff due to Japan's limited pretrial discovery rights and limited access to corporate records. Also, a supplier, processor, or importer is not liable for a product defect if the defect was not foreseeable given the level of scientific or technological knowledge in existence at the time the product was delivered [28].

19.6   Summary

To avoid legal liabilities, the most important factors for a supplier are preparation, analysis, and documentation. Companies must anticipate potential product limitations, and employees must document the concerns and notify the customers if a problem is not remedied. Equipment suppliers should educate their workforce so that employees are trained to make informed decisions. Equipment suppliers should be careful not to breach the duty of care they owe to their customers. Also, equipment suppliers should follow industry and company standards. Equipment suppliers are advised to make products as safe as possible and document the decision to use particular parts showing that such use was reasonable.

Product liability defenses are available to companies that use electronic parts outside the original manufacturer's specified temperature ranges. In order to maximize the effectiveness of these defenses, however, a company must be vigilant in seeking out ways in which it can demonstrate the viability of these defenses before the lawsuit is ever filed. Both the government contractor defense and the contract specification defense are vehicles around which a company can establish a product liability prevention program. In the final analysis, there is no substitute in product liability prevention for the training and education of employees and the endorsement by senior management of such a program.

As stated at the beginning of this chapter, the information presented here is not intended to be legal advice. All companies are advised to seek legal counsel when considering the issues presented herein. The interested reader is referred to [1], [3], [5], [8], [9], [15], [16], [21], [23], [26], [27], and [31].

19.7   References

[1]   50 App. U.S.C.A. 2091.

[2]   50 U.S.C. app. 2061-2171.

[3]   Aircraft Crash Litigation, 952 F. Supp. 1326 (1990).

[4]   *Amatulli v. Delhi Constr. Corp.*, 77 N.Y. 2d 526 (1991); *Horn v. General Motors Corp.*, 17 Cal. 3d 359 (1976); *Giordano v. Ford Motor Corp.*, 165 Ga. App. 644 (1983).

[5]   *Baker v. Monsanto Co.*, 962 F.Supp. 1143, 1151 (1995).

[6]   *Black's Law Dictionary*,1032 (6th ed. 1990).

[7]   *Boyle v. United Technologies Corp.*, 487 U.S. 500 (1988).

[8]   *Boyle v. United Technologies Corp.*, 108 S.Ct. 2510, 2518 (1988).

[9]   *Business Electronics Corp. v. Sharp Electronics Corp.*, 108 S.Ct. 1515, 485 U.S. 919 (1988).

[10]   *Carley v. Wheeled Coach*, 991 F.2d 1117 (3d Cir. 1993) (defense applied to supplier of ambulance); *Wisner v. Unisys Corp.*, 917 F. Supp. 1501 (D. Kan. 1996) (defense applied to Postal Service letter sorter machine). Hawaii Federal Asbestos Cases, 960 F.2d 806 (9th Cir. 1992) (defense did not apply to asbestos supplier).

[11]   Freedman, W., International Products Liability 105 (1995).

[12]   Freedman, Warren, *Product Liability: An International Manual of Practice France*, 46 (1990).

[13]   French Civil Code, Article 1384, p1.

[14]   Henderson, A. J. and R. N. Pearson, *The Torts Process* 338 (3rd ed., 1988).

[15]   *Hercules Inc. v. United States*, 116 S.Ct. 981, 986 (1996).

[16]   *Idaho Power Co. v. Westinghouse Electric Corp.*, 596 F.2d 924, 928 (1995).

[17]   *Kaiser Steel Corp v. Westinghouse Electric Corp.*, 55 Cal. App. 3d 939, 948 (1996).

[18] *Kennedy v. Mobay*, 84 Md. App. 397 (1990); *Aluminum Co. of America v. Alm*, 785 S.W. 2d (Tex.), cert. denied, 498 U.S. 847 (1990).

[19] *Maguire v. Hughes Aircraft Corp.*, 912 F.2d 67 (3d Cir. 1990).

[20] *Maxwell & Moore, Inc.*, 216 Va. 245 (1975); Colo. Rev. Stat. § 13-21-403(1)(a).

[21] *McLaughlin v. Sikorsky Aircraft*, 148 Cal. App. 3d 203 (1983); Turner v. Manning,

[22] *Nejad v. United States*, 924 F.Supp. 953, 955 (1989).

[23] Phillips, Products Liability 123 (3rd ed. 1988).

[24] *Ramey v. Martin-Baker Aircraft Co.*, Ltd., 874 F.2d 946 (4th Cir. 1989).

[25] *Reynolds v. United States*, 93 S.Ct. 528 (1953).

[26] Restatement (Third) of Torts: Prod. Liab. 2 (T.D. No. 1, 1994).

[27] Restatement (Third) of Torts: Prod. Liab. 6 I (T.D. NO. 2, 1995).

[28] Sako, "Japan Adopts Product Liability Law (1996)," <http://www.faegre.com/areas/area_ib3.html>.

[29] *Sanna v. National Sponge Co.*, 209 N.J. Super. 60 (1986); Tenn. Code Ann. 29-28-104.

[30] Smith, *Courts, Politics, and the Judicial Process* 38 (1995).

[31] Technology Group of Tory Tory DesLauriers & Binnington website, <http://tech group.torytory.ca/Publications/INSUR1.htm>.

[32] Wright, M., D. Humphrey and P. McCluskey, "Uprating Electronic Components for Use Outside Their Temperature Specification Limits," *IEEE Transactions on Components, Packaging, and Manufacturing Technology*, Part A, Vol. 20, No. 2, pp. 252–256, June 1997.

[33] *Woodward v. IBM*, No. 96-1000-A (E.D. Va. 1997).

Table A.1: Top twenty five franchised distributors*

Company and website	Revenue income		Product mix		Revenue by region		Number of employees	
Avnet Inc. www.avnet.com	2002	$8,920M/$664.9M	Active components	50%	North America	58%	2002	10,900
	2001	$12,814M/$15.4M	Passive components	11%	Europe	32%	2001	12,000
	2000	$9,172.2M/$145.1M	Computer products/system	39%	Asia/Pacific	10%		
Arrow Electronics Inc. www.arrow.com	2002	$7,390M/n/a	Active components	57%	North America	58%	2002	11,770
	2001	$10,127.6M/$73.8M	Passive components	16%	Europe	33%	2001	12,400
	2000	$12,959.3M/$357.9M	Computer products/system	27%	Asia/Pacific	9%		
Future Electronics www.futureelectronics.com	2002	$2,603M (est.)/n/a	n/a	n/a	n/a	n/a	2002	5,000
	2001	n/a					2001	4,500
Bell Microproducts Inc. www.bellmicro.com	2002	$2,105M(est.)/$(4M)	Active components	15%	North America	41%	2002	1,366
	2001	$2,007.1M$/$(22.1M)	Passive components	72%	Europe	46%	2001	1,476
	2000	$1,804.1M/$17.2M	Others	13%	Rest of the world	13%		
Pioneer-Standard Electronics Inc. www.pioneerstandard.com	2002	$2,323.6M/$7M	Active components	30%	North America	95%	2002	2,462
	2001	$2,901.4M/$34.5M	Passive components	15%	Rest of the world	5%	2001	2,550
	2000	$2,550.7M/$40.1M	Computer products/system	55%				

* Electronic News, "How They Rank: A Look at the Top 25 Distributors,"
<http://www.einsite.net/electronicnews/index.asp?layout=article&articleid=CA262> posted December 2, 2002, accessed December 27, 2002.

Source: Electronic News, "TOP 25 Distributor Profiles: A More In-depth Look at the Top 25,"
<http://www.einsite.net/electronicnews/index.asp?layout=article&articleid=CA262486>; posted December 2, 2002, accessed December 27, 2002.

Company and website	Revenue income		Product mix		Revenue by region		Number of employees	
Memec Group Holding Ltd. www.memec.com	2002	$1,601M (est.)/n/a	Active components	100%	North America	48%	2002	2,700
	2001	$2,100M/n/a			Europe	25%	2001	2,800
	2000	$3,700M/n/a			Asia/Pacific	25%		
					Rest of the world	2%		
Premier Farnell plc www.premierfarnell.com	2002	$1,153.2M/$55.2M	Active components	n/a	North America	56.7%	2002	5,174
	2001	$1,281M/$104.2M	Passive components	n/a	United Kingdom	26.7%	2001	5,570
	2000	$1,210.9M/$106.5M	Computer products/system	n/a	Rest of the world	16.6%		
TTI Inc. www.ttiinc.com	2002	$630M (est.)/n/a	Active components	3%	North America	84.7%	2002	1,235
	2001	$750M/n/a	Passive components	97%	Europe	13%	2001	1,270
	2000	$1,145.8M/n/a			Asia/Pacific	2%		
					Rest of the world	0.3%		
Richardson Electronics Ltd. www.rell.com	2002	$443.5M/$(11.3M)	Active components	43%	North America	56%	2002	1,111
	2001	$502.4M/$17.7M	Passive components	10%	Europe	21%	2001	1,099
	2000	$407.2M/$13.1M	Computer products/system	14%	Asia/Pacific	15%		
			Others	33%	Rest of the world	8%		
Digi-Key Corp. www.digikey.com	2002	$374M (est.)/n/a	Active components	28%	North America	97%	2002	1,140
	2001	$355.8M/n/a	Passive components	65%	Rest of the world	3%	2001	1,122
	2000	$372.4M/n/a	Others	7%				

Company and website	Revenue income		Product mix		Revenue by region		Number of employees	
All American Semiconductor Inc. www.allamerican.com	2002	$345M(est.)/n/a	Active components	88%	North America	100%	2002	600
	2001	$381.1M/$22.6M	Passive components	12%			2001	600
	2000	$522.2M/$11.2M						
Reptron Electronics Inc. www.reptron.com	2002	$320M/n/a	Active components	40%	North America	95%	2002	1,553
	2001	$398.6M/$(21.8M)	Passive components	12%	Europe	1%	2001	1,563
	2000	$576.5M/$5.7M	Contract manufacturing	48%	Asia/Pacific	4%		
Nu Horizons Electronics Corp. www.nuhorizons.com	2002	$282M/$2.2M	Active components	80%	North America	89%	2002	490
	2001	$670.6M$/$35.4M	Passive components	20%	Europe	4%	2001	471
	2000	$379.2M/$$11.7M			Asia/Pacific	7%		
DAC www.heilind.com	2002	$251.4M(est.)/n/a	Passive components	100%	North America	98%	2002	725
	2001	$273.2M/n/a			Europe	2%	2001	735
	2000	$340.4M/n/a						
Jaco Electronics Inc. www.jacoelectronics.com	2002	$194.1M/$(5M)	Active components	54%	North America	91%	2002	363
	2001	$350.2M/$9.9M	Passive components	37%	Europe	6%	2001	440
	2000	$209.3M/$6.4M	Contract manufacturing	9%	Asia/Pacific	3%		

Company and website	Revenue income		Product mix		Revenue by region		Number of employees	
Sager Electronics www.sager.com	2002	$183.9M(est.)/n/a	Active components	5%	North America	99%	2002	n/a
	2001	$200M/ n/a	Passive components	95%	Europe	1%	2001	430
	2000	$373.2M/ n/a						
California Eastern Laboratories Inc. www.cel.com	2002	$146.5M/n/a	Active components	100%	North America	50%	2002	190
	2001	$233.6M/n/a			Europe	10%	2001	206
	2000	$147.7M/n/a			Asia/Pacific	35%		
					Rest of the world	5%		
Allied Electronics www.alliedelec.com	2002	$158M/n/a	Active components	8.5%	North America	99%	2002	500
	2001	$220M/n/a	Passive components	45%	Rest of the world	1%	2001	500
	2000	$177M/n/a	Others	46.5%				
Dependable Component Supply Corp. www.dependonus.com	2002	$86M (est.)/n/a	Active components	60%	North America	93%	2002	81
	2001	$78M/n/a	Passive components	40%	Europe	2%	2001	86
	2000	$105M/n/a			Asia/Pacific	2%		
					Rest of the world	3%		
A.E. Petsche Co. Inc. www.aepetsche.com	2002	$79M(est.)/n/a	Passive components	100%	North America	84%	2002	108
	2001	$77M/n/a			Europe	13%	2001	108
	2000	$70M/n/a			Asia/Pacific	1%		
					Rest of the world	2%		

Company and website	Revenue income		Product mix		Revenue by region		Number of employees	
Master Distributors www.masterdistributors. com	2002	$71.6M/$2.3M	Active components	4%	North America	92%	2002	153
	2001	$75.3M/$2.9M	Passive components	86%	Europe	3%	2001	164
	2000	$91.4M/$3.1M	Computer products/syst.	1%	Asia/Pacific	3%		
			Contract manufacturing	8%	Rest of the world	2%		
			Other	1%				
Powell Electronics Inc. www.powell.com	2002	$72M(est.)/n/a	Passive components	100%	North America	100%	2002	220
	2001	$72M/n/a					2001	220
	2000	$80M/n/a						
RS Electronics www.rselectronics.com	2002	$71M(est.)/n/a	Active components	2%	North America	100%	2002	163
	2001	$80.1M/n/a	Passive components	75%			2001	158
	2000	$85.9M/n/a	Computer products/syst.	2%				
			Others	21%				
Electro Sonic Inc. www.e-sonic.com	2002	$68M(est.)/n/a	Active components	10%	North America	100%	2002	275
	2001	$76M/n/a	Passive components	63%			2001	300
	2000	$102M/n/a	Computer products/syst.	2%				
			Contract manufacturing	11%				
			Others	14%				
ESCO LLC www.escollc.com	2002	$64.1M(est.)/n/a	Passive components	72%	North America	100%	2002	120
	2001	$72M/n/a	Contract manufacturing	18%			2001	150
	2000	$87M/n/a	Others	10%				

Table B.1: Product categories offered by major distributors

Category	All American	Allied	Arrow	Avnet	Calif East	Digi-Key	Future Elec	Jaco	Kent	Nu Horizons	Pioneer Standard	Powell Electronics	Richardson	Sager	TTI
Other															
Software	X			X											
Motor		X		X		X						X			
Discrete semiconductors	X	X	X	X	X	X	X	X		X	X		X	X	X
Displays and optoelectronics	X	X	X	X	X	X	X	X	X	X	X	X		X	
Test and measurements		X	X	X		X	X			X			X		
PC boards				X		X				X					
Enclosures		X		X		X	X						X	X	
Industrial control				X		X				X		X	X	X	
Circuit protection	X	X	X	X		X	X	X	X	X	X	X		X	X
Connectors	X	X	X	X		X	X		X		X	X	X	X	X
Power sources	X	X	X	X		X	X	X	X	X	X	X	X	X	
Sensors		X	X	X		X	X		X	X	X	X	X	X	
ICs	X	X	X	X	X	X	X	X		X	X		X	X	
Switches and relays	X	X	X	X		X	X	X	X		X	X	X	X	X
Passive components	X	X	X	X		X	X	X	X		X		X	X	X

Table B.2: Services available from major distributors

Service	All American	Allied	Arrow	Avnet	Calif East	Digi-Key	Future Elec	Jaco	Kent	Nu Horizons	Pioneer Standard	Powell Electronics	Richardson	Sager	TTI
Other			X							X					
International service			X					X	X	X	X	X	X		X
Warranty service and support			X					X			X		X	X	
Online ordering	X	X	X	X		X		X		X	X		X	X	
Just-in-time manufacturing			X	X					X		X	X	X	X	
Supply chain and inventory management	X		X	X			X	X	X	X	X	X		X	X
Marketing	X		X					X	X	X		X		X	X
Kitting	X		X	X	X		X	X	X		X	X	X	X	
Assembly			X	X				X	X		X	X		X	X
Device programming	X		X	X		X	X	X		X	X				
Private label manufacturing													X	X	
Application and design consulting	X		X	X				X	X	X	X	X	X	X	
Contract manufacturing								X		X	X				
Packaging and testing			X	X	X			X		X		X	X		

Index